BASIC FLUID POWER

Dudley A. Pease
(deceased)

John J. Pippenger, P.E.
Educational Coordinator
FLUID POWER EDUCATIONAL FOUNDATION

Second Edition
BASIC FLUID POWER

PRENTICE-HALL, INC., Englewood Cliffs, N.J. 07632

Library of Congress Cataloging-in-Publication Data

PEASE, DUDLEY A.
   Basic Fluid power.

   Includes index.
   1. Fluid power technology.   I. Pippenger, John J.
II. Title.
TJ840.P4 1987   620.1'06   86-22719
ISBN 0-13-061508-0

Editiorial/production supervision and
   interior design: Tom Aloisi
Cover design: 20/20 Services, Inc.
Manufacturing buyer: Rhett Conklin

Printed in the United States of America

10 9 8 7 6 5 4 3 2 1

ISBN: 0-13-061508-0   025

Prentice-Hall International (UK) Limited, *London*
Prentice-Hall of Australia Pty. Limited, *Sydney*
Prentice-Hall Canada Inc., *Toronto*
Prentice-Hall Hispanoamericana, S.A., *Mexico*
Prentice-Hall of India Private Limited, *New Delhi*
Prentice-Hall of Japan, Inc., *Tokyo*
Prentice-Hall of Southeast Asia Pte. Ltd., *Singapore*
Editora Prentice-Hall do Brasil, Ltda., *Rio de Janeiro*

Contents

3   SEALING DEVICES FOR HYDRAULIC POWER   41

4   THE DISTRIBUTION OF HYDRAULIC POWER   49

5   CONDITIONING POWER FLUIDS   73

6   THE SOURCE OF HYDRAULIC POWER   90

7   THE CONTROL OF HYDRAULIC POWER   128

8   ACTUATORS PROVIDE FLEXIBILITY IN THE USE OF FLUID POWER   169

9   SYSTEM COMPONENTS AND CIRCUITS   196

SECTION TWO: PNEUMATICS

Preface

Sophistication of control is the key to more effective use of both hydraulics and pneumatics for a wide range of power transmission needs in our fast-moving mechanized society. The very solid framework built by the late Dudley Pease for a fluid power educational program is updated in this second editon of *Basic Fluid Power*.

We want to acknowledge the significant contribution by Dr. William Wolansky of the Iowa State University School of Education in the review of selective areas of this text for accuracy and academic presentation format. Professor Arthur Akers of the Mechanical Engineering Department of Iowa State University has also made a significant contribution in reviewing and updating the mathematical content.

A special thank you is in order to the key professors in the Fluid Power Educational Foundation key school network and to a large number of professional educators who have pioneered programs to weld the many new fluid power and electronic disciplines into a coherent academic program. Their contribution will be evident throughout this second edition.

As in the first edition, the structure of *Basic Fluid Power* is the summation of input from many experts in both academic institutions and industry. Dudley Pease assembled, edited, polished, and welded the expertise from many sources into a keystone structure. Hopefully, I have built on the professional status established by Dudley Pease in the first edition.

The input from contemporary professionals in the fluid power industry and the educational community has been gratifying. Credit lines in captions can offer only a brief thank you. To this I add my personal thanks.

John J. Pippenger, P.E.

BASIC FLUID POWER

I

Introduction

HISTORICAL BACKGROUND

We traditionally think of hydraulics in connection with water hydrology—the use of water to produce power by means of waterwheels or turbines. Similarly, air has been used for centuries to turn windmills or to propel ships.

But traditionally, the use of water or air to produce power depended on the movement of vast quantities of fluid at relatively low pressures. Also, we depended on nature to supply the pressure. Only in the past 100 years or so have we devised pumps and compressors to develop artificial pressures vastly greater than those nature provides, and correspondingly, we need to deal in vastly lower quantities of fluid as the operating pressure is increased. And this is the heart of the matter as far as advanced fluid power systems are concerned: the higher the pressure, the less the flow required to produce power, and the system becomes increasingly efficient.

Fluid power technology began about 300 years ago with the discovery, in 1650, of *Pascal's law*. Simply stated, this law says that pressure in a fluid at rest is transmitted equally in all directions. One hundred years later Bernoulli developed his law concerning the conservation of energy in a flowing fluid. Another 100 years were to pass before these laws were applied to industry.

By 1850 the Industrial Revolution in Great Britain was well advanced. Electrical energy had not yet been developed to power the tools of industry, but fluid power became prominent as the means of powering presses, cranes, winches, and extruding machines. In fact, by the late 1860s, cities such as London and Manchester had central industrial hydraulic distribution systems similar in nature to central steam generating and distribution systems common today in some large cities. In these systems, large steam engines drove hydraulic water pumps, delivering water at relatively high pressures through pipes to factories that used the water to power machines. But the emergence of electricity in the late nineteenth century caused fluid power to be neglected. Our brain power was concentrated on refining the technology of electricity rather than on refining the application of fluid power.

However, it became apparent by 1900 that electrical systems had certain disadvantages, as did mechanical systems for transmitting power. Engineers again looked to fluid power for the answers to problems that other systems could not solve.

An important milestone in fluid power applications came in 1906, when a hydraulic system was developed to replace electrical systems for elevating and controlling guns on battleships, first installed on the USS *Virginia*. It is important also to note that by now oil was replacing water in fluid power systems.

In 1926 the *direct hydraulic system*—a type of packaged system with the pump, controls, and actuator in a self-contained unit on the machine or vehicle using it—was developed in the United States. Since then, the applications of fluid power have become virtually limitless, and include systems on machine tools, automobiles, earth-moving machines, farm equipment, ships and locomotives, airplanes, and even in space vehicles.

Fluid power is power transmitted and controlled through use of a pressurized liquid or gas. *Hydraulics* is the engineering science pertaining to liquid pressure and flow. *Pneumatics* is the engineering science pertaining to gaseous pressure and flow.

Modern fluid power performs work through direct or central power systems and is unrelated to the academic definitions common to civil engineering or to cross-country pipeline systems. Civil engineering involves a technology in hydroelectric systems using dams, weirs, etc., whereas fluid power deals with fluids as the power media operating directly with an actuating cylinder or fluid motor to perform work.

Fluid power jacks up an automobile, drills our teeth, operates computers, launches spaceships, controls submarines, mines coal and ores, moves earth, harvests crops, and, in general, makes our everyday living easier and more enjoyable.

Over 90% of all machine tools are controlled or operated with fluid power. Farm machinery is constantly improving with fluid power. Transportation, printing, materials handling, construction, chemical, mining, manufacturing, processing, aerospace, and many other industries are finding more and more uses for fluid power. Numerical control, military applications, and high-performance machines are also important to the new and fast-growing fluid power industry.

The fantastic growth of this new and exciting technology is opening many new opportunities for trained personnel in all phases of fluid power. Engineers, technicians, mechanics, sales personnel, service personnel, operators, and supervisors are badly needed.

A critical shortage of trained teachers in the field of fluid power has opened interesting opportunities in education. Qualified teachers are needed for universities, colleges, technical institutes, vocational schools, and at the high school level.

WHY FLUID POWER?

Academically, fluid power provides flexible and easy control of variable force, distance, and speed. Fluid power can be varied from a delicate touch of a few ounces to a gigantic force of 36,000 tons or more. Tape-controlled machines can provide accuracies in measurements as low as plus or minus one ten-thousandth of an inch with amazing repetition. Fluid power can provide constant torque at speeds of nearly 100

miles per hour within a few inches or give a creeping speed of a fraction of an inch per minute.

ADVANTAGES OF USING FLUID POWER

1. Fluid power provides flexibility in the control of machines.
2. Fluid power provides an efficient method of multiplying forces.
3. Fluid power provides constant torque at infinitely variable speeds in either direction with smooth reversals.
4. Fluid power is compatible with other means of control, such as electrical, electronic, or mechanical.
5. Fluid power is accurate.
6. Fluid power gives fast response to controls.
7. Small forces can be amplified to control large forces.
8. Oil fluid power provides automatic lubrication for less wear.
9. Air fluid power is clean and safe from fire hazards.
10. Fluid power provides freedom in machine design.
11. Fluid power is simple and provides ease of installation and maintenance.
12. Fluid power is economical.
13. Fluid power is efficient and dependable.
14. Fluid power gives predictable performance.
15. Fluid power is readily available.

HOW DOES FLUID POWER WORK?

A classic illustration showing fluid power in operation is the hoist used by automobile service stations to raise a car for servicing. This application of fluid power generally uses both a liquid and a gas.

The hoist operates on the principle that air under pressure is added on top of a confined quantity of oil within a long cylinder set in the ground. The cylinder is fitted with a long movable plunger that is sealed from the atmosphere. As the pressure is added to the system, the plunger moves upward, raising the automobile.

The hoist shown in Fig. I-1 works on a simple physical law that was discovered about 1650 by the French scientist Blaise Pascal. The law states: "A pressure added to a confined fluid is transmitted undiminished throughout the fluid. It acts on all surfaces in a direction at right angles to those

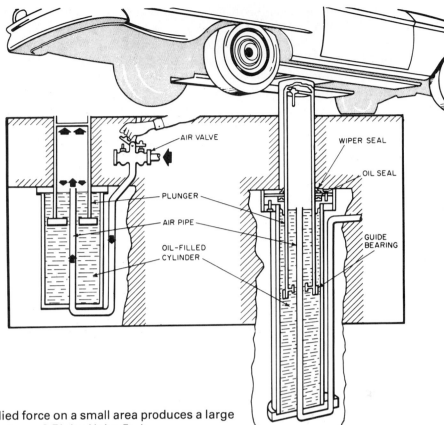

AN AIR PIPE runs through the bottom of the plunger up to the top cylinder in this type of lift. Oil flows freely into the plunger and between its outer wall and the cylinder proper. An air valve is the only control. When it is held open to admit air from the compressor line, pressure builds up between the closed upper end of the plunger and the oil beneath, raising the plunger. When the air is shut off, the plunger is held at that height. To lower it, the valve is opened on the side to the atmosphere, letting the air in the plunger bleed off slowly.

FIGURE I-1 Hoist. An applied force on a small area produces a large force on a large area. (Courtesy of Globe Hoist Co.)

surfaces." Translating the meaning of this fundamental law into a mathematical expression gives the following equation:

$$\text{Pressure} = \text{force} \div \text{area}$$

$$P = \frac{F}{A}$$

In other words, the pressure added increases the force per unit of area, or force equals pressure times area, $F = P \times A$.

Thus it can be seen that as air enters the space above the oil in a hoist, it creates a force, or push, against the area of the large plunger, causing the plunger to move upward. If the hoist plunger has a cross-sectional area of 60 in.^{2}, a pressure of 100 psi of air would create a lifting force of 6000 lb.

$$F = P \times A$$

$$\text{Force} = 100 \text{ psi} \times 60 \text{ in.}^2$$

$$= 6000 \text{ lb (3 tons)}$$

When a service person wants to lower the hoist, he or she merely bleeds air from the space above the oil, which lowers both the pressure and the hoist (see Fig. I-2).

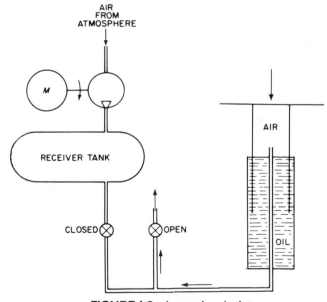

FIGURE I-2 Lowering hoist.

COMPONENTS OF THE SYSTEM

1. An air compressor to increase the pressure of air taken from the atmosphere (usually to about 120 psi)

2. An electric motor to drive the air compressor (usually about 2 hp)

3. A receiver tank to store the air energy until it is needed to raise the hoist (this permits a smaller compressor and drive motor)

4. Pipelines to transmit the air power

5. Control valves to let air into the hoist or out of the hoist

6. A hoist unit consisting of the cylinder, plunger, seals, and other fittings

7. Oil to lubricate the moving parts of the hoist and to help transmit the power to raise the hoist

All fluid power systems use the principle of Pascal's law to create a force that can provide linear and rotary motion to perform work.

A 50-TON FORCE RAISES UTILITY POLES

The compact power unit shown in Fig. I-3 is portable and can be used in areas inaccessible to vehicle-mounted devices. It is generally used to replace poles. The unit consists

FIGURE I-3 A 50-ton force raises utility poles. (Courtesy of Owatonna Tool Co.)

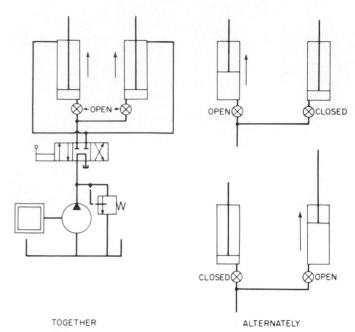

FIGURE I-4 Controls allow cylinders to operate together or alternately to rock pole.

of a 3-hp gas engine, hydraulic pump, control valves, and two small cylinders. The pump operates at 10,000 psi, and the system is capable of creating 50 tons of force. The controls allow cylinders to operate together or, alternatively, to rock the pole (see Fig. I-4).

It is apparent from this sample application that fluid power can be small and light in weight, yet provide tremendous forces. Portable fluid power is doing many jobs on the farm, in lumbering camps, on construction projects, and so on.

A NEW CONCEPT IN FARM MACHINERY

Figure I-5 shows a fluid power combine with design flexibility for placing the power components. Hydraulic drives provide full power for the wheels at low speed. It is also possible to obtain infinitely variable speeds independently of wheel drives.

FLUID POWER PROVIDES REMOTE CONTROL FOR GIANT EARTH MOVERS

The giant tandem earth movers shown in Fig. I-6 use more than a dozen hydraulic circuits from a central hydraulic power source. From the driver's seat, one operator can control all functions of the machine. Front and rear scrapers, bowls, aprons, and ejectors are all operated independently with remote controls. Because hydraulic fluid is nearly incompressible, at the pressures used here the controls operate

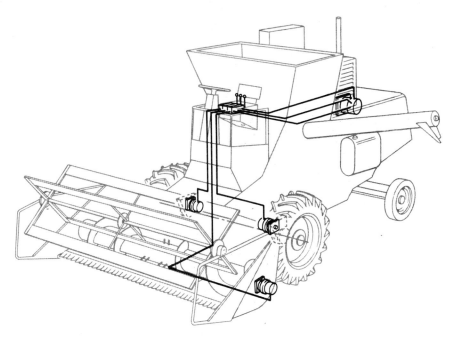

FIGURE I-5 Fluid power combine. (Courtesy of Dynex/Rivett Inc., Pewaukee, Wis.)

FIGURE I-6 Tandem. (Courtesy of Euclid.)

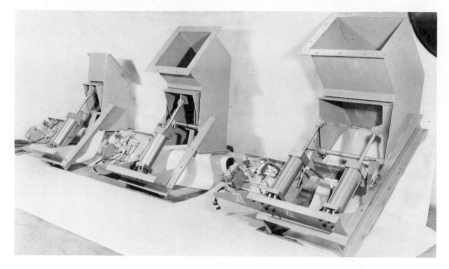

FIGURE I-7   Bin gates. (Courtesy of Parker-Hannifin Corporation, Cleveland, Ohio.)

smoothly with quick response. The machine is powered with a 444-hp engine and carries 94,000 lb of earth. Both bowls can be loaded to capacity in 1.5 minutes.

FLUID POWER USES AIR TO CONTROL CONCRETE MIXING

Fluid power plays a tremendous role in the handling of materials. Figure I-7 shows three bin gates which are used in a large concrete mixing plant. Air power is used because of the precise control needed to inch the gates open and shut and to meter the flow of aggregate onto a conveyor belt. Air power is also an excellent source of power for emergency operations in case of electrical failure.

HIGH-SPEED AIR POWER RIVETS AUTOMOBILE HEADLIGHTS

The riveting of headlight assemblies is made simple by using eight air circuits for fast-speed operation. Fast-cycling air valves operate six work cylinders and two ejector cylinders. Fluid power using air can provide rapid cycling for high-production-volume machines. Most air circuits operate on plant air pressure which is limited to approximately 100 psi; this limits air power to the lower-force applications. Fluid power helps to provide higher production with less cost of manufacture.

FLUID POWER STEERING AND BRAKES

Fluid power provides a safe and dependable power source for vehicle steering and braking systems. A pump operates off the engine and keeps the system charged. A special valving arrangement provides fluid power for both steering

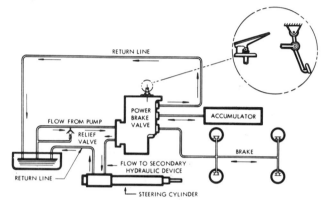

FIGURE I-8   Full-power hydraulic brakes. (Courtesy of the Bendix Corp.)

and braking and keeps the accumulator charged. (The accumulator stores pressure energy that is available for use when the engine is off.)

All the hydraulic fluid from the pump flows through the brake valve. When the accumulator pressure drops to its low limit, the brake valve diverts a small amount of fluid from the system to charge the accumulator. The brake valve automatically drives the fluid back to the system when the accumulator is fully charged (see Fig. I-8).

FLUID POWER FOR NUMERICAL CONTROL

The three-axis tape-controlled milling machine shown in Fig. I-9 is designed for numerical control and provides both profile and contour milling. Dimensions from drawings and machining data are coded and put on a control tape. In contour milling where complex curves are involved, the numerous calculations require the use of an electronic computer. As the machining commands punched in the tape are fed into the machine, they are translated into motion and force for hydraulic control operations. Hydraulic actuation

FIGURE I-9 Numerical control. (Courtesy of Mobil Oil Corporation.)

provides a very suitable link between the control tape signals and the actual machining motions because of ease of control, rapid response, variable speed, and amplification of force.

FLUID POWER BUILDS TRACTOR ENGINES

The automatic transfer machine shown in Fig. I-10 handles over 70 tractor engine blocks per hour. Castings entering the machine transfer automatically from station to station for clamping, milling, drilling, reaming, tapping, and inspection operations. Fluid power provides efficient and dependable control. Emphasis is placed on good design, lu-

brication, and preventative maintenance to keep loss of production to a minimum. The failure of a single machine component might shut down a production line or even an entire plant's operation. Hydraulic power plays a very important part in keeping machines operating smoothly and accurately for maximum production.

FLUID POWER BOTTLES OR CANS BEER

Figure I-11 shows an important use of fluid power in the making of beverages. Highly carbonated beverages must be handled gently. Throttling the flow of beer into filling ma-

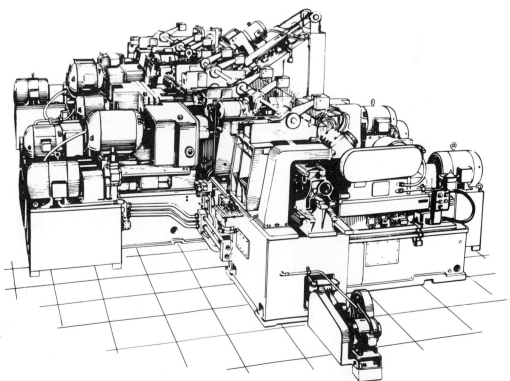

FIGURE I-10 Transfer machine. (Courtesy of Mobil Oil Corporation.)

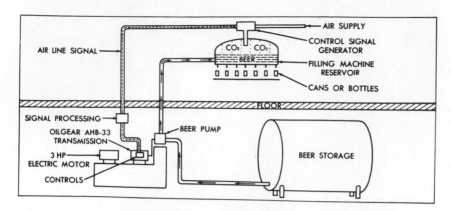

AIR LINE SIGNAL

AIR SUPPLY

CONTROL SIGNAL GENERATOR

CO₂ CO₂

BEER

FILLING MACHINE RESERVOIR

CANS OR BOTTLES

FLOOR

SIGNAL PROCESSING

OILGEAR AHB-33 TRANSMISSION

BEER PUMP

3 HP ELECTRIC MOTOR

BEER STORAGE

CONTROLS

FIGURE I-11 Stabilizing beer filling level. (Courtesy of The Oilgear Company, Milwaukee, Wis.)

chine reservoirs causes foam and irregularity of liquid height, making accurate filling of cans and bottles difficult. Fluid power provides the solution. Shown "on the job" in Fig. I-12 are six hydraulic two-way, "Any-Speed" transmissions driven by 3-hp electric motors, mounted on standard reservoir bases. Transmissions drive beer pumps through dual V belts. Beer is pumped from storage tanks to bottle or can-filling machine reservoirs on the floor above. The liquid level in the filling reservoirs is controlled automatically by a low-pressure CO_2 signal: as the level rises, the pressure rises; as the level falls, the pressure falls. The CO_2 signal is relayed to controls on the transmissions, which automatically, and instantly, increase, decrease, reverse, or stop the beer pumps. Where former throttling systems caused foam and irregular levels, this system holds the reservoir level to $\pm \frac{1}{16}$ in. under continuous operation, assuring accurate filling of each can or bottle. Beer pumps are flushed under water pressure without disconnecting drives. Several years of continuous service have proven drives to be extremely dependable under all conditions, with little or no attention.

WORLD'S LARGEST SELF-PROPELLED MACHINE USES FLUID POWER

Figures I-13, I-14, and I-15 show the rear crawlers of Peabody Coal Company's 9000-ton, 115-yd³ B-E 3850-B Stripping Shovel. It is 90 ft taller than the Statue of Liberty, wider than an eight-lane highway, and has a 420-ft reach. *A* shows two of four single-acting cylinders that support, automatically level, and equalize loads. The cylinders are 54 in. in diameter, have an 80-in. stroke, and operate at 3000 psi. Each cylinder weighs 65 tons. *B* shows single-acting cylinders 19 in. in diameter with a stroke of 90 in.; they control the 400 tons of force required to turn the crawlers through a 30° arc. They also operate at 3000 psi. *C* indicates the steering tie-rods that link crawlers. *D* is the lower frame of the machine, which is automatically held level within 2°. The shovel can move at an average speed of $\frac{1}{4}$ mile per hour. Notice the size of just one crawler compared to the man pictured in the lower right-hand corner of the photo.

FIGURE I-12 Six power units. (Courtesy of The Oilgear Company, Milwaukee, Wis.)

FIGURE I-13 Stripping shovel. (Courtesy of The Oilgear Company, Milwaukee, Wis.)

FIGURE I-14 Power unit. (Courtesy of The Oilgear Company, Milwaukee, Wis.)

FIGURE I-15 Power unit. (Courtesy of The Oilgear Company, Milwaukee, Wis.)

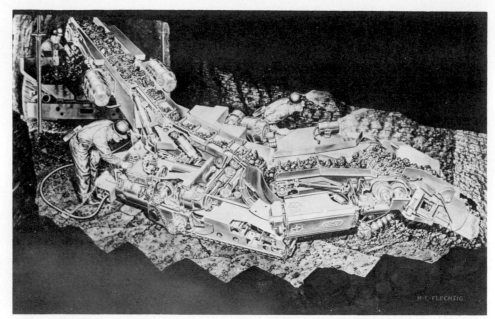

FIGURE I-16 Coal mining machine. (Courtesy of Mobil Oil Corporation.)

FLUID POWER MINES COAL

With the aid of several hydraulic cylinders and jacks, the continuous coal mining machine shown in Fig. I-16 digs and loads coal at the rate of 2 tons per minute. The combination of a ripping head and loading equipment marks a long step in coal mine mechanization.

The ripping head and rear conveyor can be swung from side to side through an arc of about 45°, which makes it possible for the machine to get around corners when moving from one position to another. During operation, the ripping head is driven into the base of a coal seam and then swung upward. Hydraulic cylinders provide the necessary force. The hydraulic controls can be actuated from one location to operate the machine. Fluid power is responsible for better, safer, and more efficient methods of mining coal.

FLUID POWER PRODUCES TOOTHBRUSH HANDLES, POCKET COMBS, AND WASHING MACHINE TUBS

Smaller machines of the type shown in Fig. I-17 turn out such items as toothbrush handles, pocket combs, and other plastic products. The large injection molding machine shown here produces larger items, such as washing machine tubs.

Granular plastic material is fed from the hopper (upper right) into a cylinder, where it is softened by heating. The huge dies are closed by the small cylinders above and below the large main ram shown in the center. As the dies close, the large ram's cylinder is prefilled with fluid from the reservoir and then locked closed with hydraulic force from a small holding pump. Flow from the main pump is then diverted to the injection-plunger cylinder (extreme right), which forces the plastic material into the die cavity with tremendous force. After curing, the dies are opened by means of the small auxiliary cylinders above and below the main ram.

A HYDRAULICALLY CONTROLLED GOLIATH

The high-pressure hydraulic forging press shown in Fig. I-18 is one of the largest self-contained presses in existence. It develops 2000 tons of force and provides high-speed operation with 20 hydraulic pumps, each having a capacity of about 35 gpm. The pumps are turned by 10 double-end shaft electric motors rated at 150 hp each. Figure I-19 shows the power units, which are separate from the press itself.

COMPARISON OF METHODS OF POWER TRANSMISSION

The three general methods of transmitting power are electrical, mechanical, and fluid power. *Transmitting power* refers to the time rate of doing work, and *work* may be defined as overcoming resistance through any distance. See Table I-1 for the components used in the various systems.

Electrical power uses nuclear energy, water turbines, or steam turbines to produce voltage and current. Voltage is the intensity factor, and current is the action factor.

FIGURE I-17 Plastic injection molding machine. (Courtesy of Mobil Oil Corporation.)

FIGURE I-18 2000-ton self-contained forging press. (Courtesy of Abex Corporation, Denison Division, Columbus, Ohio.)

Mechanical power uses gravity, heat, combustion, steam, electricity, or fluid power to produce an intensity factor, which is torque, and an action factor, speed.

Fluid power uses compressors, pumps, or intensifiers to produce an intensity factor, pressure, and an action factor, cubic feet of gas per minute or gallons of liquid per minute (cfm or gpm).

Horsepower for the three methods can be stated as follows:*

$$\text{Electrical horsepower} = \frac{\text{voltage} \times \text{current}}{746 \text{ watts}}$$

$$\text{Mechanical horsepower} = \frac{\text{torque} \times \text{speed}}{5252}$$

$$\text{Fluid power} \begin{cases} \text{Hydraulic horsepower} \\ = \dfrac{\text{pressure} \times \text{gpm}}{1714} \\ \text{Pneumatic horsepower} \\ = \dfrac{\text{pressure} \times \text{cfm}}{\text{constant}} \end{cases}$$

*The watt (W) is a unit of electrical power, and 746 W equals 1 hp; 5252 and 1714 are derived from a unit of one mechanical horsepower, which equals 33,000 ft-lb/min; the constant is derived from 33,000 ft-lb/min, but varies with temperature conditions.

TABLE I-1
COMPONENTS USED

	Electrical	Mechanical	Fluid
Distribution or transmission	Wires, cables	Wheels, shafts, gears, levers, cams, wedges, screw, pulleys, beams, belts, links, clutches, etc.	Pipes, tubes, hoses
Intensity control	Switches, transformers, regulators, relays, distributors	The form of parts transmitting torque depends on the space, weight, size, shape, etc.	Pressure valves, directional valves, flow control valves
Action device	Solenoids, motors	Wheels, pistons, levers, gears, linkage, etc.	Fluid motors, rams, and cylinders
Energy reserve	Batteries and condensers	Flywheels	Accumulators and receivers
Safety devices	Fuses, circuit breakers	Safety latches, strength of part	Relief valves, pressure switches, blowout disk

FIGURE I-19 Power units. (Courtesy of Abex Corporation, Denision Division, Columbus, Ohio.)

In each of the equations above, increasing either the intensity factor or the action factor can vary the resulting power.

MATHEMATICS AND PRINCIPLES OF PHYSICS SHOULD BE REVIEWED

The fluid power technology is somewhat related to all systems of power transmission, because it is flexible to apply and compatible with other means of control. It is therefore essential that students of fluid power understand the general laws of mechanics and the units of measurements involving forces, lengths or distances, and time.

1. *Force:* measured in pounds, or tons.
2. *Length:* measured in inches, feet, yards, rods, miles, etc.
3. *Time:* measured in seconds, minutes, hours, etc.

Other quantities can be expressed in terms of these three. For example, velocity is defined as distance (length) divided by time. Thus velocity may be miles/hour, feet/minute, feet/second, inches/day, and so on. Mathematically stated, the equation reads

$$V = \frac{d}{t}$$

where V = velocity

d = distance

t = time

Work and energy have the same units.

$$w = \text{energy} = F \times d$$

where w = work

F = force on the load in travel direction

d = distance the load moves

Notice that if the load does not move when a force is applied, $d = 0$, $F \times 0 = 0$, so no work is done. In order to have work done, the load must move (see Fig. I-20).

Power is expressed as

$$\frac{w}{t} = \frac{\text{work}}{\text{time}} = \frac{F \times d}{t}$$

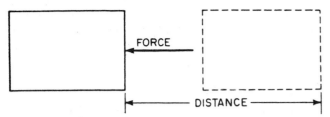

FIGURE I-20 Work = force × distance.

Notice that d/t = velocity. We could write the power equation as $P = FV$. The units of power are most commonly ft-lb/min or ft-lb/sec.

Before the internal combustion engine and electric motor were widely used, a test was performed with a horse as the power source. It was found that a good draft horse could do 33,000 ft-lb of work in 1 minute or 550 ft-lb of work in 1 second—thus originated the term *horsepower*.

$$1 \text{ hp (power)} = 33,000 \frac{\text{ft-lb}}{\text{min}} = 550 \frac{\text{ft-lb}}{\text{sec}}$$

(see Fig. I-21).

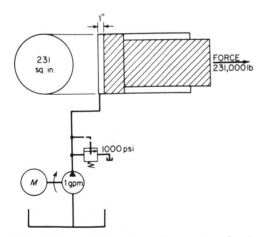

FIGURE I-21 Derivation of equation for hp formula (time of travel of 1 in. is necessary.)

EXAMPLE

A worker unloads steel stock from a flatbed to a cart, and raises the load through a distance of 4 ft. He unloads 1 ton of steel in 1 hour.

$$\text{Work done} = F \times d = 2000 \text{ lb} \times 4 \text{ ft} = 8000 \text{ ft-lb}$$

$$\text{Power used} = \frac{w}{t} = \frac{F \times d}{t} = \frac{2000 \text{ lb} \times 4 \text{ ft}}{1 \text{ hr}}$$

$$= 8000 \frac{\text{ft-lb}}{\text{hr}}$$

$$\text{Power used} = \frac{P}{33,000 \frac{\text{ft-lb}}{\text{min}}}$$

$$= \frac{8000 \frac{\text{ft-lb}}{\text{hr}}}{33,000 \frac{\text{ft-lb}}{\text{min}}}$$

To obtain the unit of power in horsepower, change hours to minutes:

$$\frac{8000 \frac{\text{ft-lb}}{\text{hr}} \times \frac{1 \text{ hr}}{60 \text{ min}}}{33,000 \frac{\text{ft-lb}}{\text{min}}} = \frac{\frac{8000}{60} \frac{\text{ft-lb}}{\text{min}}}{33,000 \frac{\text{ft-lb}}{\text{min}}}$$

$$= \frac{8000/60}{33,000} \text{ hp} = 0.004 \text{ hp}$$

We have just seen that raising an object through a vertical distance requires work to be performed on it. Oil has weight; therefore, to move it, work must be done. When work is done, power is consumed.

$$P = \frac{F \times d}{t}$$

The amount of power consumed depends on the quantity of oil, the force required to move it, and the time.

$$P = \frac{\text{gal}}{\text{min}} \times \frac{\text{lb}}{\text{in.}^2}$$

Changing gal to in.³ and in. to ft to get units of power, we have

$$P = \frac{\text{gal}}{\text{min}} \left(\frac{231 \text{ in.}^3}{\text{gal}} \right) \times \frac{\text{lb}}{\text{in.}^2} \left(\frac{1 \text{ ft}}{12 \text{ in.}} \right)$$

$$= \frac{231 \text{ ft-lb}}{12 \text{ min}}$$

$$\text{hp} = \frac{\frac{231 \text{ ft-lb}}{12 \text{ min}}}{33,000 \frac{\text{ft-lb}}{\text{min}}} = 0.000583 \text{ hp}$$

The hydraulic horsepower formula becomes

$$\text{hp} = \text{gpm} \times \text{psi} \times 0.000583$$

EXAMPLE

Oil being moved in a system is 12 gpm at a pressure of 1500 psi. What horsepower is being consumed?

Solution:

$$hp = gpm \times psi \times 0.000583$$

$$= 12 \times 1500 \times 0.000583 = 10.5 \text{ hp}$$

Another useful measurement is torque, which is used in relation to turning or rotary motion. Torque is measured in lb-ft or lb-in. Refer to Fig. I-22:

$$\text{Torque} = \text{force} \times \text{lever arm or radius}$$

$$T = F \times d$$

$$P = \frac{2\pi RTN}{33,000 \times 12}$$

where N is measured in revolutions per minute. Clearing the fraction gives

$$hp \times 33,000 \times 12 = 2\pi RTN$$

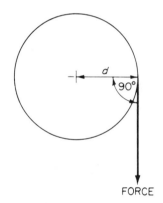

FIGURE I-22   Torque = force × distance.

Equating for unity, we have

$$1 \times 33,000 \times 12 = 2 \times 3.1416 \times 1 \times T \times 1$$

$$T = \frac{33,000 \times 12 \times 1}{2 \times 3.1416 \times 1 \times 1} = \frac{396,000 \times 1}{62,832 \times 1}$$

$$= \frac{63,025 \times hp}{1}$$

Formula for torque:

$$T = \frac{63,025 \times hp}{rpm}$$

Notice in Fig. I-22 that the factors F and d are the same as for the work formula ($W = F \times d$). The difference is that the d for the work formula is the distance the load moves while the force is being applied, whereas the d

in the torque formula represents the length of the lever arm on which the force is acting. To have work done, there must be motion. Torque requires only that a force be applied to *try* to cause a turning effect.

EXAMPLE

A mechanic tries unsuccessfully to loosen a nut. He exerts a force of 70 lb on the wrench handle, which is 18 in. long. What torque was developed?

Solution:

$$T = F \times d = 70 \text{ lb} \times 18 \text{ in.}$$

$$= 1260 \text{ lb-in.}$$

or, change inches to feet:

$$= 1260 \text{ lb-in.} \left(\frac{1\text{ft}}{12\text{in.}}\right)$$

$$= 105 \text{ lb-ft}$$

It is not possible to get out of a machine all the energy put into it, because of friction of moving parts, heat losses, leakage, and other factors. If 100 units of energy are put into a machine and 75 units of energy are available from the machine, it is 75% efficient.
Mathematically,

$$\frac{w_o}{w_i} \times 100\% = \text{efficiency}$$

where w_o = work output

w_i = work input

Also, $w_i = w_o + w_l$, where w_l is the work lost due to heat, friction, and so on. Because power is a function of work, we can also say:

$$\frac{P_o}{P_i} \times 100\% = \% \text{ efficiency}$$

where P_o = power output

P_i = power input

$P_i = P_o + P_l$

where P is the power lost.

A review of the basic rules of algebra and trigonometry will be helpful toward a better understanding of applying fluid power.

Motion accomplished by fluid power may be linear or nonlinear. This means that loads may move in a straight line in relation to an actuator, or the work may take a circulator or irregular path of travel. When it is essential to compute the effect of loads moving on an inclined plane or when levers and linkage are used, trigonometry is a useful tool.

BASIC ALGEBRAIC FORMULAS

A plus 0 equals A.

$A \times 1$ equals A.

A plus B equals B plus A.

$A \times B$ equals $B \times A$.

A plus (B plus C) equals (A plus B) plus C.

$A (BC)$ equals $(AB) C$.

$A (B$ plus $C)$ equals AB plus AC.

A^n equals $A \times A \times A \times A \times \cdots \times A$ n times.

If $^n\sqrt{A}$ equals Y, then Y^n equals A.

A^{-n} equals I/A_n; $A^{m/n}$ equals $^n\sqrt{A^m}$.

$A^m \times A^n$ equals $A^{m \text{ plus } n}$; $(A^m)^n$ equals A^{mn}.

$A^n \times B^n$ equals $(AB)^n$.

A/B plus C/D equals DA/DB plus BC/DB equals $\dfrac{DA \text{ plus } BC}{DB}$.

AB/AC equals B/C.

BASIC TRIGONOMETRIC FORMULAS (see Fig. I-23)

$\sin \phi = a/c$ opposite side/hypotenuse

$\cos \phi = b/c$ adjacent side/hypotenuse

$\tan \phi = a/b$ opposite side/adjacent side

$\sec \phi = c/b$ hypotenuse/adjacent side

$\csc \phi = c/a$ hypotenuse/opposite side

$\cot \phi = b/a$ adjacent side/opposite side

Reciprocal relations:

$\sin = 1/\csc$

$\cos = 1/\sec$

$\tan = 1/\cot$

$\sec = 1/\cos$

$\csc = 1/\sin$

$\cot = 1/\tan$

Angle	sin	cos	tan	sec	csc	cot
45°	$1/\sqrt{2}$	$1/\sqrt{2}$	1	$\sqrt{2}$	$\sqrt{2}$	1
30°	$1/2$	$\sqrt{3}/2$	$1/\sqrt{3}$	$2/\sqrt{3}$	2	$\sqrt{3}$
60°	$\sqrt{3}/2$	$1/2$	$\sqrt{3}$	2	$2/\sqrt{3}$	$1/\sqrt{3}$

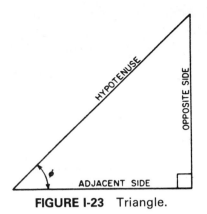

FIGURE I-23 Triangle.

EXAMPLE

Find the amount of force necessary to move 7000 lb from a standing stop up an incline of 15° (0.04 coefficient of friction) (see Fig. I-24).

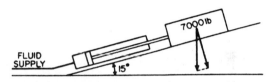

FIGURE I-24 Diagram for example.

Solution:

Force = (weight $\times \sin \theta$) + (friction $\times$ weight

     $\times \cos \theta$)

     = $(7000 \times 0.259) + (0.04 \times 7000 \times 0.966)$

     = 2083.48 lb

Many good textbooks are available for a review of basic mathematics and laws of physics. The foregoing discussion merely highlights some of the important tools helpful in the field of fluid power.

QUESTIONS

1. What is the basic law that is important in applying fluid power, when was it discovered, and by whom?

2. What theorem pertains to velocity energy, pressure energy, and friction energy?

3. How is fluid power used by the military?

4. List 10 fields of application where fluid power can be used more effectively than other power sources.

5. Compare the use of fluid power to a mechanical system, listing the advantages and disadvantages of each.

6. Explain in your own words how the hoist works to lift your car for servicing in a gas station.

7. What advantage of using fluid power is indicated by the application of raising utility poles?

8. What advantages does fluid power give to the design of farm machinery?

9. Generalize on the difference between using pneumatic fluid power and using hydraulic fluid power.

10. What effect has fluid power had on automation?

11. Why is hydraulic power especially useful with heavy work?

12. Express horsepower mathematically for the three general methods of transmitting power.

1

Basic Principles of Hydraulics

The term *hydraulics* as treated in this chapter pertains to *power transmitted and controlled through the use of a pressurized liquid. Because liquids and gases are both fluids,* it should be understood that *fluid power* is the general term for both hydraulics and pneumatics. According to general dictionary definitions, hydraulics is the branch of physics having to do with the mechanical properties of water and other liquids and the application of these properties in engineering.

Modern hydraulic power systems generally use petroleum oils or other liquids that provide automatic lubrication for all integral moving parts of the components making up the system. Hydraulic systems may use water, petroleum oils, synthetic fluids, blends and mixtures, or even liquid metals for the media in transmitting fluid power. All liquids used for hydraulic power have certain inherent physical properties which are important for an understanding of hydraulic machines (see Fig. 1-1).

Liquids have characteristics similar to those of solids, except that liquids can be pumped and transmitted through pipes and tubes to perform the work. Actually, liquids possess qualities that are common to *both* gases and solids and may be thought of as "fluid solids" from the viewpoint of transmitting fluid power. A comparison of gases, liquids, and solids and their molecular structure is shown in Fig. 1-2.

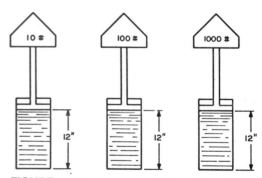

FIGURE 1-1 Incompressibility of liquids.

The molecules in gases possess more kinetic energy (the energy of movement) because they are constantly in motion at high velocity. The gas molecules must be compressed together more closely before forces are transmitted. The transmittal of these forces is in all directions.

The molecules in liquids possess less kinetic energy than those in gases. They are more closely packed, like the molecules in solids. The liquid molecules transmit forces with quick response because they are almost incompressible. The transmittal of these forces is in all directions.

The molecules in solids possess less kinetic energy than those in liquids. They stay in a closely packed condition, and hence act in unison to transmit a force in *only one direction.*

The illustrations show the main advantage of using

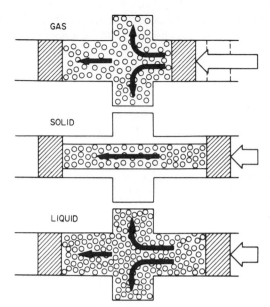

FIGURE 1-2 Molecular structure of gases, liquids, and solids for comparison.

FIGURE 1-3 Hydraulic press. (Courtesy of Abex Corporation, Denison Division, Columbus, Ohio.)

fluid power over mechanical systems—the ability of fluids (either gases or liquids) to transmit forces in all directions. Thus it is apparent that fluids can be transmitted through pipes, tubes, and controls to end up as energy used to perform useful work.

Molecules in liquids are flexible but rigid enough to maintain their fixed distance between each other, except when acted upon by extremely large external forces. A large metal press is an example of a hydraulic machine where the compressibility factor is important. Big presses such as that shown in Fig. 1-3 have large quantities of liquid under pressures of 5000 psi or higher. This makes it imperative to decompress the liquid before releasing the press.

The accepted compressibility factor for most hydraulic power systems is one-half of 1% of the liquid volume under pressure for each 1000 psi. This is illustrated in Fig. 1-4. Removing the weight and reducing the pressure will cause the oil volume to return to its original value almost instantly.

When liquids are subjected only to the force of gravity, they seek their own level regardless of the shape of the container (see Fig. 1-5). This important physical property allows hydraulic fluid to fill the cavities in all components of a system, regardless of their shape or size.

Fluid density is defined as mass per unit volume. It is, however, sometimes more convenient to use the concept of specific weight.

Thus, specific weight $= \dfrac{\text{weight}}{\text{volume}}$ or $\rho = \dfrac{W}{V}$

The density of water $= 62.4$ lb/ft^3. In other words, if a cubic foot of water at 39°F or 4°C were accurately weighed with the weight of the container subtracted, it would

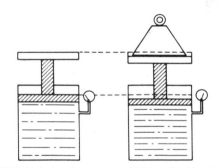

FIGURE 1-4 Compressibility of a liquid.

FIGURE 1-5 Various container shapes.

weigh 62.4 lb. Its specific weight would be expressed as 62.4 lb/ft^3.

EXAMPLE

What is the specific weight of a quart of hydraulic oil weighing 1.5 lb and occupying a volume of 57.8 in.^{3}?

Solution:

$$\rho = \frac{W}{V} = \frac{1.5 \text{ lb}}{57.8 \text{ in.}^3} = 0.0259 \text{ lb/in.}^3$$

Changing in.^{3} to ft^3 gives us

$$\frac{0.0259 \text{ lb}}{\text{in.}^3} \left(\frac{1728 \text{ in.}^3}{\text{ft}^3} \right) = \frac{44.8 \text{ lb}}{\text{ft}^3}$$

The specific gravity (sp. gr.) is a comparison of the weight of an oil or other liquid to the weight of an equal volume of water. The specific gravity of water as a standard equals 1.

$$\text{sp. gr.} = \frac{\text{weight of liquid}}{\text{weight of equal volume of water}}$$

$$= \frac{\rho \text{ of liquid}}{\rho \text{ of water}}$$

EXAMPLE

If a cubic foot of hydraulic oil weighs 45 lb, what is its specific gravity?

Solution:

$$\text{sp. gr.} = \frac{\rho \text{ of liquid}}{\rho \text{ of water}}$$

$$= \frac{45 \text{ lb/ft}^3}{62.4 \text{ lb/ft}^3} = 0.723$$

Notice that all the units cancel out, making specific gravity a dimensionless number. The oil in this case has a specific gravity of 0.723 compared to that of water, which is equal to 1.

Knowing the specific gravity of any hydraulic fluid, one can easily compute the weight or density.

$$\text{Specific weight (lb/ft}^3) = \text{sp. gr.} \times 62.4 \text{ lb/ft}^3$$

EXAMPLE

If a hydraulic oil has a specific gravity of 0.90, what is its specific gravity in lb/ft^3?

Solution:

$$\text{specific gravity (lb/ft}^3) = 0.90 \times 62.4 \text{ lb/ft}^3$$

$$\rho = 56.16 \text{ lb/ft}^3$$

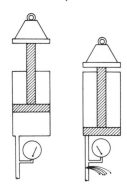

FIGURE 1-6 Fluid taking path of least resistance (power peak).

When a force is applied to the confined liquid of a hydraulic system, the liquid will take the flow path of least resistance. Figure 1-6 shows that input power is lost when leakage occurs in the hydraulic system piping.

RELATIONSHIP OF FORCE AND PRESSURE IN A HYDRAULIC SYSTEM

Knowing the difference between *force* and *pressure* is one of the most important steps toward an understanding of how hydraulic systems work. Hydraulic cylinders are used to create a force that can push or pull a load, thus providing linear motion. Hydraulic motors are used to create a force in a circular path, thus providing rotary motion (see Fig. 1-7).

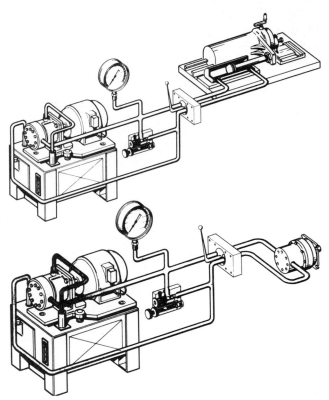

FIGURE 1-7 Linear and rotary motion.

In both linear and rotary motion the forces created may be measured in pounds or tons. The unit of force in each case is associated with the work being done by the cylinder or hydraulic motor.

It should also be understood that the *output forces are being transmitted through the use of a pressurized liquid* which is being supplied to the cylinder and motor by the hydraulic pump.

The principle of all hydraulic systems is based on *Pascal's law,* which states: "Pressure exerted on a confined liquid is transmitted undiminished in all directions and acts with equal force on all equal areas" (see Fig. 1-8).

Hydraulic liquids transmit an applied force or pressure

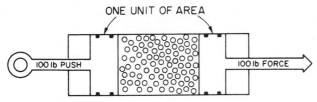

FIGURE 1-8   Principle of Pascal's law.

undiminished. A force applied on one unit of area of a confined liquid will be transmitted equally and undiminished to each unit of area of the confining container regardless of the shape of the container. Figure 1-8 shows how an applied force can be transmitted through the use of hydraulic fluid confined between two movable pistons.

In Fig. 1-9 the area of the second piston has been increased to twice the area of the first piston. With the fluid contained between the two movable pistons, the same 100 lb of applied force is now doubled. Through the application of this basic principle, a small child can easily lift an automobile with a hydraulic jack.

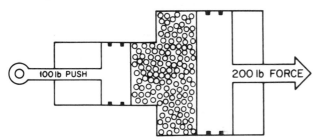

FIGURE 1-9   Pascal's law illustrated with two pistons.

It is apparent from Fig. 1-10 that the greater the area acting on a piston, the more force the piston will exert with the same applied force. In this simple system, each unit of area in the vessel is under an applied force of 100 lb. This could also be stated as: the pressure in the system is 100 lb per unit of area. Thus *pressure* may be defined as force per unit of area. Mathematically, pressure can be expressed as

$$\text{Pressure} = \frac{\text{force}}{\text{area}} \quad \text{or} \quad P = \frac{F}{A}$$

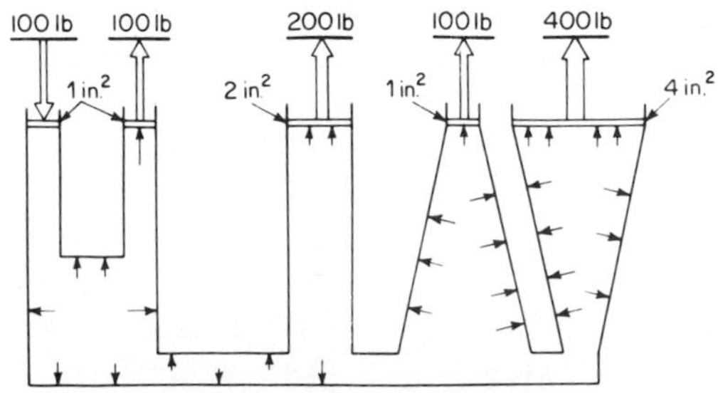

FIGURE 1-10   Four cylinders.

The units of area may be square inches or square feet. Generally, square inches are used for most calculations concerning fluid power.

The units of force are pounds or tons, with pounds being the most commonly used. Stated again, *pressure* (pounds per square inch, or psi) is equal to *force* (pounds) *divided* by *area* (square inches), or $P = F/A$.

This is the most important law in the application of fluid power. It is necessary at all times to distinguish between force and pressure. The following explanation will further emphasize this important difference.

A cubic foot of water (see Fig. 1-11) weighs 62.4 lb. It could be stated that a cubic foot of water exerts a force of 62.4 lb on the bottom of the container. This could also be stated as: the weight of 1 cubic foot of water exerts a pressure of 62.4 lb per square foot. To find pressure in pounds per square inch, it is necessary to divide the number of square inches in the base ($12 \times 12 = 144$) into the total weight or force.

$$P = \frac{\text{force}}{\text{area}} = \frac{62.4 \text{ lb}}{144 \text{ in.}^2}$$

Thus

$$\text{Pressure (psi)} = 0.433 \frac{\text{lb}}{\text{in.}^2}$$

To calculate pressure, the force, either applied or transmitted, must be known, and the area on which the force is acting must also be known.

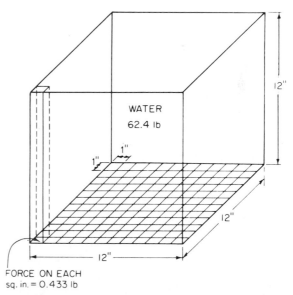

FIGURE 1-11   Cubic feet of water including 1-in.² column.

EXAMPLE 1

A force of 200 lb is being exerted by a hydraulic cylinder (see Fig. 1-12). The area of the piston is 2 in.². What is the pressure in the system?

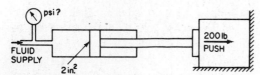

FIGURE 1-12   Force diagram for Example 1.

Solution: Known: Force = 200 lb, area = 2 in.²

Find: Pressure (psi)

The formula states that $P = F/A$. When the known values are substituted into the formula, it becomes

$$P = \frac{F}{A} = \frac{200 \text{ lb}}{2 \text{ in.}^2}$$

Therefore, the pressure in the system is 200 lb per 2 in.² or 100 lb per square inch (psi).

EXAMPLE 2

A hydraulic cylinder (see Fig. 1-13) exerts a force of 500 lb. If the cylinder has a bore of 3 in., what is the pressure in the system?

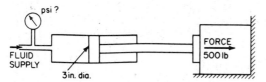

FIGURE 1-13   Force diagram for Example 2.

Solution: To find the pressure, use the formula

$$P = \frac{F}{A}$$

The force is given as 500 lb, but the area is not given. However, the diameter of the bore (also the diameter of the piston) is given as 3 in. From geometry, $A = \pi r^2$. This formula will yield the area. However, one must divide the diameter by 2 in order to use the formula. In hydraulics the diameter, or bore size, is most easily found, so it would be advantageous to have a formula for area expressed in terms of the diameter:

$$A = \pi r^2 \quad [r \text{ (radius) is equal to } \tfrac{1}{2} \text{ of the diameter}]$$

$$r = \frac{D}{2}$$

When writing the formula, instead of putting in r, put in what r is equal to, namely $D/2$. The formula now becomes

$$A = \pi \left(\frac{D}{2}\right)^2 = \frac{\pi D^2}{4}$$

and π is a constant whose value is 3.1416, so

$$A = \frac{3.1416 \, D^2}{4} = 0.7854 \, D^2$$

This provides a formula for area expressed in terms of the diameter:

$$A = 0.7854 \, D^2 = 0.7854(3 \text{ in.})^2$$

$$= 0.7854(9 \text{ in.}^2) = 7.0686 \text{ in.}^2$$

$$P = \frac{F}{A} = \frac{500 \text{ lb}}{7.0686 \text{ in.}^2} = 70.74 \frac{\text{lb}}{\text{in.}^2}$$

Because it is a mathematical expression, the equation for pressure can be stated in other forms:

$$P = \frac{F}{A} \qquad F = PA \qquad \text{and} \qquad A = \frac{F}{P}$$

EXAMPLE 3

If it were required to find the area of a cylinder exerting 200 lb of force under a pressure of 100 psi (see Fig. 1-14), the formula $A = F/P$ could be used.

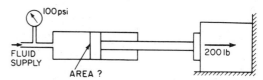

FIGURE 1-14   Diagram for Example 3.

Solution:

$$A = \frac{F}{P} = \frac{200 \text{ lb}}{100 \text{ lb/in.}^2} = \frac{2}{1/\text{in.}^2} = 2 \text{ in.}^2$$

EXAMPLE 4

Given a cylinder with a piston area of 2 in.² operating at a pressure of 200 psi (see Fig. 1-15), what force could the cylinder exert?

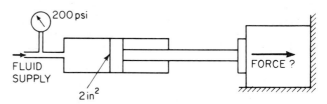

FIGURE 1-15   Diagram for Example 4.

Solution:

$$F = PA$$

$$= 200 \frac{\text{lb}}{\text{in.}^2}(2 \text{ in.}^2) = 400 \text{ lb}$$

In using Pascal's law, it is important to remember the units of each expression:

$$\text{Force} = F = \text{lb}$$

$$\text{Pressure} = P = \text{lb/in.}^2$$

$$\text{Area} = A = \text{in.}^2$$

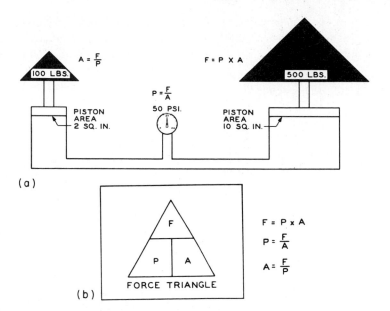

FIGURE 1-16 Force triangles.

It may be an aid in remembering the three forms of Pascal's law to use the *force triangle method* illustrated in Fig. 1-16. To use it, simply put your thumb on the factor you wish to find. Do the indicated operation with the remaining factors. For example, to find pressure, cover *P*. The remaining factors appear as *F/A,* so divide the force by the area, and pressure will be the result. To find force, cover *F*. The remaining factors appear as *PA,* so multiply pressure by area to obtain the force.

Figure 1-17 uses the following symbols: EM, the electric motor which drives the pump; P, the hydraulic pump which delivers the fluid to the system; T, the tank for storing, supplying, and conditioning the fluid; R, the relief valve to protect the system from overpressure; G, pressure gauges to determine psi at different points of the system; C, actuating cylinders which perform linear motion; and S, shutoff valve. Study the diagram with the following questions in mind:

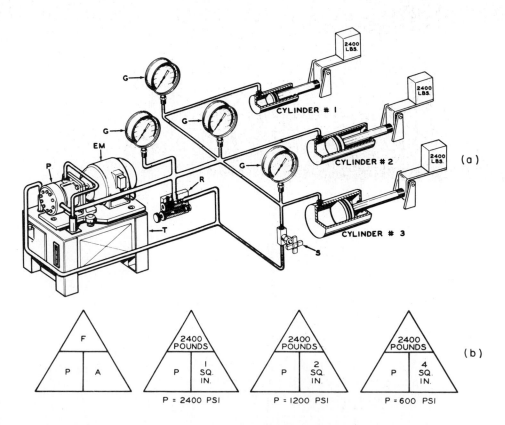

FIGURE 1-17 Three cylinders.

1. With the pump in operation, which cylinder extends first, and at what pressure? Which extends second? Which extends third?

2. What would happen if valve S were opened before the cylinders were fully extended?

There is a great deal of hydraulic knowledge that can be obtained from this illustration, so study it carefully.

To answer the first question, the force triangle (Fig. 1-17b) provides the relationship. (Disregard any frictional losses.) It is easy to see that when a pressure of 600 psi is attained, cylinder 3 will extend. Cylinders 2 and 3 will follow in turn at pressures of 1200 and 2400 psi, respectively.

If valve S were opened before cylinder 1 was fully extended, pressure energy would be lost, because the fluid would follow the path of least resistance. Consequently, the work being done by the cylinder would stop.

Hydraulic fluids under pressure can also be used to multiply forces, as shown in Fig. 1-18.

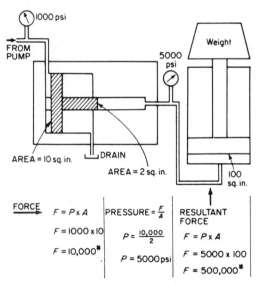

FIGURE 1-18   Intensification.

Force: $F = P \times A$ $= \dfrac{F}{A}$

$= 1000 \times 10 \text{ in.}^2$ $= \dfrac{10,000 \text{ lb}}{2 \text{ in.}^2}$

$= 10,000 \text{ lb}$ $= 5000 \text{ psi}$

Resultant force: $F = P \times A$
$= 5000 \times 100 \text{ in.}^2$
$= 500,000 \text{ lb}$

Figure 1-18 shows that 1000 psi working against an area of 10 in.² produces a force of 10,000 lb. Using this force of 10,000 lb against a smaller area of 2 in.², we obtain

a resultant pressure of 5000 psi. Using the 5000 psi in turn against a large area of 100 in.² produces a resulting force of 500,000 lb (250 tons).

This example shows the important differences between force and pressure. Further, it points up the fact that through the use of hydraulics a force of any magnitude may be obtained.

IMPORTANT CONCEPTS AND PRINCIPLES AFFECTING FLOW IN HYDRAULIC SYSTEMS

Pressure energy in a hydraulic system is a form of potential energy and is therefore essentially static in character. This form of energy is referred to as *hydrostatic pressure* or *pressure head*. Hydrostatic pressure is the pressure energy in a liquid that is at rest or relatively motionless. This pressure energy may be developed by the weight of the liquid under pull of gravity, or it may be caused by the action of a hydraulic pump or other mechanical device.

Potential energy = weight × height

EXAMPLE

What is the pressure at the bottom of a tank 40 ft high filled with oil having a specific gravity of 0.93? (See Fig. 1-19a.)

Solution:

$$\text{sp. gr.} = \frac{\text{weight (oil)}}{\text{weight (water)}}$$

$$0.93 = \frac{x}{62.4 \text{ lb/ft}^3}$$

$$W \text{ (oil)} = \text{sp. gr.} \times W \text{ (water)}$$

$$= 0.93 \times 62.4 \text{ lb/ft}^3$$

$$= 58.03 \text{ lb/ft}^3$$

Since a cubic foot of the oil in question weighs 58.03 lb, the pressure developed by 1 ft of height would be calculated as follows:

$$\text{Pressure (psi)} = \frac{\text{lb force}}{\text{area (in.}^2)}$$

The number of square inches in the base of the cube in Fig. 1-19b is 12 × 12 = 144 in.². Dividing this area into the force due to the weight of the oil equals 0.4 psi.

$$P = \frac{58.03 \text{ lb/ft}^3}{144 \text{ in.}^2} = 0.403$$

$$= 0.4 \text{ psi}$$

Now that the pressure has been found for 1 ft of height, the total pressure head (P_h) can be found by multiplying 0.4 psi × total height.

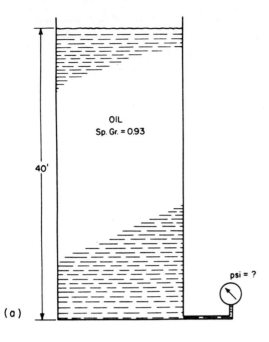

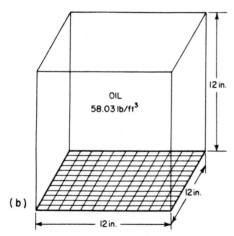

FIGURE 1-19 (a) Specific gravity; (b) cubic feet of oil.

P_h = height × 0.4 psi

= 40 × 0.4 psi

= 16 psi (pressure due to head of liquid if height is 40 ft)

The pressure energy due to the height of the oil in the tank could now be used to perform work. Suppose that the cylinder illustrated in Fig. 1-20 were fitted with a movable piston having 24 in.². How many pounds of force can be exerted with 16 psi?

$$F = P \times A$$

$$= 16 \times 24$$

$$= 384 \text{ lb}$$

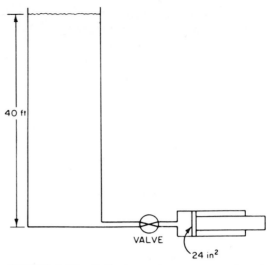

FIGURE 1-20 Cylinder using pressure head.

However, it should be pointed out that the height of the oil in the tank would have to be maintained for the full stroke of the piston. In addition, there could be no losses due to friction or other factors if the work done per foot movement of piston is 384 ft-lb. (Work = force × distance.)

It is apparent from this discussion that a power pump would be much more efficient to maintain the pressure energy to do the work. A power pump would also take care of any losses that occur, such as those caused by friction, if the proper pump is selected to do the job.

There is another phase of potential energy that is important to the design of hydraulic machinery. Weighted members of any machine that may supply energy because of their relative position must also be considered. A vertical press, whose platen and dies may weigh several hundred pounds, is an example.

Conversion formulas:

$$\text{Pressure (psi)} = \frac{\text{head (ft)} \times \text{sp. gr.}}{2.31}$$

$$\text{Head (ft)} = \frac{\text{pressure (psi)} \times 2.31}{\text{sp. gr.}}$$

$$\text{Inches of mercury vacuum} = \text{suction lift (ft)}$$
$$\times \ 0.883 \times \text{sp. gr.}$$

Kinetic energy is another form of energy that involves hydraulic machinery in motion. Kinetic energy of an incompressible mass is given by the following equation:

$$\text{Kinetic energy} = \frac{\text{weight} \times \text{velocity}^2}{2 \times \text{gravity (32 ft/sec}^2)}$$

Hydraulic fluid in motion is a result of pressure energy's being transformed into kinetic energy. It has also been discovered that many factors enter into the design of hydraulic systems because of the behavior of fluid in motion.

BERNOULLI'S EQUATION

In general, a study by Daniel Bernoulli of fluids in motion disclosed that when fluids flow through pipes or tubes *under steady-state conditions,* where the velocity is high, the pressure decreases; where the velocity is low, the pressure increases. This concept considers a perfect, frictionless fluid and is illustrated in Fig. 1-21. If no work is done on or by a flowing frictionless fluid, its energy due to pressure and velocity remains constant at all points along the streamline.

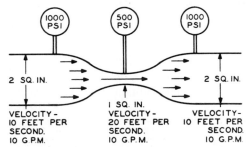

FIGURE 1-21   Bernoulli's equation.

Assume that the pump is delivering 10 gpm through the tube, and the pressure is 1000 psi. Also assume the fluid to have a velocity of 10 ft/sec. When the fluid enters the narrow section of the tube, where the area is only one-half that of the larger section, the velocity increases. This happens because the 10 gallons of oil per minute must get through this narrow section. In other words, the same number of molecules of oil must pass through an area only one-half as large. Thus the fluid must travel twice as fast in order to get all the molecules of the oil through this narrow section in the same increment of time. If there are no frictional factors involved, the velocity of flow doubles in the narrow section and the pressure energy decreases by one-half.

Steady-state flow means that the energy contained by the liquid remains constant throughout the tube and that as velocity increases, pressure decreases, or as velocity decreases, pressure increases.

EXAMPLE

The reservoir for a hydraulic system shown in Fig. 1-22 is air supercharged at 15 psi gauge. The inlet line to the pump is 6 ft below the oil level. The flow into the inlet port of the pump (point 2) is 30 gpm. Calculate the pressure at point 2 if:
  (a) It is assumed that there is no frictional head loss from point 1 to point 2.
  (b) There is a 20-ft frictional head loss from point 1 to point 2. Assume that $\rho_1 = \rho_2 = 53.1$ lb/ft^3.

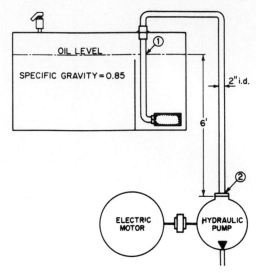

FIGURE 1-22   Bernoulli's theorem, Example 1.

Note: We have previously used the symbol ρ to designate specific weight. We referred to it as density. The use of this unit is not scientific.

Solution: *Bernoulli's law*—If no work is done on or by a flowing frictionless liquid, its energy due to pressure and velocity remains constant at all points along the streamline. In equation form,

$$Z_1 + \frac{144P_1}{\rho_1} + \frac{V_1^2}{2g} = Z_2 + \frac{144P_2}{\rho_2} + \frac{V_2^2}{2g}$$

where Z = head or reference level above an arbitrary
        base reference level (ft)

   P = pressure of fluid (psi)

   ρ = density of fluid (lb/ft^3)

   V = mean velocity of flow (ft/sec)

   g = acceleration due to gravity (32.2 ft/sec^2)

(a) $Z_1 + \dfrac{144P_1}{\rho_1} + \dfrac{V_1^2}{2g} = Z_2 + \dfrac{144P_2}{\rho_2} + \dfrac{V_2^2}{2g}$

$$6 + \frac{(144)(15)}{53.1} + 0 = 0 + \frac{(144)P_2}{53.1} + \frac{(3.07)^2}{(2)(32.2)}$$

$$6 + 40.7 = 2.715P_2 + 0.1463$$

$$\frac{46.6}{2.715} = P_2$$

$$17.15 \text{ psig} = P_2$$

(b) $\dfrac{26.554}{2.715} = P_2$

$$9.78 \text{ psig} = P_2$$

CONTINUITY EQUATION

Matter such as hydraulic oil cannot be created or destroyed, and since it is incompressible the mass rate of flow into any fixed space is equal to the mass flow out. Thus the mass flow rate of fluid past all cross sections of a tube is equal.

EXAMPLE

In Fig. 1-23, the mean oil velocity at point 1 = 5 ft/sec. Determine the minimum permissible diameter at D_2 that will limit the oil velocity to 15 ft/sec. Assume that the fluid is noncompressible.

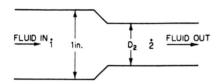

FIGURE 1-23 Continuity equation, Example 1.

Solution: *Continuity equation.* In steady flow, the mass of fluid passing all cross sections of a tube per unit time is the same.

$$\rho_1 A_1 V_1 = \rho_2 A_2 V_2$$

where ρ = density of fluid (slug/ft^3)

A = area (ft^2)

V = velocity (ft/sec)

$$(\rho_1)(0.785)(5) = (\rho_2)(0.785\, D^2)(15)$$

$$D^2 = \frac{(0.785)(5)}{(0.785)(15)} = \frac{1}{3}$$

$$D = 0.577 \text{ in.}$$

Assume that $\rho_1 = \rho_2$. Since

$$\text{Quantity 1} = \text{quantity 2}$$

then

$$\text{Velocity}_1 \times \text{area}_1 = \text{velocity}_2 \times \text{area}_2$$

EXAMPLE 2

Consider Fig. 1-24, where 231 in.^{3} of oil enters the cylinder every second. When the space is completely filled with oil the piston moves out, and in 3 seconds 693 in.^{3} of oil will enter the cylinder. The distance the cylinder moves is found by the following:

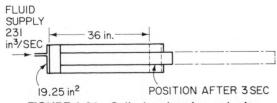

19.25 in^2 POSITION AFTER 3 SEC
FIGURE 1-24 Cylinder showing velocity.

$$\text{Distance} = \frac{\text{volume}}{\text{piston area}} = \frac{693 \text{ in.}^3}{19.25 \text{ in.}^2} = 36 \text{ in.}$$

Since the oil is entering at a constant rate, the speed with which the cylinder piston moves is given as follows:

$$\text{Velocity} = \frac{\text{distance}}{\text{time}} = \frac{36}{3} = 12 \text{ in./sec}$$

The general flow equation gives the answer more easily:

$$Q = AV$$

where Q = rate of flow (in.^{3}/sec)

A = area of cylinder (in.^{2})

V = velocity (in./sec)

Transposing the equation, we have

$$V = \frac{Q}{A}$$

$$= \frac{231 \text{ in.}^3/\text{sec}}{19.25 \text{ in.}^2}$$

$$= 12 \text{ in./sec}$$

TORICELLI'S THEOREM

The liquid velocity at an outlet discharging into the free atmosphere is proportional to the square root of the head (see Fig. 1-25).

$$V = \sqrt{2gh}$$

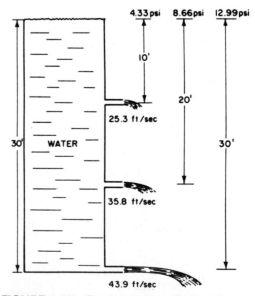

FIGURE 1-25 Tank showing Toricelli's theorem.

where V = velocity (ft/sec)

g = acceleration of gravity (32 ft/sec^2)

h = head (ft)

Total pressure head = 0.433 psi/h (ft)

= 0.433 psi × 30 ft

(0.433 psi for water)

= 13.0 psi

10 ft	20 ft	30 ft

$V = \sqrt{64 \times 10}$ $V = \sqrt{64 \times 20}$ $V = \sqrt{64 \times 30}$

$= \sqrt{640}$ $= \sqrt{1280}$ $= \sqrt{1920}$

$= 25.3$ ft/sec $= 35.8$ ft/sec $= 43.8$ ft/sec

LAMINAR AND TURBULENT FLOW

Conditions due to the construction or operation of hydraulic machinery may cause irregular flow characteristics of the fluid in the system. In addition, the type of fluid used, the size of the tubes or pipes, the smoothness of the internal surfaces of the tubes and pipes, the temperature of the fluid, and the kinds of restrictions, such as valves, fittings, and other components, all have a bearing on the nature of flow within a system.

Laminar flow or streamline flow occurs when the liquid particles flow smoothly in even layers and frictional *losses are at a minimum*. Notice that in Fig. 1-26, which shows laminar flow, there is a parabolic distribution of velocity.

LAMINAR FLOW

FIGURE 1-26   Velocity profile.

So far in our discussion of basic hydraulics, losses resulting from friction have not been considered. They are, however, extremely important to the design of hydraulic systems, and the flow pattern in the diagram shows the effect of friction between the fluid and the walls of the tubing. There would also be some friction loss within the fluid itself because of viscosity. *Viscosity* is the term for expressing a fluid's resistance to flow and is expressed in absolute units, usually centipoise, or Saybolt Universal Seconds (SUS). (This subject is covered in Chapter 2.) It is easily understood that some liquids, such as water or gasoline, flow much more readily than do heavier liquids, such as oil or glycerine.

Experiments have shown that viscosity or resistance to flow in liquids decreases when the liquids are heated. Conversely, as liquids are cooled they offer more resistance to flow; that is, their viscosity increases.

Another small loss would occur because of the kinetic or velocity energy in the fluid. Because the molecules of liquid are flowing, energy due to motion causes some friction, and this is given off in the form of heat. There would be some pressure drop in the system.

It is extremely important to keep all losses in a hydraulic system to a minimum by selecting the proper size of tubing, fittings, and other components that make up the system.

HAGEN–POISEUILLE LAW FOR LAMINAR FLOW

EXAMPLE

Oil is flowing through a 50-ft straight section of pipe at the rate of 20 gpm. The inside diameter of the pipe is 2 in. Calculate the head loss and pressure drop that will result. The sp. gr. of oil is 0.85. Assume that laminar flow oil viscosity is 150 SUS (as shown in Chapter 2).

Solution:

$$h_f = \frac{\mu L V}{46.5 D^2 g \rho}$$

or

$$\Delta P = (0.00000464)\frac{\mu L V}{D^2}$$

where ΔP = pressure drop (psi)

h_f = head loss (ft)

L = length of pipe (ft)

ρ = density (lb/ft^3)

V = oil velocity (ft/sec)

g = acceleration of gravity (32.2 ft/sec^2)

D = diameter of pipe (ft)

μ = absolute viscosity (centipoise)

$h_f = \dfrac{\mu L V}{46.5 D^2 g \rho}$ 150 SUS = 27 centipoise

$= \dfrac{(27)(50)(2.04)}{(46.5)(0.1667)^2(32.2)(53)}$

$= 1.25$ ft

$\Delta P = (0.00000464)\dfrac{\mu L V}{D^2}$

$$= (0.00000464) \left[\frac{(27)(50)(2.04)}{(0.1667)^2} \right]$$

$$= 0.46 \text{ psi pressure loss}$$

Turbulent flow occurs when the liquid particles flow in a random or erratic pattern, creating inefficiencies, high losses due to friction, and pressure drop. Notice in Fig. 1-27 that frictional losses due to erratic flow are all through the liquid. This means that greater pressure drop would occur because of heat buildup in the system. Turbulent flow occurs when tubes, pipes, fittings, and other components are too small or not properly selected for the job.

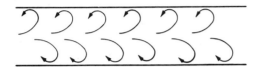

FIGURE 1-27 Turbulent flow.

REYNOLDS NUMBER

The Reynolds number is a numerical ratio of the dynamic forces of mass flow to the shear force due to viscosity. Flow usually changes from laminar to turbulent between Reynolds numbers of 2000 and 4000.

REYNOLDS' EQUATION

EXAMPLE

Oil is flowing through 1-in.-inside diameter pipe at the rate of 100 gpm. The sp. gr. of oil is 0.85. Is the flow turbulent or laminar? When 10 gpm is flowing through the same pipe, is flow turbulent or laminar?

Solution: If Re < 2000, flow is laminar; if Re > 4000, flow is turbulent.

$$\text{Re} = \frac{DV\rho}{\mu e}$$

where   Re = Reynolds number

D = pipe diameter (ft)

V = velocity (ft/sec)

ρ = density (slug/ft^3)

$\mu e = \dfrac{\mu}{1490}$ = absolute viscosity (lb/sec/ft)

μ = absolute viscosity (centipoise)

Note: 150 SUS = 27μ centipoise.

$$\text{Re} = \frac{DV\rho}{\mu e} = \frac{DV\rho(1490)}{\mu}$$

$$= \frac{(0.0833)(40.9)(53)(1490)}{27}$$

$$= 9960$$

Therefore, flow is turbulent for 100 gpm.

$$\text{Re} = \frac{(0.0833)(4.09)(53)(1490)}{(27)}$$

$$= 996$$

Therefore, flow is laminar for 10 gpm.

DARCY'S FORMULA

Darcy's formula is used to determine the pressure drop to flow friction through a tube or pipe.

$$h_f = \frac{fLV^2}{D2g}$$

where h_f = head loss (ft)

f = friction factor

L = length of tube or pipe (ft)

V = mean velocity of flow (ft/sec)

g = gravity (32 ft/sec^2)

D = internal diameter of tube or pipe (ft)

In any tube or pipe through which a fluid is flowing, there is a constant loss of pressure.

EXAMPLE

Oil is flowing through an 80-ft straight section of pipe at the rate of 60 gpm. The inside diameter of the pipe 0.75 in. Assume a friction factor of 0.05f. Calculate the head loss and pressure drop that will result. The sp. gr. of oil is 0.85. Assume turbulent flow.

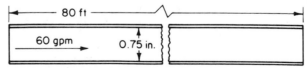

FIGURE 1-28 Darcy's formula.

Solution: Use Darcy's formula: (see Fig. 1-28).

$$h_f = \frac{fLV^2}{D2g}$$

$$\Delta P = \frac{\rho fLV^2}{144D2g} \quad \text{(in terms of psi)}$$

where Δp = pressure drop (psi)

ρ = density (slug/ft^3)

f = friction factor

L = length of pipe (ft)

V = oil velocity (ft/sec)

D = diameter of pipe (ft)

g = acceleration of gravity (32.2 ft/sec^2)

$$\Delta P = \frac{(62.4)(0.85)(0.05)(80)(43.6)^2}{(144)(0.0625)(2)(32.2)}$$

$$= 695.8 \text{ psi pressure loss}$$

$$h_f = \frac{fLV^2}{D2g}$$

$$= \frac{(0.05)(80)(43.6)^2}{(0.0625)(2)(32.2)}$$

$$= 1890 \text{ ft head loss}$$

The basic principles and concepts discussed in this section will be utilized throughout the book in their practical aspect. A great deal of the complexity involving laws and formulas has been removed by the formulation of charts, nomographs, and tables providing engineering data for the design of hydraulic systems. Much of this information will be found in its pertinent section of the book or in the appendix.

QUESTIONS

1. Define *fluid power*.

2. Name the important characteristics of a fluid used to transmit power.

3. Which state of matter contains the least kinetic energy? Which state of matter contains the most kinetic energy?

4. What is the definition of a fluid?

5. What is the density of a hydraulic oil that has a specific gravity of 0.86?

6. Explain why liquids will follow the path of least resistance.

7. What is the difference between force and pressure?

8. What hydraulic device is used to create a force that can push or pull a load?

9. What hydraulic device is used to create a force in a circular path?

10. What force would be developed by a cylinder operating at 600 psi if the area is 7 in.^{2}?

11. What is the hydrostatic principle? What is the hydrodynamic principle?

12. What is pressure head?

13. Explain the meaning of Bernoulli's equation and how it affects fluid flow in a hydraulic system.

14. What is the continuity equation, and what implication does it have for fluid flow?

15. Explain laminar flow. Explain turbulent flow.

16. What causes friction when fluid is flowing through pipes and orifices?

2
Fluids for Hydraulic Power

The primary function of the liquid in a hydraulic system is to transmit power to perform useful work. The hydraulic fluid must transmit an applied force from one part of the system to another and must respond quickly to reproduce any change in magnitude or direction of the applied force. To perform this primary function, the fluid must be relatively incompressible and have flowability.

In addition to this primary function of the hydraulic fluid, there are secondary factors that are extremely important to the successful operation of the hydraulic system. These important factors include: film strength to seal close clearances between moving parts against leakage; the minimization of wear and friction by providing adequate lubrication in bearings and between sliding surfaces in pumps, valves, cylinders, and other components of the system; resistance to physical and chemical change and prevention of rusting or corrosion; rapid settling and separation of insoluble contaminants that may enter the system; and removal of heat from the system.

Any number of liquids would fulfill the primary function of a hydraulic fluid. For instance, water is used as the fluid medium in hydraulic systems for coastal oil-drilling rigs, forging presses, and elevators. This is ideal from the standpoint of availability and cost, but there are also numerous disadvantages. Water promotes rust, freezes and boils at frequently encountered operating temperatures, and has very little lubricating value. Through the addition of soluble oils or other materials, water can be made suitable for special applications where large quantities of fluid are essential or for working hot materials where a fire hazard exists.

PETROLEUM-BASE FLUIDS

Modern advances in the refining and blending of petroleum-base oils have been responsible for the production of satisfactory fluids for hydraulic systems. The petroleum refiner is capable of producing oils of almost any viscosity. Additives also provide other desirable characteristics which prolong the life expectancy of hydraulic machines and increase their efficiencies.

The ability of a fluid to be pumped and transmitted through the system is most important. This ability to flow is determined by the oil's viscosity. Viscosity is the measure of an oil's resistance to flow or its internal resistance to shear. Viscosity is affected by temperature and must be expressed as a certain value at a certain temperature.

The engineer is sometimes faced with design problems where absolute units of measurement and the metric system are involved. Fluid friction is caused by the movement of molecules with respect to each other; this movement is termed *viscosity*. Figure 2-1 shows the concept of this movement.

As a thin moving plate is drawn over a film of fluid,

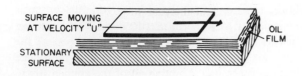

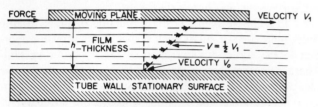

FIGURE 2-1 Concept of viscosity. (Courtesy of Mobil Oil Corporation.)

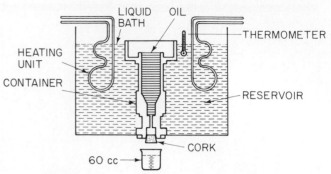

FIGURE 2-2 Viscosity tesing device using a fixed orifice. *Definition*: The viscosity of petroleum products and lubricants is the duration of time that is required to measure 60 ml through a standard orifice at a predetermined temperature (viscosity is a measure of the flowability at this definite temperature). (From J.J. Pippenger and T.G. Hicks, *Industrial Hydraulics*, 3rd ed., ©1979. Reproduced with the permission of the Gregg Division of McGraw-Hill, Inc.)

the fluid adheres to both the moving plate and the stationary surface. Fluid in contact with the moving plate moves at the same velocity as that of the moving plate, whereas the fluid in contact with the stationary surface is at a velocity of zero. Between V_1 and V_0 the fluid flows as imaginary layers, with each at a velocity that is proportional to the distance above the stationary surface. The force pushing the moving plate is proportional to the viscosity of the fluid.

The measurement of viscosity in the metric system would be as follows: If the moving plane surface is 1 cm^2 in area, the film thickness h is 1 cm, the velocity V_1 is 1 cm/sec, and the force is 1 dyne, the viscosity of the fluid would be 1 poise. The poise is the standard unit of absolute viscosity in the centimeter-gram-second system. It is defined as the ratio of the shearing stress to the shear rate of a fluid, and is expressed in dyne-seconds per square centimeter; 1 centipoise equals 0.01 poise. The poise can be converted into other common units of measurement for viscosity.

The method used most commonly as a measure of viscosity is SUS, or Saybolt Universal Seconds. The method of determination is shown in Fig. 2-2. The formula for changing SUS to viscosity in centipoise is

$$\mu = \text{sp. gr.} \left(0.22 \text{ SUS} - \frac{135}{\text{SUS}} \right)$$

where μ = viscosity (centipoise)
sp. gr. = specific gravity of fluid
SUS = Saybolt Universal Seconds

(see Table 2-1).

Viscosity is generally considered to be one of the most important physical properties of a hydraulic fluid and is the starting point in the selection of a fluid. If the fluid does not have the proper viscosity, it cannot perform, regardless of other superior characteristics.

If the viscosity is too low, the following results can be expected (lightweight oils):

TABLE 2-1
VISCOSITY CLASSIFICATION

SAE number	Viscosity Range, SUS (Equivalent Centistokes in Parentheses)			
	At 0°F		At 210°F	
	Minimum	Maximum	Minimum	Maximum
5W		4,000(880)		
10W	6,000(1,320)*	12,000(2,640)		
20W	12,000(2,640)†	48,000(10,500)		
20			45(5.75)	58(9.7)
30			58(9.7)	70(13.0)
40			70(13.0)	85(16.85)
50			85(16.85)	110(22.75)

*Minimum viscosity of 10W at 0°F can be waived if viscosity at 210°F is not below 40 SUS (4.2 cS).

†Minimum viscosity of 20W at 0°F can be waived if viscosity at 210°F is not below 45 SUS (5.75) cS).

$$\frac{\text{Centipoises}}{\text{Specific gravity}} = \text{centistokes (cS)}$$

Source: SAE (Society of Automotive Engineers).

1. Less film strength, and thus more wear on moving parts
2. More leakage
3. More pressure loss
4. Lower volumetric efficiencies in pumps and motors
5. Less precision control and slower responses
6. Lower overall efficiency

If the viscosity is too high, the following results can be expected (heavyweight oils):

1. Higher pressure drop due to friction
2. Excessive heat generation
3. Sluggish operation
4. More power consumption
5. Lower mechanical efficiency
6. Starvation of the pump inlet, causing cavitation (*cavitation* is starvation of the pump inlet)

VISCOSITY INDEX

The viscosity index (V.I.) is a measure of the relative change in viscosity for a given change in temperature. An oil with a high viscosity index shows less change in viscosity for a given change in temperature than does an oil with a low viscosity index (Fig. 2-3).

The viscosity index has little significance when the operating temperature range is small. This is the reason for selecting a viscosity of an oil at a specific operating temperature range. However, in hydraulic systems where the temperature range is difficult to control, the viscosity index is very important: for example, a lift truck working both indoors and outdoors during extreme temperature conditions. Oil having a V.I. rating above 90 is usually selected for hydraulic systems. The viscosity index nomograph shown in Fig. 2-3 and the table in Fig. 2-4 can be used to select oils with proper viscosity indexes.

STABILITY OR RESISTANCE TO OXIDATION

Another important property of a good hydraulic oil is stability, or resistance to chemical change. The three main accelerators of oxidation are high operating temperatures, high rate of mixing with air, and the catalytic effects of metals and contaminants, such as dirt and water.

The normal operating temperature range for most hydraulic systems is between 100 and 130°F. The rate of oxidation doubles for approximately every 20°F rise in temperature. A high rate of oxidation cannot be tolerated in a hydraulic system because of the varnishes and sludges produced. A high rate of oxidation also increases the acidity content of the fluid, causing further contamination and damage to system components.

The temperature of the oil in the reservoir does not give the true operating temperature of the system. It is better to use the pump-discharge side or other points wherever the temperature is the greatest.

It should also be recognized that pressure increases viscosity, which, in turn, will develop more frictional heat, thus raising the operating temperature of a system and increasing the rate of oxidation.

The amount of air that can be held in solution by an oil increases rapidly with increase in pressure. At atmospheric pressure an oil may contain 10% by volume of dissolved air. Air provides the oxygen necessary to promote oxidation.

Contaminants such as cutting oils, greases, dirt, moisture, paint, pipe compound, and insoluble oxidation products themselves accelerate the rate of oxidation.

Metals such as copper produce a catalytic action for oxidation, especially in the presence of water. Iron and aluminum also provide some catalytic action, increasing oxidation.

Premium-grade hydraulic oils contain inhibitors which tend to decrease the factors causing oxidation. They may be of the type that breaks the chain of reactions, thus preventing oxidation, or they may reduce the catalytic effect of the metals.

The ASTM (American Society for Testing and Materials) Oxidation Test (ASTM D-943) is important as a research tool in determining which oils have proven to be resistant to oxidation.

RUST PREVENTION

Rust occurs in a hydraulic system because of the presence of water. The source of moisture may be leakage from an oil cooler, or may be condensation from the air present in the system. Air may enter through the reservoir breather at ambient temperatures above 72°F. When the system is shut down and the air cools, moisture is given off. Water can also enter the system through the use of coolants during machinery operations.

Suitable inhibitors have been provided to control rust in a hydraulic system. High-grade hydraulic fluids contain an additive material, called *rust inhibitor,* which has an affinity for metal surfaces. It adheres to the metal surfaces, forming a protective coating that resists water, thus preventing a corrosive action between the metal and water.

RESISTANCE TO FOAMING

In addition to entrained air in hydraulic oil because of contact with the atmosphere, high-velocity mixing of air and oil also takes place in a hydraulic system. This occurs when oil is being discharged over a relief valve and back to the reservoir. Another source of air into a hydraulic system is through the pump inlet line, the pressure of which is generally less than atmospheric pressure.

It is important to bleed air out of all lines and components when starting up a system. Also, it is important to fill all lines and components when a hydraulic machine is being serviced. Good chemical compounds are available to break out

Viscosity Index

A.S.T.M. D-567
A.S.A. Z11.45

WHAT IS VISCOSITY INDEX?

The Viscosity Index is an empirical number indicating the rate of change in viscosity of an oil within a given temperature range. A low viscosity index signifies a relatively large change in viscosity with temperature, while a high viscosity index shows a relatively small change in viscosity with temperature. Viscosity Index cannot be used to measure any other quality of an oil.

Viscosity Index is calculated as follows:

$$VI = \frac{L-U}{L-H} \times 100$$

U = viscosity at 100F of the oil whose viscosity index is to be calculated.

L = viscosity at 100F of an oil of 0 viscosity index having the same viscosity at 210F as the oil whose viscosity index is to be calculated.

H = viscosity at 100F of an oil of 100 viscosity index having the same viscosity at 210F as the oil whose viscosity index is to be calculated.

THE SPREAD MAKES THE DIFFERENCE IN VISCOSITY INDEX

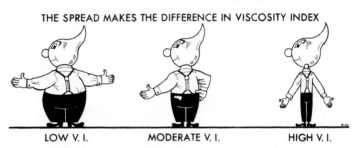

LOW V. I.   MODERATE V. I.   HIGH V. I.

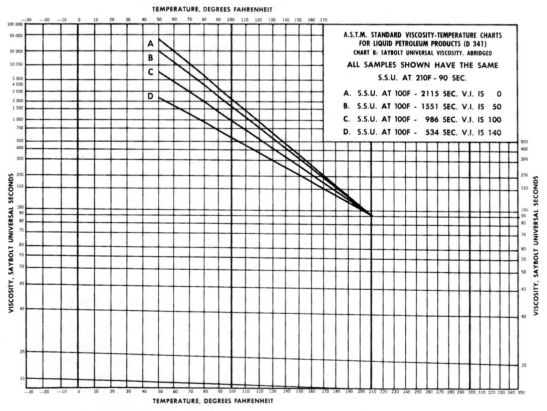

A.S.T.M. STANDARD VISCOSITY-TEMPERATURE CHARTS
FOR LIQUID PETROLEUM PRODUCTS (D 341)
CHART B: SAYBOLT UNIVERSAL VISCOSITY. ABRIDGED

ALL SAMPLES SHOWN HAVE THE SAME
S.S.U. AT 210F - 90 SEC.

A. S.S.U. AT 100F - 2115 SEC. V.I. IS   0
B. S.S.U. AT 100F - 1551 SEC. V.I. IS   50
C. S.S.U. AT 100F - 986 SEC. V.I. IS 100
D. S.S.U. AT 100F - 534 SEC. V.I. IS 140

FIGURE 2-3 Viscosity index. (Courtesy of United states Steel Corp.)

VISCOSITY INDEX

TABLE 100F

V.I.	40	45	50	55	60	65	70	75	80	85	90	95	100	105	110	115	120	125	130	135	140	145	150	155	V.I.
0	138	265	422	596	781	976	1182	1399	1627	1865	2115	2375	2646	2928	3220	3524	3838	4163	4498	4845	5202	5570	5959	6339	0
5	136	261	414	584	763	953	1153	1364	1585	1816	2059	2311	2573	2846	3129	3423	3727	4042	4366	4701	5046	5402	5768	6145	5
10	135	256	405	570	745	930	1124	1329	1543	1767	2002	2246	2500	2765	3038	3323	3616	3920	4233	4557	4890	5234	5587	5950	10
15	133	252	397	557	727	907	1095	1294	1502	1718	1946	2182	2427	2683	2947	3222	3505	3799	4101	4413	4734	5065	5406	5756	15
20	132	247	389	545	710	884	1066	1259	1460	1670	1889	2117	2355	2601	2856	3121	3394	3677	3968	4269	4578	4897	5225	5562	20
25	130	243	380	532	692	861	1038	1224	1418	1621	1833	2053	2282	2520	2765	3021	3284	3556	3836	4125	4423	4729	5044	5368	25
30	129	238	372	519	674	837	1009	1188	1376	1572	1776	1989	2209	2438	2674	2920	3173	3434	3703	3981	4267	4561	4863	5173	30
35	127	234	364	506	656	814	980	1153	1334	1523	1720	1924	2136	2356	2583	2819	3062	3313	3571	3837	4111	4392	4682	4979	35
40	126	230	355	493	638	791	951	1118	1293	1474	1663	1860	2063	2274	2492	2718	2951	3191	3438	3693	3955	4224	4501	4785	40
45	124	225	347	480	621	768	922	1083	1251	1425	1607	1795	1990	2193	2401	2618	2840	3070	3306	3549	3799	4056	4320	4590	45
50	123	221	339	468	603	745	893	1048	1209	1377	1551	1731	1918	2111	2311	2517	2729	2948	3173	3405	3643	3888	4139	4396	50
55	121	216	330	455	585	722	864	1013	1167	1328	1494	1667	1845	2029	2220	2416	2618	2827	3041	3261	3487	3719	3957	4202	55
60	119	212	322	442	568	699	835	978	1125	1279	1438	1602	1772	1948	2129	2316	2507	2705	2908	3117	3331	3551	3776	4007	60
65	118	207	314	429	550	676	806	943	1084	1230	1381	1538	1699	1866	2038	2215	2396	2584	2776	2973	3175	3383	3595	3813	65
70	116	203	305	416	532	653	777	908	1042	1181	1325	1473	1626	1788	1947	2114	2285	2462	2643	2829	3019	3215	3414	3619	70
75	115	199	297	403	514	630	749	873	1000	1132	1268	1409	1553	1703	1856	2014	2175	2341	2511	2685	2864	3046	3233	3425	75
80	113	194	288	391	497	606	720	837	958	1083	1212	1345	1480	1621	1765	1913	2064	2219	2378	2541	2708	2878	3052	3230	80
85	112	190	280	378	479	583	691	802	916	1035	1155	1280	1408	1539	1674	1812	1953	2098	2246	2397	2551	2710	2871	3036	85
90	110	185	272	365	461	560	662	767	875	986	1099	1216	1335	1457	1583	1711	1842	1976	2113	2253	2396	2542	2690	2842	90
95	109	181	263	352	443	537	633	732	833	937	1042	1151	1262	1376	1492	1611	1731	1855	1981	2109	2240	2373	2509	2647	95
100	107	176	255	339	426	514	604	697	791	888	986	1087	1189	1294	1401	1510	1620	1733	1848	1965	2084	2205	2328	2453	100
105	106	172	247	326	408	491	575	662	749	839	930	1023	1116	1212	1310	1409	1509	1612	1716	1821	1928	2037	2147	2259	105
110	104	167	238	314	390	468	546	627	707	790	873	958	1043	1131	1219	1309	1398	1490	1583	1677	1772	1869	1966	2064	110
115	103	163	230	301	372	445	517	592	666	741	817	894	970	1049	1128	1208	1287	1369	1451	1533	1616	1700	1785	1870	115
120	101	159	222	288	355	422	488	557	624	693	760	829	898	967	1037	1107	1176	1247	1318	1389	1460	1532	1604	1676	120
125	99	154	213	275	337	399	460	522	582	644	704	765	825	886	946	1007	1066	1126	1186	1245	1305	1364	1423	1482	125
130	98	150	205	262	319	375	431	486	540	595	647	701	752	804	855	906	955	1004	1053	1101	1149	1196	1242	1287	130
135	96	145	197	249	301	352	402	451	498	546	591	636	679	722	764	805	844	883	921	957	993	1027	1061	1093	135
140	95	141	188	236	284	329	373	416	457	497	534	572	606	640	673	704	733	761	788	813	806	859	880	899	140

FIGURE 2-4 Viscosity index table.

33

entrained air, quickly separating air from oil. Premium-grade hydraulic oils are available with foam depressants.

POUR POINT

Pour point is the characteristic of a fluid that indicates its ability to flow at low temperatures. It is important to consider the extreme temperature conditions under which a system must function, rather than normal conditions. For most installations, the pour point should be at least 20°F below the lowest expected operating temperature.

ANTIWEAR PROPERTIES

In addition to its many other functions, a good hydraulic oil must lubricate all integral moving parts of the system. Petroleum-base fluids provide excellent lubricating qualities.

Antiwear tests are conducted with a vane type hydraulic pump. Results indicate that premium-grade hydraulic oils provide excellent lubrication and usually, extended working life.

NEUTRALIZATION NUMBER

According to ASTM D-974-54T, the neutralization number is a measure of the acidity or alkalinity of a hydraulic fluid. This is referred to as the *pH factor* of a fluid. Fluids with a low neutralization number are recommended to prevent harmful chemical reaction. High acidity causes the oxidation rate in an oil to increase rapidly. The best method of determining the pH factor of an oil is to send a sample of the oil to the supplier in a properly prepared bottle. Most suppliers and some universities have modern laboratory facilities for making the various tests necessary to determine the condition of a fluid.

FIRE-RESISTANT FLUIDS

In many applications of fluid power, hazardous conditions and safety requirements dictate the use of a fire-resistant fluid. Hazardous operating conditions found in coal mines or in hot-metalworking processes are good examples. Safety requirements regarding fluid power systems for aircraft and marine uses call for maximum safety for the prevention of fire.

Manufacturers of fluids have done an excellent job in research and development to keep abreast of the increasing demands in all industries for better and dependable fire-resistant hydraulic fluids.

A fire-resistant fluid is a fluid, difficult to ignite, which shows little tendency to propagate flame. Such fluids should be used where any hazard exists that may endanger human lives or destroy valuable property.

In general, fluid-powered equipment is designed for use with petroleum fluids. The manufacturers of the equipment and of the fire-resistant fluid should be consulted for detailed consideration of any installation or conversion problem. When fire-resistant fluids are used, the suppliers are the best source of important information concerning design features, test requirements, operational techniques, maintenance procedures, equipment life, or other pertinent data, such as compatibility with system materials.

GENERAL INDUSTRIAL TYPES OF FIRE-RESISTANT FLUIDS

The general types of fire-resistant fluids include water-glycol, straight synthetics, and water-in-oil emulsions. The manufacturers of these various types of fire-resistant fluids have designed their products to be used in the same standard atmospheric conditions as petroleum oils. Whenever there are extreme atmospheric conditions, such as in foundries, where smoke and dirt may be drawn into the hydraulic system, the fluid composition could change. This might also happen in chemical and plating plants, where highly corrosive conditions exist. The change in composition could, over a period of time, be harmful to the hydraulic system. Always consult the suppliers when adverse environmental conditions must be met.

WATER-GLYCOL FLUIDS

Short-term vapor exposure or handling of water-glycol fluids in their complete form has no pronounced toxic effect. Leakage or spillage of these fluids can easily be cleaned up by flushing or mopping up with water.

Viscosity Control

Viscosity of the water-glycol base fluids depends on the water content, which ranges from 30 to 50%. It is good practice to check the fluid in a system at regular intervals to determine the average rate of water evaporation so that proper viscosity can be maintained. Loss of water could also seriously reduce the effectiveness of fire resistance.

Pure water, such as distilled or clean condensate or deionized water, should be used for makeup. This avoids the possibility of introducing harmful iron, lime, salts, or other foreign materials into the system.

Evaporation may also cause the loss of important inhibitors. These should be replaced by consulting the fluid supplier.

Operating Temperature

The operating temperature of the fluid should not exceed 120°F because of excessive evaporation rate and loss of water and inhibitors. High temperatures could also cause separation of water from the glycol base. The separated fluids may or may not go back into solution as the temperature is reduced to normal. It is essential to maintain good temperature control to avoid serious damage to pumps and other system components.

Excessive temperature may also cause physical change of the inhibitors, resulting in a gummy substance which does not go back into solution. This may cause loss of protection against fire and may also contaminate the hydraulic system.

Foaming and Aeration

Air retention in water-glycol fluids is more prominent than with petroleum oils. Reservoirs should be selected large enough, with adequate vents and baffles to help eliminate air from the fluid. The fluid returning to the reservoir from the system should be at a maximum distance from the inlet line.

Wear-Resistance Characteristics

Wear-resistance characteristics of water-glycol fluids have been found to be adequate in systems where components are carefully selected to ensure compatibility with the fluid. Antifriction bearings and other parts are definitely limited at high-speed, high-load capacities when lubricated with these fluids.

Hydraulic equipment manufacturers, particularly pump manufacturers, should be consulted concerning the use of their equipment with water-glycol fluids, since some modification may be desirable for wearing parts.

Specific Gravity and Pump Inlet

The higher specific gravity of water-glycols can aggravate pump inlet conditions. Inlet strainers or filters should be amply sized. Fine filtration should be restricted to the discharge side of the pump. Pump inlet lines must also be large enough to prevent excessive velocities of the fluid to the pump. Lines that are too small may starve the pump, causing damage.

Compatibility Factors

Water-glycol fluids attack zinc, cadmium, and magnesium, forming sticky or gummy substances. This residue may cause plugging of strainers or orifices, sticky valves and pump parts, and deterioration or destruction of parts plated or alloyed with these elements. Check with the fluid supplier for compatibility with the following:

1. All major components
2. Galvanized pipe
3. Strainers that have internal parts plated with zinc or cadmium
4. Die-cast reservoir fittings such as vents, fillers, and oil-level gauges
5. Instrumentation components such as thermometer bulbs, gauge bulbs, diaphragms, fittings, level indicators, heater coils, and tubing

These fluids also have a solvent action on most paints, enamels, and varnishes. (Some coatings are now available which resist this action and may be used for internal and external use.) Softening coatings on the inside of sumps or reservoirs must be removed before water-glycols are put into the system.

Experience to date indicates that the Buna N elastomers and neoprene seals and packings, which are normally used with petroleum oils, are satisfactory for use with water-glycol fluids. Silicones, butyl rubbers, nylon, Teflon, Viton, Fluorel, and Kel-f are also suitable materials. (These are trade names for compounds used in sealing devices.)

Asbestos, leather, and cork-impregnated materials should be avoided in rotary seals, since these materials have a tendency to swell because of their water-absorption properties.

CHANGING FLUIDS IN A SYSTEM

Thoroughness and carefulness should be prime considerations when one is changing fluids in a hydraulic system. Sufficient time, thought, and care can often mean the difference between good, successful operation or complete shutdown and major overhaul. The importance of working closely with fluid suppliers' recommendations cannot be overstressed. Representatives of fluid manufacturers are readily available and are willing to provide recommendations and assistance.

Changing from Petroleum Oil to Water-Glycol Fluid

1. The system should be completely drained and cleaned.
2. Drain and blow out the pipe lines. Manually clean the valves and reservoirs.
3. Pumps, accumulators, and cylinders should be actuated to ensure removal of all fluid.

4. Strainers should be dismantled, washed, and cleaned. Disposable cartridges should be replaced.

5. Drain and clean the filter housings. Replace or clean the filter elements.

6. Check the component functions before the system is closed and water-glycol fluid is added.

7. Close the system and flush with a small amount of water-glycol fluid or a suitable flushing compound recommended by the fluid supplier. Do not use carbon tetrachloride or other chlorinated solvents, as they can react with water by hydrolysis and form corrosive hydrochloric acid. Carbon tetrachloride produces toxic vapors when it comes in contact with flame.

8. Check the system carefully after it is in operation. Leakage at seals is appreciably more expensive with water-glycols than it is with petroleum oils.

Changing from Water-Glycol to Petroleum Oil

1. The system should be completely drained and cleaned. If all the water-glycol is not removed, a sufficient amount of water may remain to cause harmful effects to the replacement oil. As little as 1% water in petroleum oil in a high-pressure hydraulic system can be very detrimental.

2. Pipelines should be opened and thoroughly drained. Valves should be opened or removed and drained if there is a possibility of fluid in recesses. Reservoirs should be opened for thorough washing and cleaning.

3. Pumps, accumulators, and cylinders should be actuated to ensure removal of all fluid.

4. Strainers should be dismantled, washed, and cleaned. Disposable cartridges should be replaced.

5. Drain and clean the filter housings. Replace or clean the filter elements.

6. Coolers should be thoroughly drained and cleaned. A thin insulating film may adhere to cooler tubes. This film should be removed to maintain original cooler capacity.

The use of steam, a high-pressure water jet stream, or the wiping of surfaces with a brush, cloth, or other suitable material, and water are effective methods of cleaning surfaces that have been in contact with water-glycol fluid. Care must be taken to prevent rust or corrosive action if steam or water is used.

Do not attempt to use carbon tetrachloride or commercial chlorinated solvents, since they may form corrosive reaction products that do not dissolve with water-base fluids. Usually, a "sticky" or "tacky" gummed surface film results, which makes cleaning with water more difficult. Many common water-detergents speed cleaning action, especially in reservoirs. A clear-water rinse should follow detergent cleaning.

Circulate a minimum amount of petroleum oil through the system and flush out as much as possible without reopening the system. This oil should not be reused.

Trade Names and Manufacturers of Water-Glycol Fluids

Ucon Hydrolube, Union Carbide Chemicals Company

Houghto-Safe (600 Series), E. F. Houghton & Company

Mobil Nyvac Fluid, Mobil Oil Company

Celluguard, Celanese Chemical Company

Fyre Safe 225, Nalco Oil Company

Safety Fluid 200, 300, Texaco, Inc.

Quintolubric, Quaker Chemical Corp.

STRAIGHT SYNTHETIC PHOSPHATE ESTER AND PHOSPHATE ESTER–BASE FLUID

These fluids ordinarily do not present any irritating or sensitizing action to the skin. Continued exposure tends to dry the skin. Skin tends to be more sensitive to the chlorine-containing fluids than to the phosphate esters. Ordinary care, avoidance of continual skin contact, and normal personal hygiene is advised.

Vapors from the fluid *at recommended temperatures* are not considered harmful. Irritating fumes may be developed when fluid is sprayed on high-temperature metals. Leakage or spillage, such as that of petroleum oils, can be cleaned up with the use of cleaning solvents.

Viscosity Control

Generally, no additives are necessary to maintain original viscosity characteristics. Synthetic fluids are compounds and by nature are inherently stable. Changing, draining, or periodic adjustments are not normally required. Synthetic fluids have a lower compressibility factor than petroleum oils, which contributes to faster circuit response.

Operating Temperature

Normal operating temperatures between 120 and 130°F are recommended. Specific temperature limitations should be obtained from fluid suppliers.

If the viscosity index is low, fluid will thin out more

rapidly with increased temperatures. This causes an increase in leakage and can also change flow characteristics through control valves.

If the viscosity index is high, the fluid may contain "viscosity index improvers," which may "shear down" with use in pumps and valves, causing the viscosity to lower. Consult the fluid supplier to determine if the fluid will "shear down."

Foaming and Aeration

Air retention in straight synthetics may be more prominent than with petroleum oils. The reservoir, vents, and effective baffles should be adequate, and the pump inlet line should be as far as possible from the system return line. This will provide maximum deaeration.

Wear-Resistance Characteristics

Wear-resistance characteristics of straight synthetic fluids appear to be similar to those of good grades of petroleum oils.

Specific Gravity

Straight synthetic fluids have a high specific gravity and are not water soluble. Water tends to go to the top when the fluid is not agitated. Because of the higher specific gravity, pump inlet conditions may be affected. Lower-micron filtration should be restricted to the discharge side of the pump. Inlet strainers and pump inlet lines must be of adequate size to maintain proper inlet conditions at the pump. It is possible that Fuller's earth active-type depth filters can be used to advantage if the synthetic fluids are of the *nonadditive* type.

Compatibility Factors

No corrosive action is known to occur when straight synthetic fluids are used with the metals normally used in the components of a hydraulic system. Corrosion can occur if water contaminates the system.

The straight synthetic fluids have a strong solvent action on most paints, enamels, and varnishes. Nylon-base paints, cured phenolic, and epoxy paints are compatible with these fluids. Obtain specific information from the fluid supplier.

Sealing materials normally used with petroleum oils are *not* suitable for use with straight synthetic fluids. Seals should be changed when the system is being converted to this fluid. Suitable sealing materials are butyl rubbers, some silicones, Teflon, Kel-f, nylon, Viton, and Fluorel.

Although neoprene swells in this type of fluid, hoses have been used satisfactorily. However, butyl-lined hoses are recommended.

Changing from Petroleum Oil to Straight Synthetic Fluid

1. The system should be completely drained and cleaned. Very small amounts of residual oils usually do not interfere with the performance of phosphate ester–base and chlorinated hydrocarbon–base fluids. Resistance to fire is reduced by the amount of oil left in the system.

2. Drain and blow out pipe lines. Manually clean valves and reservoirs.

3. Valves should be opened or removed and drained. Replace gaskets, packings, and seals with those compatible with straight synthetic fluids. It is particularly important to change dynamic seals. Static seals do not necessarily have to be changed if there is no objection to replacing them each time that hydraulic components are serviced.

4. Pumps, accumulators, and cylinders should be removed if necessary for draining purposes. Replace packings, seals, gaskets, and bladders with those compatible with synthetic fluids. Consult seal and packing manufacturers for recommended materials.

5. Check the component functions before the system is closed and straight synthetic fluid is added.

6. Close the system and flush with a small amount of straight synthetic fluid. Do not continue to use the flushing fluid unless the manufacturer approves.

7. Check the system carefully after it is in operation. Leakage at seals is appreciably more expensive with straight synthetic fluids than with petroleum oils.

Changing from Straight Synthetic Fluid to Petroleum Oil

1. The system should be completely drained and cleaned.

2. Pipelines should be opened and thoroughly drained.

3. Valves should be opened or removed and drained. Replace gaskets, packings, and seals to those compatible with oil.

4. Pumps, accumulators, and cylinders should be removed, if necessary, for draining purposes. Replace packings, seals, gaskets, and bladders with compatible materials. Substitute rubbers recommended for straight synthetic fluids are, in general, not resistant to the solvent or softening action of conventional petroleum base oils.

5. Reservoirs should be opened so that fluid can be thoroughly removed and surfaces wiped clean.

6. Strainers should be dismantled and completely cleaned.

7. Drain and clean filter housings. Replace or clean filter elements.

8. Coolers should be thoroughly drained.

9. Ordinary commercial cleaning solvents will "cut" the straight synthetic fluids and can be used to clean reservoirs or to wipe parts.

10. Circulate a small amount of oil through the system and flush out as much as possible without reopening the system. The flushing oil should not be reused.

Trade Names and Manufacturers of Straight Synthetic Fire-Resistant Fluids: Phosphate Ester and Phosphate Ester Base

Cellulube, Celanese Corporation

Pydraul F9, 150, 625, 60, Monsanto Chemical Company

Skydrol (aircraft and ground support equipment), Monsanto Chemical Company

Mobil Pyrogard 42, 43, 53, 55, Mobil Oil Company

Houghto-Safe (1000 Series), E. F. Houghton & Company

Shell, SFR B, C, D, Shell Oil Company

Fyre Safe 1090, Nalco Oil Company

Chlorinated hydrocarbon base:

Pydraul A-200, Monsanto Chemical Company

WATER-IN-OIL EMULSIONS

Short-term vapor exposure or handling of emulsion fluids in their complete form has no pronounced toxic effect. Normal precautions for handling and cleaning any spillage are the same as with other hydraulic fluids.

Water-in-oil emulsions are premixed or finished emulsions ready for use in hydraulic systems. The amount of water in various brands usually ranges around 40% and should be maintained according to the fluid manufacturer's recommendation.

A number of manufacturers have concentrates available consisting of the base oil plus emulsifiers and additives. Finished emulsions can be prepared from these concentrates by adding water.

Viscosity Control

Viscosity of the water-in-oil emulsion decreases as the water content is reduced. It is good practice to check the fluid at regular intervals after fluid is first put into operation in order to determine the average water evaporation rate. Check with the supplier to determine proper testing procedures, because loss of water increases fire hazard. Pure or distilled water free of harmful iron, lime, salts, or other foreign matter should be used for makeup.

Operating Temperature

Fluid operating temperatures should not exceed 120°F. Higher temperatures may cause difficulty because of excessive evaporation of water or oxidation. Products of oxidation can contribute to emulsion instability and the separation of free water from the fluid.

The performance of emulsion fluids depends on their ability to resist phase separation. Emulsions which are trapped in stagnant areas, such as single-acting cylinders or accumulators, will, in time, suffer some oil and water separation. In such cases, reemulsification can be established by circulating through the pump or across relief valves.

Repeated freezing and thawing of emulsions in unprotected storage can cause phase separation. Storage above 100°F can also result in phase separation.

Foaming and Aeration

The foaming and air-entrainment characteristics of emulsion fluids are comparable with those of conventional petroleum oils.

Wear Resistance

The qualities of wear resistance of many emulsion fluids have been found to be adequate in systems where components are carefully selected to insure compatibility with the fluid. However, any water separation or reversion of the fluid may greatly reduce wear resistance. Emulsions may exhibit a reduction in viscosity when subjected to the high rates of shear existing in most hydraulic pumps. Consequently, a viscosity level somewhat higher than that of petroleum oils is manufactured into the fluid purposely, and the reduction in viscosity occurs when the emulsion is sheared.

Pump inlet conditions are especially critical with these fluids of higher viscosity, and precautions should be taken to assure proper installation. Antifriction bearings and other parts are definitely limited at high-speed, high-load capacities when lubricated with these fluids. Consult hydraulic equipment manufacturers concerning any modification necessary to use this type of fluid.

Specific Gravity

The higher specific gravity of these fluids can affect the pump inlet conditions. Fine filtration should be restricted to the discharge side of the pump. The degree of filtration allowed depends on the fluid used, and the supplier should be consulted.

Filter materials, such as metal screens or disks and sintered metal filters, are suitable for use with emulsion-type fluids. Emulsion-type fluids tend to hold dirt and fine metallic particles in suspension more readily than petroleum oils. Filtration should include magnets to remove ferrous materials.

Compatibility Factors

Emulsion-type fluids are not corrosive to the metals normally encountered in industrial hydraulic systems. Emulsion-type fluids may have a solvent action on some paints, enamels, or varnishes. However, many of the coatings used with petroleum oils are satisfactory for use with water-in-oil emulsion fluids. Fluid suppliers should be consulted for the proper protective coating to use.

Experience to date indicates that Buna N-type elastomers and neoprene seals and packings, which are normally used for petroleum oils, are satisfactory for use with emulsion fluids. Silicones, Teflon, Viton, Fluorel, and Kel-f are also suitable materials.

It is generally not necessary to replace static seals until they are disturbed during overhaul. Butyl dynamic seals should be replaced immediately when the system is charged over for use with emulsion fluids.

Changing from Petroleum Oil to Emulsion Fluid

1. The system should be completely drained and cleaned. Very small amounts of residual oils usually do not interfere with the performance of emulsion fluids.

2. Drain and blow out the pipelines. Manually clean the valves and reservoirs.

3. Pumps, accumulators, and cylinders should be actuated to ensure removal of all fluid.

4. Strainers should be dismantled, washed, and cleaned.

5. Drain and clean the filter housings. Replace or clean the filter elements.

6. Check the component functions before the system is closed and emulsion fluid is added.

7. Close the system, fill with emulsion fluid, and circulate for at least 8 hours.

8. Drain the flushing charge, blow down the system, and install operating charge of emulsion fluid.

Changing from Emulsion Fluid to Petroleum Oil

1. The system should be completely drained and cleaned.

2. Drain and blow out the pipelines, and manually clean the lines and reservoirs.

3. Pumps, accumulators, and cylinders should be actuated to ensure removal of the fluid.

4. Strainers should be dismantled and completely washed and cleaned.

5. Drain and clean the filter housings. Replace or clean the filter elements.

6. Close the system and install a minimum quantity of petroleum oil. Circulate for 16 to 24 hours.

7. Drain and blow down the system and install the operating charge of petroleum oil.

Trade Names and Manufacturers of Water-in-Oil-Emulsion Fire-Resistant Fluids

Gulf FR Fluid, Gulf Oil Company

Houghto-Safe (5000 Series), E. F. Houghton & Company

Pyrogard C, D, Mobil Oil Company

Sinclair Duro FR Fluid, Sinclair Refining Company

Shell Irus Fluid 902, Shell Oil Company

Sunsafe, Sun Oil Company

Hul-E-Mul, Hulbert Oil & Grease Company

Imol, Humble Oil Company

Dasco, D. A. Stuart Oil Company

Texaco TL 4-625, Texaco, Inc.

Puro RF, Pure Oil Company

Permanul, American Oil Company

Fyre Safe w/o, Nalco Oil Company

LSE-1256, Atlantic Oil Company

Pacemaker, Cities Service

Quintolubric, Quaker Chemical Corp.

QUESTIONS

1. What is the primary function of a liquid in a hydraulic system?

2. What type of fluid is most generally used to transmit power in a hydraulic system?

3. What is viscosity, and how is it measured?

4. What effect does temperature have on viscosity?

5. What happens when the viscosity of an oil is too high?

6. Why should normal operating temperatures be controlled below 130°F in most hydraulic systems?

7. What causes catalytic action in a hydraulic oil?

8. How does moisture get into a hydraulic system?

9. To what does the neutralization number of a fluid refer?

10. When are fire-resistant fluids used in a hydraulic system?

11. What is compatibility?

12. What effect does a higher specific gravity have on the inlet of a pump?

13. Why is it necessary to use precautions when changing from a petroleum-base fluid to a fire-resistant fluid?

3
Sealing Devices for Hydraulic Power

The success of applying fluid power to any application depends largely on the ability of the sealing devices to prevent both internal and external leakages in the system. Around the year 1650 Blaise Pascal discovered the fundamental laws of physics by which all modern fluid power systems are governed. Unfortunately, his discovered knowledge was not put to practical use until about 150 years later, when Joseph Bramah developed the first hydraulically operated press (see Fig. 3-1). The success of this first press, which used water for the transmission of power, is accredited to the invention of the cup packing. Cup packing was the sealing device used to prevent leakage within the cylinder or ram.

Modern fluid power systems depend more and more on the development of sealing materials that will operate with special fluids and at conditions of high pressures and temperatures. The manufacturers of sealing devices have kept pace with the needs of all industry through excellent research and development programs.

Because of this progress in the chemistry of dealing with sealing materials, many new trade names associated with products are constantly being introduced. The manufacturers of sealing devices are the experts in this highly specialized field, and their recommendations are always important in any application.

Fluid power design engineers work very closely with the suppliers of sealing devices to ensure proper sealing for rotating shafts, reciprocating rods and stems, and for stationary mating parts of fluid power equipment.

Losses of fluid from hydraulic machinery caused by leakage can be very costly not only in terms of the cost of the fluid but also in production. External leaks in a system can be quickly discovered and fixed, but internal leakage is not so easily detected. Internal leaks slow up machines, causing more rejects of parts, as well as loss in production. Often, lost production from one machine may slow up the entire operation of a plant.

Good maintenance procedures with regard to external leaks are essential to minimize the safety hazards created by oil leaking on the floor and around the machines. Also, oil gets on machined parts, causing an extra cleaning operation. Experience indicates that where maintenance procedures have been established that minimize excessive leakage, increased production has been the result.

CONDITIONS AFFECTING THE SELECTION OF SEALING DEVICES

Each application for a seal must be treated as a specific problem because of slight variations in minor conditions of operation. The entire set of conditions encountered in the operation of a component, such as a pump, fluid motor, cylinder, or control valve, must be considered collectively. For example, a given type of seal design might have the

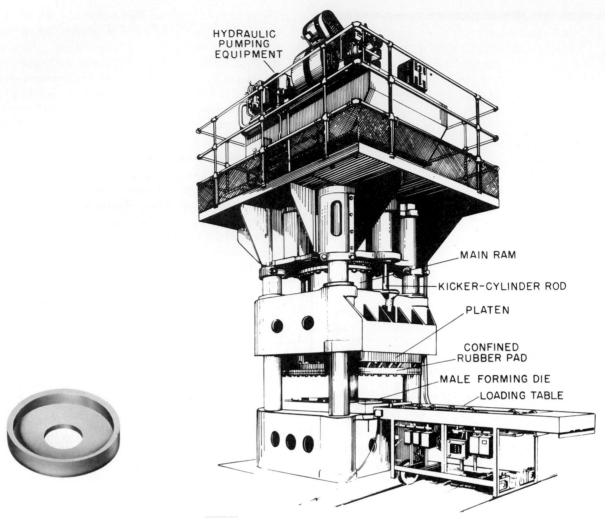

HYDRAULIC
PUMPING
EQUIPMENT

MAIN RAM

KICKER-CYLINDER ROD

PLATEN

CONFINED
RUBBER PAD

MALE FORMING DIE

LOADING TABLE

FIGURE 3-1   (First press)—cup packing. A 25,000 ton press made successful through cup packing. (Courtesy of Mobil Oil Corporation).

ability to stand a given pressure, or it might be able to stand a given speed, but it might not be able to withstand the same given pressure and speed occurring together.

1. Speed is an important factor in determining frictional temperature buildup, which results in wear of both the seal and the shaft. Therefore, surface speed or peripheral speed and shaft size are important considerations. Seals may also be affected by frequency, depending upon the revolutions per minute. This is important when a seal must operate satisfactorily under eccentricity, wobble, and whip conditions of rotation. The same is true concerning end play of shafts. For example, a 1-in.-diameter shaft having a speed of 4000 rpm presents a condition more serious than a 4-in.-diameter shaft rotating at 1000 rpm, with the same

eccentricity, although the surface speed is the same in both cases.

2. Pressure in the system in contact with a sealing device may increase the contact force. This would increase friction, causing heat buildup and more wear. In some cases, seals may be pressure balanced, which would eliminate increased friction. Pulsating pressures are troublesome because they set up vibrations in the sealing surface which interfere with the sealing contact.

3. Temperature at the point of seal depends on the materials in contact, the shaft speed, pressure, amount of lubrication, heat from bearings, heat from fluid, or other sources. The sealing device selected must be unaffected by frictional temperature or the fluid being sealed.

4. Compatibility between sealing devices, the fluid, and other materials that make up the complete system is extremely important. Selecting the wrong sealing materials may result in costly shutdowns and loss of production. Foreign materials, corrosion, excessive heat, and many other conditions can affect sealing operations.

In addition to the factors listed above, there are many machine operating conditions which may also affect good sealing practices. Fast reversals, intermittent operation, long intervals between operations, or any erratic motions are a few examples.

The designer who specifies the type of sealing device for any given application should study all of the problems carefully, using the supplier for consultation and help in making the final decision.

SEALING MATERIALS

Sealing materials may be classed into three general categories: leather, fabricated rubber, and homogeneous (see Table 3-1).

Leather is the oldest material used for sealing devices and is still very popular for many applications. Leather impregnated with synthetic rubber may be operated safely at temperatures up to 180°F. Leather seals have low frictional properties and relatively high tensile strength which resists extrusion into clearance spaces. Thus leather is used for applications where higher system operating pressures are used. It is also used as antiextrusion rings to bridge clearances for using O-rings and thus preventing their extrusion into the clearances (see Fig. 3-2).

Leather does not score metal surfaces. However, rough metal surfaces have a tendency to shorten the life of any

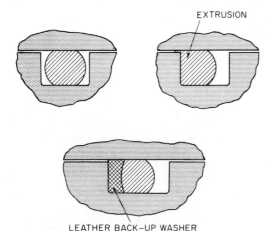

FIGURE 3-2 Anti-extrusion ring for "O" ring. (Courtesy of Mobil Oil Corporation).

TABLE 3-1
COMPATIBILITY CHART

Sealing material	Fluid compatibility	Remarks
Natural rubber (NR)	Water-base fluids	Operating temperature range low to 200°F. Compounded to suit fluid requirements
Leather (Treated)	Petroleum-base fluids	Operating temperature range −65 to 265°F
(Impregnated with thiokol)	Some synthetics and phosphate ester	Multiple seals good for high pressure
Butyl rubber (11R)	Water-base fluids and phosphate ester	Operates from −65 to 250°F.
Nitrile rubber (Buna-N) (Hycar)	Petroleum-base fluids Water-base fluids and some synthetics	Operates from −65 to 250°F; does not have corrosive tendencies, but swells in aromatic oils and some synthetics
Silicone rubbers	Water-base fluids: Fair with petroleum base; good with some synthetics	Operating temperature range from −120 to 500°F; not suitable for silicone fluids
Fluoro-plastics (Teflon) (Kel-F)	Good with almost all fluids used	Operating temperature range from −320 to 500°F; low friction with good chemical resistance
Fluoro-elastomers (Viton A)	Good with water-base, petroleum-base, and most synthetic fluids	Operating temperature range from −20 to 500°F
Polyacrylates (Hycar 4021) (Vyram)	Good with chlorinated hydrocarbons	Not recommended for low temperatures; high temperature about 350°F
Polysulphides (Thiokol)	Good with water-base, petroleum-base, and most synthetic fluids	Operating temperature range from 10 to 160°F; vaporizes above 160°F
Styrene (SBR)	Good for water-base fluids	Operating temperature range slightly higher than natural rubber

material used for sealing. A metal surface between 8 and 40 rms (root mean square) (microfinish) is recommended.

Leather sealing devices are manufactured in U, V, cup, and flange shapes (see Fig. 3-3), which are in accordance with the Fluid Power Standards.

Fabricated rubber sealing devices are composed of synthetic rubber compounds and fabrics. The fabrics rein-

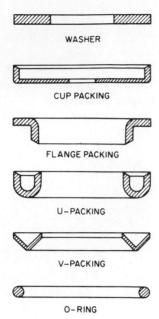

WASHER

CUP PACKING

FLANGE PACKING

U-PACKING

V-PACKING

O-RING

FIGURE 3-3   "U" "V" cup and flange shapes of seals.

force the synthetic rubber to give sealing devices resistance to extrusion. There are numerous combinations of synthetic rubbers and fabrics.

Cotton duck, asbestos, and nylon are three common kinds of fabrics in use. Duck is used for normal operating temperatures, asbestos for high-temperature operation, and nylon for greater strength and flexibility.

The synthetic rubbers or polymers used in fabricated seals depend on the type of fluid, temperature, and other operating conditions. Some of the common synthetic rubber bases include chloroprene, Buna N, Buna S, Butyl, and Viton. Chloroprene and Buna N are suited for petroleum oils, Buna S for water systems, Butyl for straight synthetic fire-resistant fluids, and Viton for high-temperature operations.

Fabricated seals have a wider operating temperature range than leather, because the base polymer can be varied. For example, a combination of asbestos and Viton would have an operating temperature range of -65 to $400°F$. The metal finish recommended for fabricated seals is between 16 and 32 rms. Fabricated sealing devices are made in U, V, cup, and flange in Fluid Power Standard sizes for interchangeability with other material seals.

Homogeneous sealing devices are compounded from many different base polymers of synthetic rubber. They are made in many hardnesses, depending on the shape, application, and intended operating pressure.

The compounding of materials with elastic qualities in a synthetic rubber for a specific fluid or operating condition is complex. Resilient compounds have intricate, semi-plastic, colloidal structures which are capable of almost infinite variation because of the hundreds of ingredients from which they are produced.

A hardness range from 8 Shore durometer (see Fig. 3-4) to 40 durometer measured on the A scale are commonly available. Other hardnesses are available for special installations.

Homogeneous seals operate over a wide temperature range, similar to that of fabricated seals (-65 to $400°F$). When operating pressures in the system exceed 1500 psi,

FIGURE 3-4   Durometer. (Courtesy of Chicago Rawhide Mfg. Co.)

antiextrusion rings or minimum metal clearances must be used. The metal surface finish for homogeneous seals should be between 8 and 40 rms to minimize friction. Homogeneous seals are made in V, U, cup, and O-ring shapes, according to Fluid Power Standards for interchangeability.

SELECTING SHAPE OF SEAL

Selecting the seal that is properly shaped for the job depends on how the seal must be applied and how the part to be sealed is constructed. Figure 3-5 shows common uses of the various seals.

Multiple Vs may be used for an extreme range in operating pressures. They are also adaptable for rotational or reciprocating motion with pistons or shafts. When V seals are installed, the gland nuts should be torqued just enough to seal the seal and then backed off to a point where the sealing is still maintained. Damage may occur from excessive friction if the seals are kept too tight. Often a spring is used to maintain the correct force on the seals at all times. A value of 10 lb/in. of mean diameter is generally used for selecting spring force. V seals are considered a balanced seal.

The O-ring (see Fig. 3-6a and b) is placed in a groove

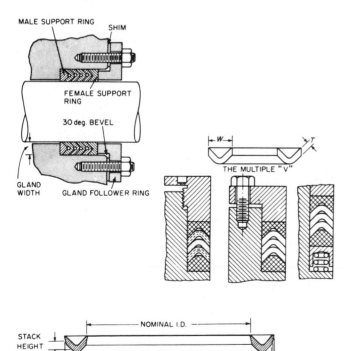

FIGURE 3-5 "V" seals. (Courtesy of Mobil Oil Corporation.)

in such a way that there is enough contact pressure, because of direct compressive stress in the material, to accomplish sealing at low pressures. When greater hydraulic pressure is applied, the material has a tendency to reshape itself in the space away from the pressure, which increases contact pressure.

O-rings are used as stationary seals or for reciprocating motion on piston or shaft applications. At higher pressures, antiextrusion rings or backup washers should be used to prevent the seal from extruding into the clearance space. O-ring seals are not generally used for rotary applications, except for very low speeds and pressures. They are very popular as static seals, but should not be used where vibration or motion is excessive. O-rings should be replaced when components are disassembled and reassembled during service or repair. O-rings are considered a balanced seal.

The U seal (see Fig. 3-7) is a form of pressure-actuated seal that depends on the hydraulic pressure in the system to act against the lip to create contact pressure for sealing. In order for the lip to move freely, the groove space should be slightly greater than the seal. This type of seal is more effective when the groove is properly proportioned to allow lip flexing with little solid compression. When used at operating pressures over 400 psi, it is advantageous to use a leather or fabricated backup washer to prevent extrusion of material into the clearance space. Seals of this type are commonly used for pressures as high as 3000 psi for cylinder pistons and rods. The U seal is considered a balanced type of seal.

The flange seal (see Fig. 3-8) is a pressure-actuated seal that effects sealing through the hydraulic pressure in the system. Some of these seals are made with cylindrical sides, or lips, the same dimension as the cylinder rod, but it is considered better practice to make the lip tapered toward the member to be sealed so that the flange seal is somewhat under the rod size. Flange seals are generally applied in the lower pressure ranges. The flange seal is considered an unbalanced seal.

Cup seals (see Fig. 3-9) are used primarily for sealing cylinder pistons. This seal is also pressure-actuated, with the sealing accomplished by the contact pressure outward against the cylinder bore. It is essential that these seals be clamped in place, and the method of clamping is important. The lip of the seal cannot be too long or compressed too tightly against the cylinder walls. If the lip is too long, repeated stroking of the cylinder causes weakening and the seal breaks down and bits of it extrude into the clearance spaces. Eventually, the pressure breaks the cup all the way through. It is therefore important that the cup lip be no higher than is necessary to follow the side play.

This type of seal is considered an unbalanced seal, and its operating pressure depends on the material and conditions under which it operates.

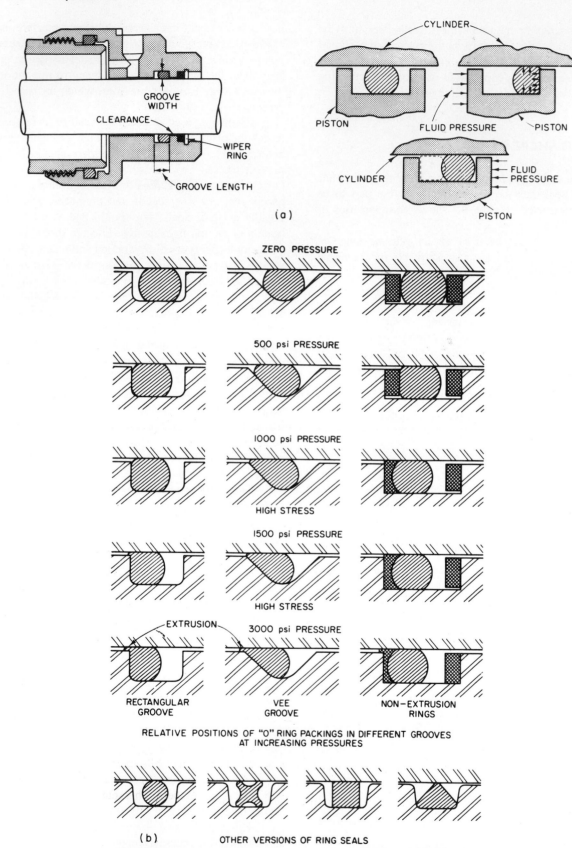

RELATIVE POSITIONS OF "O" RING PACKINGS IN DIFFERENT GROOVES
AT INCREASING PRESSURES

(b)   OTHER VERSIONS OF RING SEALS

FIGURE 3-6 (a) & (b) "O" rings. (Courtesy of Mobil Oil Corporation
and E. F. Houghton Co.)

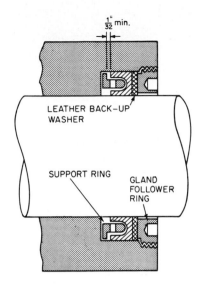

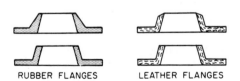

"V" AND "U" SEALS

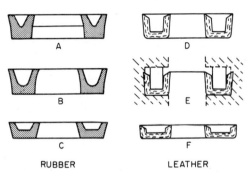

FIGURE 3-7 "U" seals. (Courtesy of Mobil Oil Corporation.)

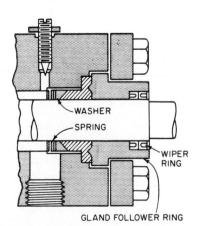

FIGURE 3-8 Flange seals. (Courtesy of Mobil Oil Corporation.)

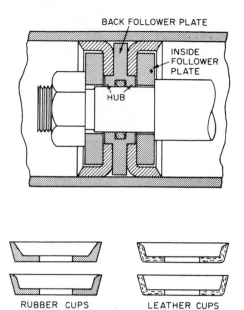

FIGURE 3-9 Cup seals. (Courtesy of Mobil Oil Corporation.)

BALANCED MECHANICAL FACE SEALS

To prevent leakage between a rotating shaft and a stationary housing through which the shaft passes often requires a low friction seal.

Mechanical face seals have this capability. They are especially well suited for applications where minimal leakage of the contained fluid is critical and a significant design consideration.

When properly specified, designed, and assembled, mechanical seals can withstand high operating pressures, temperatures, and speeds. They offer long life, low leakage, and require little maintenance (Fig. 3-10).

This type of face seal can function under both static and dynamic operating conditions, withstand large pressure changes, and be compatible with many different types of fluids. They adapt to bidirectional shaft rotation, as well as fluctuations of pressure, speed, and temperature. Within allowable tolerances, face seals can withstand some misalignment and eccentricity of shaft and bore. Because face seals do not induce wear, shaft metallurgy, surface finish, hardness, and roundness are much less critical.

Basic Design

Major sealing interface in a face seal is between the rotating and stationary members. This sealing plane is perpendicular to the axis of the shaft. The sealing area is a relatively narrow ring where the two sealing elements interface. Common mating materials are carbon and hardened steel. For some seals a hard stainless may be involved. Low friction plastics may also be used.

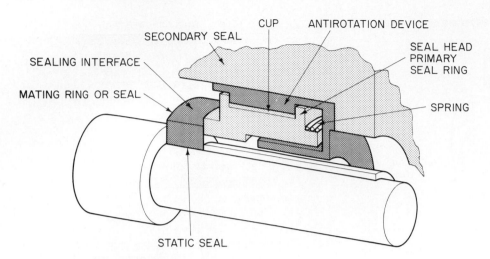

SECONDARY SEAL

CUP   ANTIROTATION DEVICE

SEALING INTERFACE

MATING RING OR SEAL

SEAL HEAD
PRIMARY
SEAL RING

SPRING

STATIC SEAL

FIGURE 3-10 Mechanical face seals.

The two major elements of a mechanical seal are a seal head which is usually spring loaded and a mating ring or seat, Fig. 3-10. A seal head consists of axially movable parts which accomodate linear space variations in the installed length of the seal assembly. The seal head compensates for any wear of the opposing flat surface or faces and adjusts to misalignment and longitudinal movement in the unit being sealed.

For design convenience the seal head can be rotary or stationary. If the seal head rotates, the seat on the mating ring remains stationary or vice versa. Rotating seal heads are most commonly used with shafts machined to close tolerances and designed to rotate at relatively slow speeds. Stationary seal heads, on the other hand, are more suitable for higher-speed applications. The dynamic balancing is less difficult. Only pressure balancing is needed for the head and rotational balancing for the seat.

To perform its primary or dynamic sealing function, the seal head must maintain some sliding contact with the mating ring mounted on the shaft, or in the housing through which the shaft passes, depending on the seal design.

Two static, secondary seals are also needed between the head-and-shaft and the mating-ring-and-housing, or vice versa, depending on design.

QUESTIONS

1. What results when a hydraulic system has internal leaks?

2. What effect does the speed of an actuating cylinder or fluid motor have on the selection of sealing devices?

3. What effect does pulsating pressure have on seals?

4. What effect does temperature have on sealing materials?

5. What happens when the sealing material and the fluid are not compatible?

6. Which type of sealing materials is generally used for systems operating at higher pressures?

7. What is an antiextrusion ring?

8. Which types of sealing materials are used for higher-temperature operations?

9. Sketch various shapes of sealing devices normally used with hydraulic systems.

10. What is a static seal?

4
The Distribution of Hydraulic Power

The distribution of hydraulic power involves pipe, tubing, hose assemblies, manifolds, and the various fittings used to connect the components of a system. The choice of pipe, tube, or hose depends largely on the system's operating pressure and flow. Other important factors to consider include environmental conditions, type of fluid and operating temperature, shock loads, relative motion between connected parts, practicality, and compliance with certain standards.

All materials used to convey fluid power are commonly classified as *conductors,* and the various fittings for connecting components are classified as *connectors*.

Fabricating a hydraulic system requires skill, patience, and pride of workmanship. Personnel engaged in this type of work are required to have a good understanding of basic fluid power and to know how to read circuit diagrams. Skills in the art of measuring, bending, cutting, flaring, or threading of conductors are acquired mainly through experience or practice.

Good tools are also essential for the proper fabrication of a hydraulic system. Using proper tools for flaring tubing, or threading pipe prevents leakage and aids in the prevention of detrimental contamination to the system. Small particles of metal from ragged threads or flares can cause serious damage to such components as pumps, motors, cylinders, and valves. Making the proper bends and connections can also help to minimize friction and reduce power losses.

This entire area of distributing fluid power is very significant in terms of the successful operation of any hydraulic equipment.

OPERATING PRESSURE AND FACTOR OF SAFETY

Conductors and connectors used for hydraulic circuits are selected mainly to operate at the highest pressure level that is reached during a cycle of the equipment operation. The materials used must have a continuous operating pressure rating which will withstand working pressures and provide a margin of safety for short-lived pressure peaks resulting from hydraulic shock.

Hydraulic shock occurs when control valves operate suddenly, stopping or reversing flow which is backed up by large mass forces or high velocities. This condition produces a shock wave due to the kinetic energy in the fluid. Hydraulic shock waves also result from the sudden deceleration, stopping, or reversing of heavy work loads. In either case, the produced shock waves travel through the conductors on the downstream side into various branches at the speed of sound for the fluid, and then bounce back toward the point of origin. This process repeats, with each bounce or reaction traveling a lesser distance than the preceding, until the kinetic energy is dissipated into friction and heat.

Pressure ratings of the various conductors depend mainly on the tensile strength of the materials used and the wall thicknesses of the conductor. The wall thicknesses and safety factors recommended by the Fluid Power Standards are based on calculations using *Barlow's formula:*

Minimum wall thickness (in.)

$$= \frac{\text{maximum pressure (psi)} \times \text{OD of conductor (in.)}}{2 \times \text{tensile strength (psi)}}$$

where OD is the outside diameter.

EXAMPLE

(a) What would be the maximum pressure for SAE_{1010} Dead Soft cold-drawn-steel tubing with an outside diameter of $\frac{3}{4}$ in., tensile strength of 55,000 psi, and wall thickness of 0.065 in.?

(b) What would be a safe working pressure for this tubing if the factor of safety must be at least a 4:1 ratio?

Solution:

(a) Max. psi
$$= \frac{\text{wall thickness (0.065)} \times 2 \times 55,000 \text{ psi T.S.}}{\text{OD } 0.750 \text{ in.}}$$

$$= 9533 \text{ psi}$$

where T.S. is the tensile strength.

(b) Working pressure (psi) $= \dfrac{\text{max. pressure (psi)}}{\text{factor of safety}}$

$$= \frac{9533 \text{ psi}}{4}$$

$$= 2383 \text{ psi}$$

The formula is adequate for practical purposes in selecting conductor wall thicknesses to withstand the maximum rate of surge peak pressures at the frequencies developed by the cycling of the hydraulic equipment operation. Surge pressures that may be encountered within a system are rarely known and seldom appreciated in their full potential strength. Actual experience has shown that surge peaks of two to four times that of design working pressure are not uncommon and often not suspected by the system designer. When detailed information on surge peaks is not readily available, the importance of a safety factor becomes appreciated, and it becomes a prime consideration in the circuit design.

In addition to normal working pressures and surge pressure peaks, there may be mechanical stresses produced by thermal expansion, abuse, and environmental factors. Evaluation of all phenomena is difficult, so the proper selection of conductors and connectors is usually a compromise drawing heavily on what has been experienced as most successful. Suppliers are most cooperative in helping to solve problems, and their experience may serve well to avoid trouble and provide excellent performance with maximum safety.

However, pressure surges and general operating characteristics within a hydraulic system can be accurately measured by the use of instruments such as a hydrauliscope (see Fig. 4-1) or with an oscilloscope with power supply and proper accessories. Instruments such as these use a transducer or type of strain gauge that changes the pressure surge into an electronic pulse, which, in turn, is transmitted onto a screen for visual reading. The screen is graduated into divisions that give accurate measurement of the pressure surges within the circuit.

When excessive shock loads are discovered through instrumentation, it is possible to add relief valves, hydraulic fuses, or shock absorbers to prevent serious damage to the hydraulic equipment.

Pressure relief valves appropriately installed are of prime importance for the protection of all system components, including pumps, fluid motors, cylinders, control devices, conductors, connectors, and auxiliary equipment. The relief or safety valve helps to protect a hydraulic system during overpressure conditions that may be impossible to predict. The person fabricating a hydraulic system has a responsibility to recognize the need for additional pressure relief protection that may be caused by a particular installation. The following may serve as a guide in providing safety from overpressures:

1. Provide ample protection for all parts of the system, using one or more relief valves of proper capacity.

2. Install relief valves to provide priority protection for the pump.

3. Provide an additional relief valve or hydraulic fuse to protect extra-long conduit runs.

4. Provide an additional relief valve or hydraulic fuse to protect against trapped fluid which may be subject to shock loads.

5. Use relief valves of the design type best suited to the specific application.

6. Apply shock absorbers where the intensity of surges is beyond the damping characteristics of common relief valves.

Conductors and connectors are available for working pressures to 10,000 psi and higher. The choice of pipe, tubing, or hose is usually a matter of economics. Tubing provides greater flexibility, cleaner appearance, fewer connectors, less leakage, and better reusability. However, pipe is readily available and is generally less in cost. Pipe also has advan-

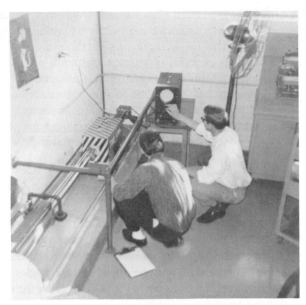

FIGURE 4-1 Hydrauliscope in action. Kenosha Technical Institute students are shown using a hydrauliscope to measure the pressure surges in a shock absorber when a 700-lb weight is accelerated to 15 mph and stopped within a couple of inches. The shock absorber absorbs 72,000 lb of kinetic energy in stopping the weight. Maximum pressure surge recorded nearly 8000 psi.

tages for low-pressure applications, where larger conduits are needed. Figure 4-2 shows a typical installation using modern techniques to connect system components.

PRESSURE DROP

Pressure drop is essentially a result of a change in the hydraulic-energy form. Hydraulic pressure energy starting at the pump with an initial potential value has the ability to perform a specific amount of work. However, as the fluid flows through the conductors or around the bends and turns through the various connectors and components of the system, pressure energy is exchanged for velocity energy and heat. This means that some of the hydraulic energy is lost along the way, because of resistance and friction. Factors that influence loss of energy or pressure include sudden contractions, sudden enlargements, number and kinds of bends, size and smoothness of conductors, as well as the temperature and properties of the fluid itself.

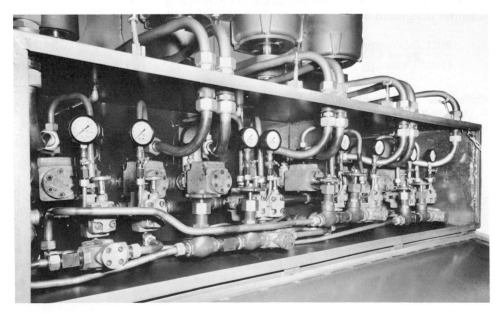

FIGURE 4-2 Distribution lines. (Courtesy of Flodar.)

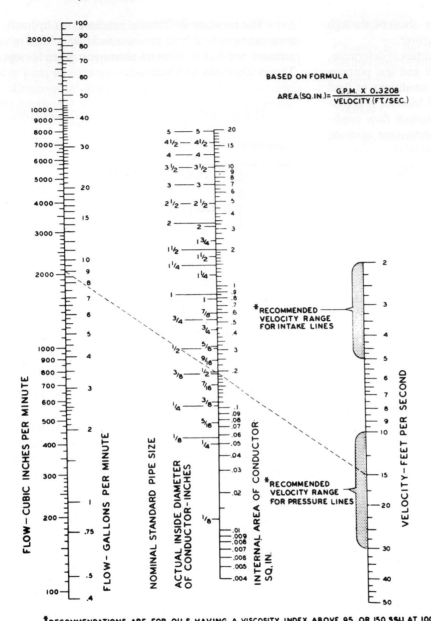

BASED ON FORMULA

$$\text{AREA (SQ.IN.)} = \frac{\text{G.P.M.} \times 0.3208}{\text{VELOCITY (FT/SEC.)}}$$

*RECOMMENDED VELOCITY RANGE FOR INTAKE LINES

*RECOMMENDED VELOCITY RANGE FOR PRESSURE LINES

FLOW—CUBIC INCHES PER MINUTE

FLOW—GALLONS PER MINUTE

NOMINAL STANDARD PIPE SIZE

ACTUAL INSIDE DIAMETER OF CONDUCTOR—INCHES

INTERNAL AREA OF CONDUCTOR SQ.IN.

VELOCITY—FEET PER SECOND

*RECOMMENDATIONS ARE FOR OILS HAVING A VISCOSITY INDEX ABOVE 95 OR 150 SSU AT 100°F AND 50 SSU AT 210°F (ANALINE POINT 190° TO 210°). ASTM #2 GRADE

FIGURE 4-4 Flow capacities of conductors at recommended flow velocities.

To find the inside diameter, use the following equation:

$$D = \sqrt{\frac{\text{area}}{0.7854}}$$

$$= \sqrt{\frac{0.1950}{0.7854}}$$

$$D^2 = \frac{0.1950}{0.7854}$$

$$D = \tfrac{1}{2} \text{ in.} = 0.1964 \text{ in.}^2 \text{ (consult the nomograph)}$$

Check the solution by using the formula

$$\text{Area (in.}^2) = \frac{\text{gpm} \times 0.3208}{V}$$

$$= \frac{9 \times 0.3208}{15}$$

$$= \frac{2.88}{15}$$

$$= 0.192 \text{ in.}^2$$

When selecting pipe, hose, or tubing as the conductor, it is general practice to use the nearest larger standard size.

The velocity of the hydraulic fluid should not exceed the ranges recommended in the right-hand column for maintaining laminar flow.

VIBRATION

Another important design and installation consideration is the effect of vibration on a system. Excessive and continuous vibrations cause conductors, connectors, etc., to fracture with fatigue. Vibration can be absorbed effectively if conductors are properly supported at regular intervals. Table 4-2, taken from Industry Hydraulic Standards, should serve as a guide for the placement of tie-down supports.

TABLE 4-2
PLACEMENT OF TIE-DOWN SUPPORTS

Tube (O. D. in.)	Distance between supports (feet)
$\frac{1}{4}, \frac{5}{16}, \frac{3}{8}, \frac{1}{2}$	3
$\frac{5}{8}, \frac{3}{4}, \frac{7}{8}$	4
1	5
$1\frac{1}{4}, 1\frac{1}{2}$	7

A good ductile material, such as wood, various plastics, and composition materials, can be used as tie-down blocks. Industry standards state that all piping shall be continuous from one piece of apparatus to another, with couplings used only where necessary. Piping shall not be welded to supports and must be removable without dismantling equipment components. When proper installation is made, noise due to vibrations can be minimized. This is an important factor in any circuit.

CHEMICAL REACTION

Corrosion and pitting due to moisture or chemical reactions are destructive agents that attack all components, conductors, and other system parts. Every precaution should be exercised to prevent contamination, which tends to cause malfunction, wear, and a reduction in strength of all materials that make up the system.

RECORDS AND IDENTIFICATION

Records and machinery history are important and should be maintained to have ready information available concerning circuit diagrams, source and identification of each component used in the system, spare parts and service information, type of fluid, and so on. When systems call for two or more pressure levels, branch circuits, auxiliary systems, or metered flow, and so on, the color code pattern shown in Table 4-3 may be helpful to apply for the training of operators and service personnel (see Fig. 4-5).

SELECTION OF CONDUCTORS

The prime considerations for selecting conductors for a hydraulic power system are the type of materials, capacity, and pressure rating. A number of materials, such as steel, iron, and plastic, in both pipe and tubing, are readily available to adequately handle a wide range of flow capacities.

Industry Recommendations for Systems Below 3000 psi

If the system operating pressures are below 3000 psi, the following tables, taken from the Hydraulic Standards, give

TABLE 4-3
COLOR CODE PATTERNS

Function	Definition	Diagram color	Zip-a-Tone no.	
Intensified pressure	Pressure in excess of supply pressure	Black	#7R	
Supply pressure	Power actuating fluid	Red	#177	
Charging press. Pilot pressure Reduced pressure	Inlet supercharge above atmosphere Control actuating Auxiliary pressure lower than supply	Intermittent Red	#52	
Metered flow	Fluid at controlled flow rate, other than pump delivery	Yellow	#63	
Exhaust	Return of power and control fluid to reservoir	Blue	#75	
Intake Drain	Subatmospheric pressure Clearance leakage return to reservoir	Green	#16	
Inactive	Fluid which is idle during the phase being represented	Blank	Blank	

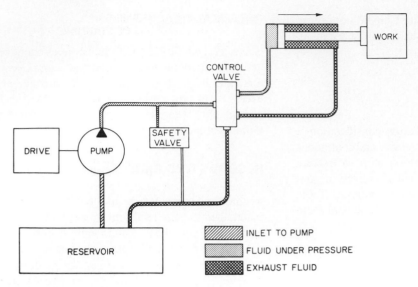

INLET TO PUMP

FLUID UNDER PRESSURE

EXHAUST FLUID

FIGURE 4-5 Circuit showing use of color coding.

pertinent data important as a guide in the selection of pipe and tubing.

Schedule Pipe

Piping was originally classified by weight as "standard," "extra heavy," and "double extra heavy." This classification has been superseded by classification according to schedule numbers. Hot or cold drawn seamless pipe is recommended for use with hydraulic systems and must be free internally from rust, scale, and dirt.

Schedule numbers run from 40 (formerly standard), to schedule 80 (formerly extra heavy), to schedule 160 (formerly double extra heavy).

Tables 4-4 and 4-5 indicate the usage of schedule pipe under the Hydraulic Standards. Tables 4-6 and 4-7 show typical hydraulic tubing data. Table 4-8 gives the dimensions of schedule 160 pipe and the rated working pressure for each wall thickness.

The Hydraulic Standards recommend a 4:1 factor of safety for systems operating above 2500 psi working pressure. (Welded flange fittings, fittings having metal-to-metal

TABLE 4-4
OPERATING PRESSURE (0 TO 1000 psi)

Flow rate (15ft sec) gpm	Valve size	Pipe schedule	Tubing O. D.	Tubing-wall thickness
1	$\frac{1}{8}$	80	$\frac{1}{4}$	0.035
1.5	$\frac{1}{8}$	80	$\frac{5}{16}$	0.035
3	$\frac{1}{4}$	80	$\frac{3}{8}$	0.035
6	$\frac{3}{8}$	80	$\frac{1}{2}$	0.042
10	$\frac{1}{2}$	80	$\frac{5}{8}$	0.049
20	$\frac{3}{4}$	80	$\frac{7}{8}$	0.072
34	1	80	$1\frac{1}{4}$	0.109
58	$1\frac{1}{4}$	80	$1\frac{1}{2}$	0.120

Safety factor 8 : 1

TABLE 4-5
OPERATING PRESSURE (1000 TO 2500 psi)

Flow rate (15ft sec) gpm	Valve size	Pipe schedule	Tubing O. D.	Tubing-wall thickness
2.5	$\frac{1}{4}$	80	$\frac{3}{8}$	0.058
6	$\frac{3}{8}$	80	$\frac{5}{8}$	0.095
10	$\frac{1}{2}$	80	$\frac{3}{4}$	0.120
18	$\frac{3}{4}$	80	1	0.148
30	1	80	$1\frac{1}{4}$	0.180
42	$1\frac{1}{4}$	160	$1\frac{1}{2}$	0.220

Safety factor 6 : 1. Above $\frac{1}{2}$ in. tubing, welded flange fittings or fittings having metal to metal seals or seals that seal with pressure are recommended.

seals, or seals that seal with pressure are recommended.) Pipe should be selected accordingly from the tables. Proper area size for a given flow at a recommended velocity can be determined by using the nomograph or its formula: $A = gpm \times 0.3208 \div velocity$.

Tubing for Higher-Pressure Systems

Experience has indicated that on a broad basis tubing is proving to be increasingly more applicable to hydraulic systems than pipe. However, each has advantages and disadvantages.

Table 4-9 has been tabulated for the selection of tubing for working pressures up to 10,000 psi with a 4:1 safety factor. Very little difficulty exists in the selection of tubing for high-pressure systems in the smaller-capacity sizes (below $\frac{1}{2}$-in.-OD tubing).

In the larger sizes of $\frac{1}{2}$ to $1\frac{1}{2}$ in. OD (graph), Table 4-10 gives the recommended wall thickness for a given working pressure. Compilations are based on AISIA 4130 steel. Table 4-11 lists standard tube radii in inches.

TABLE 4-6
DEAD SOFT COLD-DRAWN-STEEL TUBING
SAE 1010 RECOMMENDED BY INDUSTRY
STANDARDS AND COMMONLY CARRIED
IN STOCK BY METAL WAREHOUSES

O. D. (in.)	Wall thickness (in.)	Burst pressure (psi)
$\frac{1}{8}$	0.032	28,000
$\frac{3}{16}$	0.032	18,500
$\frac{1}{4}$	0.035	15,400
$\frac{5}{16}$	0.035	12,300
	0.049	17,300
$\frac{3}{8}$	0.035	10,200
	0.049	14,400
$\frac{1}{2}$	0.035	7,700
	0.049	10,700
	0.065	14,000
$\frac{5}{8}$	0.049	8,600
	0.065	11,500
	0.083	14,600
$\frac{3}{4}$	0.049	7,200
	0.065	9,400
	0.095	13,900
$\frac{7}{8}$	0.065	8,150
	0.083	10,000
1	0.049	5,400
	0.065	7,150
	0.095	10,400
$1\frac{1}{4}$	0.083	7,300
$1\frac{1}{2}$	0.095	6,950

TABLE 4-7
COMPOSITION AND PHYSICAL PROPERTIES

Tubing composition		Physical properties	
Carbon	0.08–0.18	Tensile strength—max.	55,000 ps
Manganese	0.30–0.60	Elongation in 2 in.—min.	35%*
Phosphorus	0.50 max.	Rockwell hardness	B65 max
Sulfur	0.55 max.		

*For tubes with O. D. of $\frac{3}{8}$ in. and/or wall thickness of 0.035 a minimum elongation of 30 per cent is permitted.

Plastic Conductors

Plastic hose and tubing are being applied with success to a number of hydraulic systems requiring a lightweight and flexible conductor in the lower pressure ranges.

A number of plastics are available which include nylon, polyethylene, polypropylene, and polyvinyl chloride. Each of these materials has its own advantages and disadvantages for specific applications. Some plastics have the ability to absorb water, swell, and lose strength, which may make it necessary to consult the manufacturer for this important property. Table 4-12 shows the pressure ratings for the materials mentioned above at temperature conditions.

TABLE 4-8
DIMENSIONS OF STANDARD PIPE
AND TUBING FOR THE HIGHER PRESSURE
RANGES. (ASA B36. 10-1950) SCHEDULE 160 PIPE

Size (in.)	O. D. (in.)	I. D. (in.)	Wall thickness (in.)	Internal area (in.)	Rated pressure (psi)
$\frac{1}{2}$	0.84	0.466	0.187	0.171	5010
$\frac{3}{4}$	1.05	0.614	0.218	0.208	4670
1	1.315	0.815	0.250	0.522	4280
$1\frac{1}{4}$	1.660	1.160	0.250	1.060	3390
$1\frac{1}{2}$	1.900	1.338	0.281	1.410	3330
2	2.375	1.689	0.343	2.250	3250
$2\frac{1}{2}$	2.875	2.125	0.375	3.55	2940
3	3.500	2.626	0.437	5.41	2810
4	4.500	3.438	0.531	9.32	2655
5	5.563	4.313	0.625	14.65	2530
6	6.625	5.189	0.718	21.20	2440
8	8.625	6.813	0.906	36.70	2365

Although testing for a specific purpose is recommended, a 4:1 safety factor is considered good engineering practice in the average hydraulic systems. Table 4-13 gives the standard sizes available in two popular pressure ranges.

In general, plastic tubing can be worked and installed with the ordinary tubing tools. It cuts easily and can be heated and given permanent bends. It can be used with standard metallic compression and flare fittings designed for metal tubing. Many new developments are being made in the nature of tools, fittings, quick disconnects, and other devices designed especially for plastic fabrications.

Flexible Hoses

Synthetic hose is manufactured from a large number of elastomeric or rubberlike compounds and plastics. Manufacturers are keeping up with the present trend toward more horsepower with higher pressure. Flexible hose and connectors are now available to fulfill the requirements of circuits operating at pressures up to 10,000 psi and higher. Tables 4-14 through 4-18 list important data for the selection of flexible hoses covering a broad range of working pressures with a 4:1 safety factor. As pressures increase it becomes more important to carefully assess installation factors such as minimum bend radius and positioning of the hoses so that they cannot damage personnel or machinery in the event of a failure that could allow the hose to swing free.

Important Factors for the Application of Flexible Hose

1. Hose assemblies must be of proper overall length.
2. Hose under variations in working pressure must have enough length to expand and contract.

TABLE 4-9
TUBING FOR PRESSURE UP TO 10,000 PSI, TYPE AISIA 4130 OR EQUIVALENT
RECOMMENDED FOR HIGH PRESSURE SYSTEMS
TENSILE STRENGTH 70,000 PSI (STANDARD NORMALIZED)

Size or O.D. (in.)	I.D. (in.)	Wall thickness (in.)	Internal area (sq in.)	Size or O.D. (in.)	I.D. (in.)	Wall thickness (in.)	Internal area (sq in.)
$\frac{1}{8}$	0.069	0.028	0.003739		0.482	0.134	0.1825
					0.437	0.156	0.1500
$\frac{3}{16}$	0.131	0.028	0.01348		0.375	0.188	0.1104
					0.312	0.219	0.07645
$\frac{1}{4}$	0.180	0.035	0.02545		0.250	0.250	0.04909
	0.166	0.042	0.02164	$\frac{7}{8}$	0.709	0.083	0.3948
	0.152	0.049	0.01815		0.685	0.095	0.3685
	0.134	0.058	0.01410		0.657	0.109	0.3390
	0.120	0.065	0.01131		0.635	0.120	0.3167
	0.084	0.083	0.005542		0.625	0.125	0.3068
	0.060	0.095	0.002827		0.607	0.134	0.2894
$\frac{5}{16}$	0.214	0.049	0.03597		0.562	0.156	0.2481
	0.196	0.058	0.03017		0.500	0.188	0.1963
	0.182	0.065	0.02602		0.438	0.219	0.1507
	0.146	0.083	0.01674		0.375	0.250	0.1104
	0.122	0.095	0.004072	1	0.834	0.083	0.5463
$\frac{3}{8}$	0.277	0.049	0.06026		0.810	0.095	0.5153
	0.259	0.058	0.05269		0.782	0.109	0.4803
	0.245	0.065	0.04714		0.760	0.120	0.4536
	0.209	0.083	0.03431		0.750	0.125	0.4418
	0.185	0.095	0.02688		0.732	0.134	0.4208
	0.135	0.120	0.01431		0.687	0.156	0.3707
$\frac{1}{2}$	0.402	0.049	0.1269		0.625	0.188	0.3068
	0.384	0.058	0.1158		0.562	0.219	0.2481
	0.370	0.065	0.1075		0.500	0.250	0.1963
	0.334	0.083	0.08762		0.250	0.375	0.04909
	0.310	0.095	0.07548	$1\frac{1}{4}$	1.060	0.095	0.882
	0.282	0.109	0.06246		1.010	0.120	0.801
	0.260	0.120	0.05309		0.982	0.134	0.7574
	0.232	0.134	0.04227		0.937	0.156	0.6896
	0.187	0.156	0.02746		0.920	0.165	0.6648
$\frac{5}{8}$	0.509	0.058	0.2035		0.875	0.188	0.6013
	0.495	0.065	0.1924		0.812	0.219	0.5178
	0.459	0.083	0.1655		0.750	0.250	0.4418
	0.453	0.095	0.1612		0.625	0.313	0.3068
	0.407	0.109	0.1301		0.500	0.375	0.1963
	0.385	0.120	0.1164	$1\frac{1}{2}$	1.310	0.095	1.348
	0.357	0.135	0.1001		1.260	0.120	1.247
	0.312	0.156	0.07645		1.232	0.134	1.192
	0.250	0.188	0.04909		1.187	0.156	1.108
	0.187	0.219	0.02746		1.125	0.188	0.9940
$\frac{3}{4}$	0.606	0.072	0.2884		1.062	0.219	0.8866
	0.584	0.083	0.2679		1.000	0.250	0.7854
	0.560	0.095	0.2463		0.875	0.313	0.6013
	0.532	0.109	0.2223		0.750	0.375	0.4418
	0.510	0.120	0.2043		0.500	0.500	0.1963

Tubing Composition

Carbon	0.28–0.33
Manganese	0.40–0.60
Phosphorus	0.04 max.
Sulfur	0.04 max.
Silicon	0.20–0.35
Chromium	0.80–1.10
Molybdenum	0.15–0.25

TABLE 4-10
RECOMMENDED WALL THICKNESS

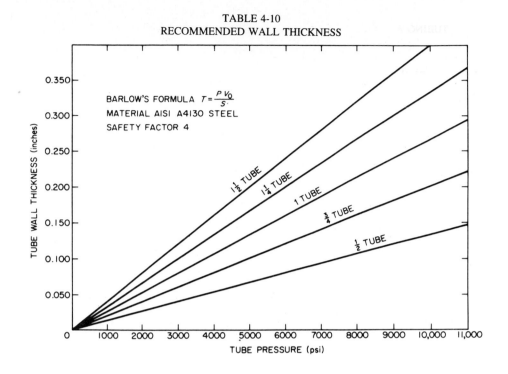

4-11
STANDARD TUBE RADII-INCHES

Number	Tube size O. D.	Radius bend centerline
2	$\frac{1}{8}$	$\frac{3}{8}$
3	$\frac{3}{16}$	$\frac{7}{16}$
4	$\frac{1}{4}$	$\frac{9}{16}$
5	$\frac{5}{16}$	$\frac{11}{16}$
6	$\frac{3}{8}$	$\frac{15}{16}$
8	$\frac{1}{2}$	$1\frac{1}{2}$
10	$\frac{5}{8}$	2
12	$\frac{3}{4}$	$2\frac{1}{2}$
14	$\frac{7}{8}$	3
16	1	$3\frac{1}{2}$

Approximately 3X the tube O. D. except for heavy-walled tube, for which the following formula may be useful.

$$\text{Minimum radius bend} = \frac{0.5 \times \text{O.D. in.}}{\% \text{ elongation in a 2 in.}}$$
length expressed in decimal

TABLE 4-12
PRESSURE RATINGS FOR PLASTIC TUBING

Operating temperature at room temperature	Nylon		Polyethylene		Polypropylene		Polyvinyl chloride	
	Burst pressure 2200	Work pressure 550	Burst *pressure 250–600	Work *pressure 75–160	Burst *pressure —	Work pressure 20–55	Burst pressure 1800	Work pressure 450
140°	1500	375	125–300	37–80	—	—	1200	300
230°	890	225	—	—	—	—	540	135
300°	310	75	—	—	—	—	90	25

*Depending on tube size and wall thickness

TABLE 4-13
STANDARD SIZES NORMALLY AVAILABLE

	Up to 250 psi *working pressure*					*Up to* 500 psi *working pressure*				
O. D.	$\frac{1}{8}$	$\frac{3}{16}$	$\frac{1}{4}$	$\frac{5}{16}$	$\frac{3}{8}$	$\frac{1}{8}$	$\frac{3}{16}$	$\frac{1}{4}$	$\frac{5}{16}$	$\frac{3}{8}$
I. D.	0.096	0.138	0.190	0.242	0.295	0.078	0.110	0.150	0.188	0.225
Minimum bend radius inches	$\frac{5}{8}$	1	$1\frac{1}{4}$	2	3	$\frac{3}{8}$	$\frac{5}{8}$	$1\frac{1}{4}$	3	$2\frac{1}{2}$

TABLE 4-14
ONE-WIRE BRAID HOSE
(SAE 100 RI)

I.D. (in.)	O.D. (in.)	Burst pressure (psi)	Work pressure (psi)	Min bend radii (in.)
$\frac{1}{4}$	$\frac{11}{16}$	11,000	2750	4
$\frac{3}{8}$	$\frac{25}{32}$	9,000	2250	5
$\frac{1}{2}$	$\frac{29}{32}$	8,000	2000	7
$\frac{3}{4}$	$1\frac{3}{16}$	5,000	1250	$9\frac{1}{2}$
1	$1\frac{1}{2}$	4,000	1000	$9\frac{1}{2}$
$1\frac{1}{4}$	$1\frac{13}{16}$	2,500	625	16
$1\frac{1}{2}$	$2\frac{1}{16}$	2,000	500	20

TABLE 4-15
TWO-WIRE BRAID HOSE
(SAE 100 R2)

I.D. (in.)	O.D. (in.)	Burst pressure (psi)	Work pressure (psi)	Min bend radii (in.)
$\frac{1}{4}$	$\frac{11}{16}$	20,000	5000	4
$\frac{3}{8}$	$\frac{27}{32}$	16,000	4000	5
$\frac{1}{2}$	$\frac{31}{32}$	14,000	3500	7
$\frac{3}{4}$	$1\frac{1}{4}$	9,000	2250	$9\frac{1}{2}$
1	$1\frac{9}{16}$	7,500	1875	$9\frac{1}{2}$
$1\frac{1}{4}$	2	6,500	1625	16
$1\frac{1}{2}$	$2\frac{1}{4}$	5,000	1250	20
2	$2\frac{3}{4}$	4,500	1125	22

TABLE 4-16
*THREE-WIRE BRAID HOSE

I.D. (in.)	O.D. (in.)	Burst pressure (psi)	Work pressure (psi)	Min bend radii (in.)
$\frac{1}{4}$	$\frac{3}{4}$	28,000	7000	5
$\frac{3}{8}$	$\frac{29}{32}$	22,000	5500	6
$\frac{1}{2}$	$1\frac{5}{64}$	20,000	5000	8
$\frac{3}{4}$	$1\frac{5}{16}$	16,000	4000	10
1	$1\frac{9}{16}$	11,000	2750	12
$1\frac{1}{4}$	$2\frac{1}{16}$	10,000	2500	17
$1\frac{1}{2}$	$2\frac{5}{16}$	8,000	2000	21
2	$2\frac{7}{8}$	8,000	2000	26

*Wire diameter is of equal importance, and the various manufacturers may have different ways of expressing wire braid or spiral wrapping.

TABLE 4-17
FOUR-WIRE BRAID HOSE

I.D. (in.)	O.D. (in.)	Burst pressure (psi)	Work pressure (psi)	Min bend radii (in.)
$\frac{1}{4}$	$\frac{13}{16}$	35,000	8750	5
$\frac{3}{8}$	$\frac{15}{16}$	30,000	7500	6
$\frac{1}{2}$	$1\frac{3}{32}$	25,000	6250	8
$\frac{3}{4}$	$1\frac{7}{16}$	20,000	5000	10
1	$1\frac{3}{4}$	16,000	4000	12
$1\frac{1}{4}$	2	12,000	3000	18
$1\frac{1}{2}$	$2\frac{1}{4}$	10,000	2500	22
2	$2\frac{25}{32}$	10,000	2500	—

*Wire diameter is of equal importance, and the various manufacturers may have different ways of expressing wire braid or spiral wrapping.

TABLE 4-18
SIX-WIRE BRAID HOSE

I.D. (in.)	O.D. (in.)	Burst pressure (psi)	Work pressure (psi)	Min bend radii (in.)
$\frac{1}{4}$	$\frac{15}{16}$	45,000	11,250	5
$\frac{3}{8}$	$1\frac{1}{16}$	40,000	10,000	6
$\frac{1}{2}$	$1\frac{7}{32}$	30,000	7,500	8
$\frac{3}{4}$	$1\frac{9}{16}$	25,000	6,250	10
1	$1\frac{29}{32}$	20,000	5,000	12
$1\frac{1}{4}$	$2\frac{5}{32}$	20,000	5,000	12
$1\frac{1}{2}$	$2\frac{13}{32}$	14,000	3,500	18
2	$2\frac{31}{32}$	12,000	3,000	22

*Wire diameter is of equal importance, and the various manufacturers may have different ways of expressing wire braid or spiral wrapping.

3. Do not clamp high- and low-pressure hoses together.

4. Avoid clamping at bends so that the bend radii can absorb change.

5. When there is relative motion between the two ends of a hose assembly, always allow for adequate length of travel.

6. To prevent twisting, hose should be bent in the same plane as the motion of the boss to which it is connected.

7. To prevent twisting in hose lines bent in two planes, clamp hose at the change of plane.

8. Use the proper hydraulic adapters to reduce the number of joints and improve performance as well as appearance.

9. Where the radius falls below the required minimum bend, an angle adapter should be used.

10. Avoid contact with sharp edges or moving parts.

11. Arrange proper positioning of hose and adapters before tightening to avoid distortion.

12. Apply clamps properly and maintain tight to prevent abrasion due to line surge.

13. Be sure to use the proper strength of hose to maintain a good margin of safety.

14. Select the proper size of hose to stay within the recommended velocity range (consult a velocity-flow nomograph; see Figure 4-4).

15. Prevent dirt, chips, or any foreign materials from entering the system during fabrication of the circuit.

16. Be sure that all materials used are compatible with the hydraulic fluid designated for the system.

CONNECTORS AND INSTALLATION INFORMATION

Hydraulic systems often require pipe, tubing, and hose in the same system. Connections must be transposed from one design configuration to another and satisfy variables such as size, thread type, adaptability, and so on.

Connectors are available in standard sizes, design types, and materials that provide excellent results for the installation of hydraulic systems covering practically all fields of applications and operation pressures. However, several primary requirements must be considered in the selecting of hydraulic system connectors.

1. The efficiency of a connector to pass fluid with a minimum pressure drop determines its performance as an important part of the hydraulic circuit.

2. Connectors shall provide leakproof joints under maximum pressures. (This important requirement, however, depends largely on the level of cleanliness in the installation.)

3. The degree of workability and ease with which installations can be made or assembled. Connectors may also be considered from a reusability factor for economy.

To help minimize variations in connectors and to establish that certain connections can be universally used, standards were set up giving recommended practices for making connections and joints. The following list is taken from the Hydraulic Standards and may serve as a guide.

1. Piping between actuating and feed-control devices should be constructed rigidly to confine a minimum volume of fluid. The purpose is to maintain constant, controlled motion and to restrict the effort of varying forces.

2. Solderless pipe connections such as flare, flareless, self-flaring, and flange should be used.

3. Performance of any type of fitting used should be equal to or exceed the requirements of MIL-F-5506.

4. All pipe threads should be Dryseal American (National) Standard Taper Pipe Threads (NPTF).

5. Fittings with straight threads and which incorporate pressure seals may be used in place of pipe thread fittings.

6. All piping connections should be designed and installed to permit quick removal and reassembly by means of hand tools.

7. Only steel should be used as fitting materials. Copper alloy fittings should not be used with petroleum-base fluids. Fittings with restricted or stepped-up passages are not recommended.

8. Whenever practicable, each piping run should be integral and continuous from one piece of apparatus to another. Piping runs should be removable without dismantling and without bending or springing the tubing in a manner that would damage it.

9. When high or extra pressure piping is used, all connections should be welded to steel flanges or connections should be used which are equal in performance and ease of assembly. Flanged connections should use sealing devices that seal with pressure.

Threads and Ports

The following thread standards have been accepted by users of hydraulic equipment including pipe, tubing, hose, and fittings. Table 4-19 coordinates the dimensions of the various conductors and connectors according to thread type and size. Table 4-20 provides data for pipe threads.

1. Dryseal Standard American Pipe Thread (NPTF)
2. Joint Industry Conference (J.I.C.) Standard Thread with 37° Male Flare
3. Joint Industry Conference (J.I.C.) Standard Thread with 37° Female Flare

TABLE 4-19
STANDARD THREAD TYPES
COORDINATING INSIDE DIAMETER OF HOSE WITH
TUBING AND PIPE

Hose I.D.	Tube O.D.	Pipe thread	J.I.C 37° flare	SAE straight thread "O" ring
$\frac{1}{4}$	$\frac{5}{16}$	$\frac{1}{4}$	$\frac{1}{2}$-20	$\frac{1}{2}$-20
$\frac{3}{8}$	$\frac{1}{2}$	$\frac{3}{8}$	$\frac{3}{4}$-16	$\frac{3}{4}$-16
$\frac{1}{2}$	$\frac{5}{8}$	$\frac{1}{2}$	$\frac{7}{8}$-14	$\frac{7}{8}$-14
$\frac{3}{4}$	$\frac{3}{4}$	$\frac{3}{4}$	$1\frac{1}{16}$-12	$1\frac{1}{16}$-12
1	1	1	$1\frac{5}{16}$-12	$1\frac{5}{16}$-12
$1\frac{1}{4}$	$1\frac{1}{4}$	$1\frac{1}{4}$	$1\frac{5}{8}$-12	$1\frac{5}{8}$-12
$1\frac{1}{2}$	$1\frac{1}{2}$	$1\frac{1}{2}$	$1\frac{7}{8}$-12	$1\frac{7}{8}$-12
2	2	2	$2\frac{1}{2}$-12	$2\frac{1}{2}$-12

TABLE 4-20
PIPE THREADS

Pipe size (in.)	O.D. (in.)	Length of thread screwed in fitting (in.)	Threads per in.	Taper
$\frac{1}{8}$	0.405	$\frac{1}{4}$	27	$\frac{3}{4}$ in. per ft
$\frac{1}{4}$	0.540	$\frac{3}{8}$	18	″
$\frac{3}{8}$	0.675	$\frac{3}{8}$	18	″
$\frac{1}{2}$	0.840	$\frac{1}{2}$	14	″
$\frac{3}{4}$	1.050	$\frac{1}{2}$	14	″
1	1.315	$\frac{9}{16}$	$11\frac{1}{2}$	″

4. Society of Automotive Engineers (SAE) Standard Male Thread with an O-ring

The Dryseal American (National) Standard Taper Pipe Threads differ from National Standard Pipe Threads by engaging roots and crests before flanks, so that spiral clearance occurring in standard threads is avoided. Special taps are used for Dryseal Threads. Table 4-21 provides a convenient comparison of areas of circles.

This section includes design illustrations of the various fittings and pointers on how to bend tubing.

Tube bending and fabrication is an art that can be learned only through practical experience. The information given is useful as a guide for students to follow, and to identify tube and pipe fittings commonly used.

Flange and Weld Connections

Flange connections for large components and on some critical applications are sometimes preferred. Sealing in a flanged joint is achieved with a static seal against a ground surface. The flanges are usually welded to the pipe or tube. In the case of flexible hose, the flange is fixed permanently or clamp-connected to hose. Vibration resistance in this type of connection is excellent and reuseability is unlimited. Restrictions of passages are avoided, since ports of both mating parts coincide. The relatively large area which is required

TABLE 4-21
AREAS OF CIRCLES

D = diameter $\hspace{6cm}$ $A = 0.7853981634\ D^2$

DIAM.	0	1	2	3	4	5	6	7	8	9	10	11	12	13	14	15	16	17	DIAM.
0		.7854	3.1416	7.0686	12.566	19.635	28.274	38.485	50.266	63.617	78.540	95.033	113.10	132.73	153.94	176.71	201.06	226.98	0
1/64	.000192	.8101	3.1909	7.1424	12.665	19.753	28.422	38.656	50.462	63.838	78.785	95.303	113.39	133.05	154.28	177.08	201.45	227.40	1/64
1/32	.000767	.8352	3.2405	7.2166	12.763	19.881	28.570	38.829	50.659	64.060	79.031	95.574	113.69	133.37	154.63	177.45	201.85	227.82	1/32
3/64	.001726	.8607	3.2906	7.2912	12.863	20.005	28.718	39.002	50.856	64.282	79.278	95.845	113.98	133.69	154.97	177.82	202.24	228.23	3/64
1/16	.003068	.8866	3.3410	7.3662	12.962	20.129	28.866	39.175	51.054	64.504	79.525	96.116	114.23	134.01	155.32	178.19	202.64	228.65	1/16
5/64	.004794	.9128	3.3918	7.4415	13.062	20.253	29.015	39.348	51.252	64.727	79.772	96.388	114.57	134.33	155.66	178.56	203.03	229.07	5/64
3/32	.006903	.9306	3.4430	7.5173	13.162	20.378	29.165	39.522	51.450	64.950	80.019	96.660	114.87	134.65	156.01	178.93	203.43	229.49	3/32
7/64	.009396	.9666	3.4946	7.5934	13.263	20.503	29.315	39.696	51.649	65.173	80.267	96.932	115.17	134.98	156.35	179.30	203.82	229.91	7/64
1/8	.01227	.9940	3.5466	7.6699	13.364	20.629	29.465	39.871	51.849	65.397	80.516	97.205	115.47	135.30	156.70	179.67	204.22	230.33	1/8
9/64	.01553	1.0218	3.5989	7.7468	13.465	20.755	29.615	40.046	52.048	65.621	80.764	97.479	115.76	135.62	157.05	180.04	204.61	230.75	9/64
5/32	.01917	1.0500	3.6516	7.8241	13.567	20.881	29.766	40.222	52.248	65.845	81.013	97.752	116.06	135.94	157.39	180.42	205.01	231.17	5/32
11/64	.02320	1.0786	3.7048	7.9017	13.669	21.008	29.917	40.398	52.448	66.070	81.263	98.026	116.36	136.27	157.74	180.79	205.40	231.59	11/64
3/16	.02761	1.1075	3.7583	7.9798	13.772	21.135	30.069	40.574	52.649	66.296	81.513	98.301	116.66	136.59	158.09	181.16	205.80	232.01	3/16
13/64	.03240	1.1369	3.8121	8.0582	13.875	21.263	30.221	40.750	52.850	66.522	81.763	98.575	116.96	136.91	158.44	181.53	206.20	232.44	13/64
7/32	.03758	1.1666	3.8664	8.1370	13.978	21.391	30.374	40.927	53.052	66.747	82.014	98.850	117.26	137.24	158.79	181.91	206.60	232.86	7/32
15/64	.04314	1.1967	3.9211	8.2162	14.082	21.519	30.526	41.105	53.254	66.974	82.265	99.126	117.56	137.56	159.14	182.28	206.99	233.28	15/64
1/4	.04909	1.2272	3.9761	8.2958	14.186	21.648	30.680	41.282	53.456	67.201	82.516	99.402	117.86	137.89	159.48	182.65	207.39	233.71	1/4
17/64	.05541	1.2577	4.0315	8.3757	14.291	21.777	30.833	41.461	53.659	67.428	82.768	99.678	118.16	138.21	159.83	183.08	207.79	234.13	17/64
9/32	.06213	1.2893	4.0873	8.4561	14.396	21.906	30.987	41.639	53.862	67.655	83.020	99.955	118.46	138.54	160.19	183.40	208.19	234.55	9/32
19/64	.06922	1.3209	4.1435	8.5368	14.501	22.036	31.141	41.818	54.065	67.883	83.272	100.232	118.76	138.86	160.54	183.78	208.59	234.98	19/64
5/16	.07670	1.3530	4.2000	8.6179	14.607	22.166	31.296	41.997	54.269	68.112	83.525	100.509	119.06	139.19	160.89	184.15	208.99	235.40	5/16
21/64	.08456	1.3854	4.2570	8.6994	14.713	22.297	31.451	42.177	54.473	68.341	83.779	100.787	119.37	139.52	161.24	184.53	209.39	235.83	21/64
11/32	.09281	1.4182	4.3143	8.7813	14.819	22.428	31.607	42.357	54.678	68.570	84.032	101.066	119.67	139.84	161.59	184.91	209.79	236.25	11/32
23/64	.1014	1.4513	4.3720	8.8636	14.926	22.559	31.763	42.537	54.883	68.799	84.286	101.344	119.97	140.17	161.94	185.28	210.20	236.68	23/64
3/8	.1104	1.4849	4.4301	8.9462	15.033	22.691	31.919	42.718	55.088	69.029	84.541	101.623	120.28	140.50	162.30	185.66	210.60	237.10	3/8
25/64	.1198	1.5188	4.4886	9.0292	15.141	22.823	32.076	42.899	55.294	69.259	84.796	101.903	120.58	140.83	162.65	186.04	211.00	237.53	25/64
13/32	.1296	1.5532	4.5475	9.1126	15.249	22.955	32.233	43.081	55.500	69.490	85.051	102.182	120.88	141.16	163.00	186.42	211.40	237.96	13/32
27/64	.1398	1.5879	4.6067	9.1964	15.357	23.088	32.390	43.263	55.707	69.721	85.306	102.462	121.19	141.49	163.36	186.79	211.80	238.39	27/64
7/16	.1503	1.6230	4.6664	9.2806	15.466	23.221	32.548	43.445	55.914	69.953	85.562	102.743	121.49	141.82	163.71	187.17	212.21	238.81	7/16
29/64	.1613	1.6584	4.7264	9.3652	15.575	23.355	32.706	43.628	56.121	70.184	85.819	103.024	121.80	142.15	164.06	187.55	212.61	239.24	29/64
15/32	.1726	1.6943	4.7868	9.4501	15.684	23.489	32.865	43.811	56.329	70.417	86.075	103.305	122.11	142.48	164.42	187.93	213.02	239.67	15/32
31/64	.1840	1.7305	4.8476	9.5354	15.794	23.623	33.024	43.995	56.537	70.649	86.332	103.587	122.43	142.81	164.77	188.31	213.42	240.10	31/64
1/2	.1963	1.7671	4.9088	9.6212	15.904	23.758	33.183	44.179	56.745	70.882	86.590	103.869	122.72	143.14	165.13	188.69	213.82	240.53	1/2
33/64	.2088	1.8042	4.9703	9.7072	16.015	23.893	33.343	44.363	56.954	71.116	86.848	104.151	123.03	143.47	165.49	189.07	214.23	240.96	33/64
17/32	.2217	1.8415	5.0322	9.7937	16.126	24.029	33.503	44.548	57.163	71.349	87.106	104.434	123.33	143.80	165.84	189.45	214.64	241.39	17/32
35/64	.2349	1.8793	5.0946	9.8806	16.237	24.165	33.663	44.733	57.373	71.583	87.365	104.717	123.64	144.13	166.20	189.83	215.04	241.82	35/64
9/16	.2485	1.9175	5.1573	9.9678	16.349	24.301	33.824	44.918	57.583	71.818	87.624	105.001	123.95	144.47	166.56	190.22	215.45	242.25	9/16
37/64	.2625	1.9560	5.2203	10.0544	16.461	24.438	33.985	45.104	57.793	72.053	87.883	105.285	124.26	144.80	166.91	190.60	215.85	242.68	37/64
19/32	.2769	1.9949	5.2838	10.1435	16.574	24.575	34.147	45.290	58.004	72.288	88.143	105.569	124.57	145.13	167.27	190.98	216.26	243.11	19/32
39/64	.2916	2.0342	5.3477	10.2318	16.687	24.713	34.309	45.477	58.215	72.524	88.404	105.804	124.88	145.47	167.63	191.36	216.67	243.54	39/64
5/8	.3068	2.0739	5.4119	10.3206	16.800	24.850	34.472	45.664	58.426	72.760	88.664	106.139	125.19	145.80	167.99	191.75	217.08	243.98	5/8
41/64	.3223	2.1140	5.4765	10.4098	16.914	24.989	34.634	45.851	58.638	72.996	88.925	106.425	125.50	146.14	168.35	192.13	217.48	244.41	41/64
21/32	.3382	2.1545	5.5415	10.4994	17.028	25.127	34.798	46.039	58.850	73.233	89.186	106.771	125.81	146.47	168.71	192.52	217.89	244.84	21/32
43/64	.3545	2.1953	5.6069	10.5893	17.142	25.266	34.961	46.227	59.063	73.470	89.448	106.997	126.12	146.81	169.07	192.90	218.30	245.28	43/64
11/16	.3712	2.2365	5.6727	10.6796	17.257	25.406	35.125	46.415	59.276	73.708	89.710	107.284	126.43	147.14	169.43	193.28	218.71	245.71	11/16
45/64	.3883	2.2782	5.7388	10.7703	17.372	25.546	35.289	46.604	59.489	73.946	89.973	107.571	126.74	147.48	169.79	193.67	219.12	246.14	45/64
23/32	.4057	2.3201	5.8054	10.8614	17.488	25.686	35.454	46.793	59.703	74.184	90.236	107.858	127.05	147.82	170.15	194.06	219.53	246.58	23/32
47/64	.4236	2.3623	5.8723	10.9528	17.604	25.826	35.619	46.983	59.917	74.423	90.499	108.146	127.36	148.15	170.51	194.44	219.94	247.01	47/64
3/4	.4418	2.4053	5.9396	11.0447	17.721	25.967	35.785	47.173	60.132	74.662	90.763	108.434	127.68	148.49	170.87	194.83	220.35	247.45	3/4
49/64	.4604	2.4484	6.0073	11.1369	17.837	26.108	35.951	47.363	60.347	74.901	91.027	108.723	127.99	148.83	171.24	195.21	220.76	247.89	49/64
25/32	.4794	2.4929	6.0753	11.2295	17.954	26.250	36.117	47.554	60.562	75.141	91.291	109.012	128.30	149.17	171.60	195.60	221.18	248.32	25/32
51/64	.4987	2.5359	6.1438	11.3236	18.072	26.392	36.283	47.745	60.778	75.382	91.556	109.301	128.62	149.51	171.96	195.99	221.59	248.76	51/64
13/16	.5185	2.5802	6.2126	11.4159	18.190	26.535	36.450	47.937	60.994	75.622	91.821	109.591	128.93	149.84	172.32	196.38	222.00	249.20	13/16
53/64	.5386	2.6248	6.2819	11.5096	18.308	26.678	36.618	48.129	61.211	75.863	92.087	109.881	129.25	150.18	172.69	196.77	222.41	249.63	53/64
27/32	.5591	2.6699	6.3515	11.6038	18.427	26.821	36.787	48.321	61.427	76.105	92.353	110.171	129.56	150.52	173.05	197.15	222.83	250.07	27/32
55/64	.5800	2.7153	6.4215	11.6983	18.546	26.964	36.954	48.514	61.645	76.346	92.619	110.462	129.88	150.86	173.42	197.54	223.24	250.51	55/64
7/8	.6013	2.7612	6.4918	11.7933	18.665	27.109	37.122	48.707	61.862	76.589	92.886	110.753	130.19	151.20	173.78	197.93	223.65	250.95	7/8
57/64	.6230	2.8074	6.5624	11.8885	18.785	27.252	37.291	48.900	62.080	76.831	93.153	111.045	130.51	151.54	174.15	198.32	224.07	251.39	57/64
29/32	.6450	2.8540	6.6337	11.9842	18.906	27.398	37.461	49.094	62.299	77.074	93.420	111.337	130.82	151.88	174.51	198.71	224.48	251.83	29/32
59/64	.6675	2.9010	6.7052	12.0803	19.026	27.543	37.630	49.288	62.518	77.317	93.688	111.630	131.14	152.22	174.88	199.10	224.90	252.26	59/64
15/16	.6923	2.9483	6.7771	12.1768	19.147	27.688	37.800	49.483	62.737	77.561	93.956	111.922	131.46	152.57	175.25	199.49	225.31	252.70	15/16
61/64	.7135	2.9961	6.8494	12.2736	19.268	27.834	37.971	49.678	62.956	77.805	94.225	112.215	131.78	152.91	175.61	199.89	225.73	253.15	61/64
31/32	.7371	3.0442	6.9221	12.3708	19.390	27.981	38.142	49.874	63.176	78.050	94.494	112.509	132.09	153.25	175.98	200.28	226.15	253.59	31/32
63/64	.7610	3.0920	6.9952	12.4684	19.512	28.127	38.313	50.069	63.396	78.295	94.763	112.803	132.41	153.59	176.35	200.67	226.56	254.03	63/64

to accommodate flange bolts somewhat restricts its use to the larger components. However, excellent results have been obtained with the flanged joint in connection with high-pressure hydraulic systems.

Types of Connectors

Connectors may be divided into two classifications, the flare type and the compression type. These two groups are further subdivided as follows:

Flare type	Compression type
37° flare	Beveled sleeve
45° flare	Tapered sleeve (bite type)
Inverted flare	Threaded sleeve
	O-ring-split sleeve
	Nut-shearable sleeve
	Flex sleeve design
	Tube-supported sleeve design

All of the designs above have specific advantages (disadvantages) and should be selected for hydraulic service on the basis of the manufacturers' specifications. Figures 4-6 through 4-14 show the physical characteristics of a few of the popularly used designs.

In view of the difficulties encountered when large or heavy-walled conductors are being bent, it is often advisable to buy prefabricated bends and other weld-type connectors for a better and neater system. The bends and fittings may generally be supplied with the proper radius and have chamfered edges. The inside and outside diameters of the corresponding tube or pipe can be welded by butt welding, resulting in a neat and streamlined appearance. Figure 4-15 illustrates a swivel joint. Figure 4-16 illustrates the use on a machine.

Hose Fittings

Hose end connectors can be obtained in permanently pressed-on designs, reusable clamp designs, and reusable screw-type designs. The reusable types of connectors require skill and time in making assembly.

It is usually preferable to have both ends of the hose be connected by union connectors which incorporate a freely turning nut. Nuts are located either in the hose end coupling or in the connector fitting. For short hoses it is possible to have one end rigidly connected; however, during assembly the entire hose must turn. Unions cost more than rigid fittings, so it is generally desirable to locate the union in the connector rather than in the hose coupling.

A considerable number of hose end connector designs are available, different in the basic principle employed and the type of flare and thread used.

FOUR EASY STEPS FOR FIGURING OFFSET BENDS

Step 1. Determine the total amount of offset required (dimension Y in diagram) and angle of offset. Wherever possible use 45° offset bends. This will enable you to figure the total amount of tubing required for a given application, as explained in the following section.

Step 2. Figure the length of tube which is needed to meet your offset requirements (X in dimension diagram) from Table 4-22. For example, say that the amount of offset you require (Y dimension, Step 1) is $2\frac{1}{2}$ in. and the offset angle is 45°. Check the 45° column and find $2\frac{1}{2}$ in. The figure next to this is the amount of tubing required for the offset bend you want (X dimension). In this case it is $3\frac{17}{32}$ in.

Step 3. Determine where you want the center of the offset bend on the tube and make a reference mark (A). Now measure off the X dimension (determined in Step 2) starting from the reference mark and make a second mark (B). You are now ready to make the bends.

Step 4. Align mark A with reference mark R on bender and proceed with first bend. Then align B with reference mark R and make a second bend in the proper direction. Figure 4-17 provides data for the use of hand benders.

Note: When the amount of offset exceeds what is listed on the table, choose an offset from the table which is a multiple of the offset you need. Look this up on the table and multiply the X dimension by the multiple you used.

EXAMPLE

For an offset of 20 in. for a 45° bend, look up the 5-in. offset on the table in the 45° column and multiply the X dimension ($7\frac{7}{16}$ in.) by 4. The resulting X dimension you would use is $28\frac{1}{4}$ in.

HOW TO FIGURE TOTAL LENGTH OF TUBING REQUIRED FOR 45°-OFFSET APPLICATIONS

Determine the X dimension required for a particular application and subtract the amount of offset from this. From the example above (Step 3), a $2\frac{1}{2}$-in. offset was required and the X dimension as determined from the table was $3\frac{17}{32}$ in. The difference between these two figures is $1\frac{1}{32}$ in. Simply add this to the vertical distance from the starting point to the finishing point (the Z dimension in the illustration above).

In summary: Figure 4-18 illustrates typical pipe data. Figure 4-19 lists typical 37° steel flare adapters. Figure 4-20 illustrates how to identify coupling ends.

TUBING CONNECTORS

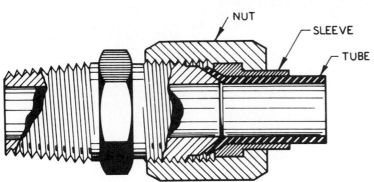

NUT — SLEEVE — TUBE

FIGURE 4-6

37° FLARE

The 37° Flare provides excellent results for connections when tubing is flarable. The free floating sleeve supports tube and dampens vibration. Approved by Underwriters Laboratories., S.A.E., J.I.C., and A.S.M.E. This type of connector is reuseable, does not twist during assembly and requires low assembly torque.

FIGURE 4-7

45° FLARE

The 45° Flare may be used with flarable tubing and will withstand pressures up to 5,000 PSI. This design is approved by Underwriters Laboratories, A.S.A., A.S.M.E., J.I.C., and S.A.E. It is reuseable and available at comparatively low cost.

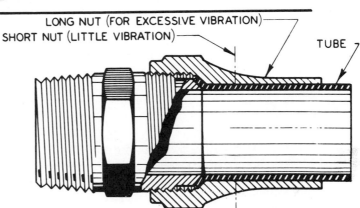

LONG NUT (FOR EXCESSIVE VIBRATION)
SHORT NUT (LITTLE VIBRATION)
TUBE

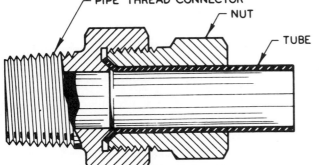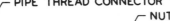

PIPE THREAD CONNECTOR — NUT — TUBE

FIGURE 4-8

45° INVERTED FLARE

The 45° inverted Flare provides protection for seat and threads which are recessed. It is reuseable, low cost, and fits in tight places. This design is also approved by U.L., A.S.A., A.S.M.E., J.I.C. and S.A.E.

FIGURE 4-9

STRAIGHT THREAD "O" RING PORT

The straight thread "O" ring design is proving to provide excellent results for high pressure hydraulic applications. This design is approved by S.A.E. and gives a tighter seal as pressure is increased. There is less chance of over-tightening as it gives a positive hit-home feel when assembled. The "O" ring used must be compatible with the fluid. This configuration covers AND 10050 and MS 16142.

This is the preferred connection for hydraulic service.

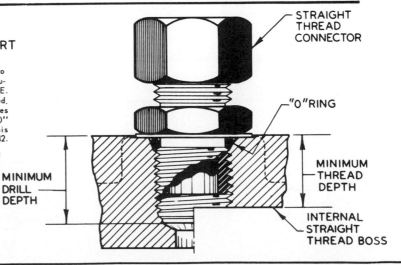

STRAIGHT THREAD CONNECTOR
"O" RING
MINIMUM DRILL DEPTH
MINIMUM THREAD DEPTH
INTERNAL STRAIGHT THREAD BOSS

FIGURE 4-10
SELF-FLARING

In this design, assembly is made by a ferrule or sleeve being pulled into the fitting body as the nut is tightened. As the ferrule rides into the body, it follows a taper which causes the sleeve to bite into the tubing and provide seal. Good results have been experienced with this design and it is JIC approved.

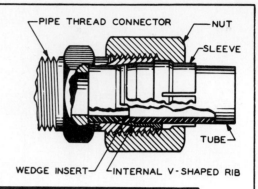

FIGURE 4-11
STRAIGHT THREAD PORT

This is another design of the straight thread port and connection. It is approved by J.I.C. and allows a metal to metal seal.

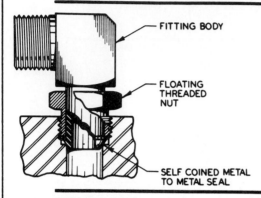

FIGURE 4-12
COMPRESSION

This design of the tapered sleeve (bite type) compression fitting may be used with heavy wall tubing and is approved by U.L., S.A.E., J.I.C., and A.S.M.E. It is reuseable. Good results have been experienced with this design for high pressure applications. However, extreme care should be given to the amount of torque in assembly.

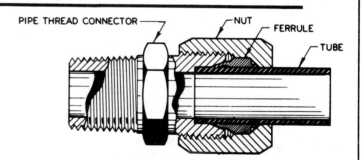

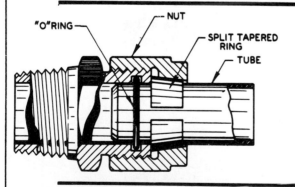

FIGURE 4-13
COMPRESSION

This design is of the "O" ring and split sleeve type and is approved by J.I.C. and S.A.E. It is reuseable and allows considerable tolerance in the length of tube cut and squareness of cut. The "O" ring must be compatible with the fluid used in the system.

FIGURE 4-14
COMPRESSION

This design is proving satisfactory for plastic tubing. It is easy to install, reuseable and comes in a variety of materials including steel and plastic.

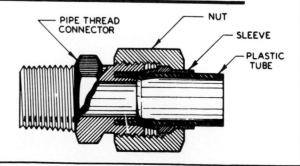

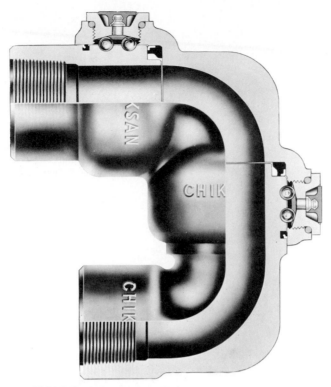

FIGURE 4-15 Swivel fittings are used when there is relative motion between actuator and machine members. This type is available for pressure up to 15,000 psi. (Courtesy of FMC Corp., Chiksan Div.)

FIGURE 4-16 Application of swivel fittings. (Courtesy of FMC Corp., Chiksan Div.)

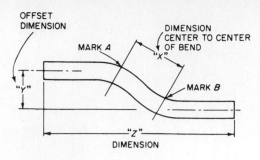

TABLE 4-22
OFFSET BEND CALCULATOR

Angle of offset 15°		Angle of offset 30°		Angle of offset 45°		Angle of offset 60°		Angle of offset 75°	
Amount of Offset (Y Dimension)	(X Dimension)	Amount of Offset (Y Dimension)	(X Dimension)	Amount of Offset (Y Dimension)	(X Dimension)	Amount of Offset (Y Dimension)	(X Dimension)	Amount of Offset (Y Dimension)	(X Dimension)
1	$3\frac{7}{8}$	1	2	1	$1\frac{13}{22}$	1	$1\frac{5}{32}$	1	$1\frac{1}{32}$
$\frac{1}{8}$	$4\frac{11}{32}$	$\frac{1}{8}$	$2\frac{1}{4}$	$\frac{1}{8}$	$1\frac{19}{32}$	$\frac{1}{8}$	$1\frac{5}{16}$	$\frac{1}{8}$	$1\frac{5}{32}$
$\frac{1}{4}$	$4\frac{27}{22}$	$\frac{1}{4}$	$2\frac{1}{2}$	$\frac{1}{4}$	$1\frac{25}{32}$	$\frac{1}{4}$	$1\frac{7}{16}$	$\frac{1}{4}$	$1\frac{9}{32}$
$\frac{3}{8}$	$5\frac{5}{16}$	$\frac{3}{8}$	$2\frac{3}{4}$	$\frac{3}{8}$	$1\frac{15}{16}$	$\frac{3}{8}$	$1\frac{19}{32}$	$\frac{3}{8}$	$1\frac{7}{16}$
$\frac{1}{2}$	$5\frac{25}{32}$	$\frac{1}{2}$	3	$\frac{1}{2}$	$2\frac{1}{8}$	$\frac{1}{2}$	$1\frac{23}{32}$	$\frac{1}{2}$	$1\frac{9}{16}$
$\frac{5}{8}$	$6\frac{9}{32}$	$\frac{5}{8}$	$3\frac{1}{4}$	$\frac{5}{8}$	$2\frac{5}{16}$	$\frac{5}{8}$	$1\frac{7}{8}$	$\frac{5}{8}$	$1\frac{11}{16}$
$\frac{3}{4}$	$6\frac{3}{4}$	$\frac{3}{4}$	$3\frac{1}{2}$	$\frac{3}{4}$	$2\frac{15}{32}$	$\frac{3}{4}$	$2\frac{1}{32}$	$\frac{3}{4}$	$1\frac{13}{16}$
$\frac{7}{8}$	$7\frac{1}{4}$	$\frac{7}{8}$	$3\frac{3}{4}$	$\frac{7}{8}$	$2\frac{21}{32}$	$\frac{7}{8}$	$2\frac{5}{32}$	$\frac{7}{8}$	$1\frac{15}{16}$
2	$7\frac{23}{32}$	2	4	2	$2\frac{13}{16}$	2	$2\frac{5}{16}$	2	$2\frac{1}{16}$
$\frac{1}{8}$	$8\frac{7}{32}$	$\frac{1}{8}$	$4\frac{1}{4}$	$\frac{1}{8}$	3	$\frac{1}{8}$	$2\frac{15}{32}$	$\frac{1}{8}$	$2\frac{3}{16}$
$\frac{1}{4}$	$8\frac{11}{16}$	$\frac{1}{4}$	$4\frac{1}{2}$	$\frac{1}{4}$	$3\frac{3}{16}$	$\frac{1}{4}$	$2\frac{19}{32}$	$\frac{1}{4}$	$2\frac{5}{16}$
$\frac{3}{8}$	$9\frac{3}{16}$	$\frac{3}{8}$	$4\frac{3}{4}$	$\frac{3}{8}$	$3\frac{11}{32}$	$\frac{3}{8}$	$2\frac{3}{4}$	$\frac{3}{8}$	$2\frac{15}{32}$
$\frac{1}{2}$	$9\frac{21}{32}$	$\frac{1}{2}$	5	$\frac{1}{2}$	$3\frac{17}{32}$	$\frac{1}{2}$	$2\frac{7}{8}$	$\frac{1}{2}$	$2\frac{19}{32}$
$\frac{5}{8}$	$10\frac{5}{32}$	$\frac{5}{8}$	$5\frac{1}{4}$	$\frac{5}{8}$	$3\frac{23}{32}$	$\frac{5}{8}$	$3\frac{1}{32}$	$\frac{5}{8}$	$2\frac{23}{32}$
$\frac{3}{4}$	$10\frac{5}{8}$	$\frac{3}{4}$	$5\frac{1}{2}$	$\frac{3}{4}$	$3\frac{7}{8}$	$\frac{3}{4}$	$3\frac{3}{16}$	$\frac{3}{4}$	$2\frac{27}{32}$
$\frac{7}{8}$	$11\frac{3}{32}$	$\frac{7}{8}$	$5\frac{3}{4}$	$\frac{7}{8}$	$4\frac{1}{16}$	$\frac{7}{8}$	$3\frac{5}{16}$	$\frac{7}{8}$	$2\frac{31}{32}$
3	$11\frac{19}{32}$	3	6	3	$4\frac{1}{4}$	3	$3\frac{15}{32}$	3	$3\frac{3}{32}$
$\frac{1}{8}$	$12\frac{1}{16}$	$\frac{1}{8}$	$6\frac{1}{4}$	$\frac{1}{8}$	$4\frac{13}{32}$	$\frac{1}{8}$	$3\frac{19}{32}$	$\frac{1}{8}$	$3\frac{7}{32}$
$\frac{1}{4}$	$12\frac{9}{16}$	$\frac{1}{4}$	$6\frac{1}{2}$	$\frac{1}{4}$	$4\frac{19}{32}$	$\frac{1}{4}$	$3\frac{3}{4}$	$\frac{1}{4}$	$3\frac{3}{8}$
$\frac{3}{8}$	$13\frac{1}{32}$	$\frac{3}{8}$	$6\frac{3}{4}$	$\frac{3}{8}$	$4\frac{25}{32}$	$\frac{3}{8}$	$3\frac{29}{32}$	$\frac{3}{8}$	$3\frac{1}{2}$
$\frac{1}{2}$	$13\frac{17}{32}$	$\frac{1}{2}$	7	$\frac{1}{2}$	$4\frac{15}{16}$	$\frac{1}{2}$	$4\frac{1}{32}$	$\frac{1}{2}$	$3\frac{5}{8}$
$\frac{5}{8}$	14	$\frac{5}{8}$	$7\frac{1}{4}$	$\frac{5}{8}$	$5\frac{1}{8}$	$\frac{5}{8}$	$4\frac{3}{16}$	$\frac{5}{8}$	$3\frac{3}{4}$
$\frac{3}{4}$	$14\frac{1}{2}$	$\frac{3}{4}$	$7\frac{1}{2}$	$\frac{3}{4}$	$5\frac{1}{16}$	$\frac{3}{4}$	$4\frac{11}{32}$	$\frac{3}{4}$	$3\frac{7}{8}$
$\frac{7}{8}$	$14\frac{31}{32}$	$\frac{7}{8}$	$7\frac{3}{4}$	$\frac{7}{8}$	$5\frac{15}{32}$	$\frac{7}{8}$	$4\frac{15}{32}$	$\frac{7}{8}$	4
4	$15\frac{1}{2}$	4	8	4	$5\frac{21}{32}$	4	$4\frac{5}{8}$	4	$4\frac{1}{8}$
$\frac{1}{8}$	$15\frac{15}{16}$	$\frac{1}{8}$	$8\frac{1}{4}$	$\frac{1}{8}$	$5\frac{27}{32}$	$\frac{1}{8}$	$4\frac{3}{4}$	$\frac{1}{8}$	$4\frac{9}{32}$
$\frac{1}{4}$	$16\frac{7}{16}$	$\frac{1}{4}$	$8\frac{1}{2}$	$\frac{1}{4}$	6	$\frac{1}{4}$	$4\frac{29}{32}$	$\frac{1}{4}$	$4\frac{13}{32}$
$\frac{3}{8}$	$16\frac{29}{32}$	$\frac{3}{8}$	$8\frac{3}{4}$	$\frac{3}{8}$	$6\frac{3}{16}$	$\frac{3}{8}$	$5\frac{1}{16}$	$\frac{3}{8}$	$4\frac{17}{32}$
$\frac{1}{2}$	$17\frac{3}{8}$	$\frac{1}{2}$	9	$\frac{1}{2}$	$6\frac{3}{8}$	$\frac{1}{2}$	$5\frac{3}{16}$	$\frac{1}{2}$	$4\frac{21}{32}$
$\frac{5}{8}$	$17\frac{7}{8}$	$\frac{5}{8}$	$9\frac{1}{4}$	$\frac{5}{8}$	$6\frac{17}{32}$	$\frac{5}{8}$	$5\frac{11}{32}$	$\frac{5}{8}$	$4\frac{25}{32}$
$\frac{3}{4}$	$18\frac{11}{32}$	$\frac{3}{4}$	$9\frac{1}{2}$	$\frac{3}{4}$	$6\frac{23}{32}$	$\frac{3}{4}$	$5\frac{15}{32}$	$\frac{3}{4}$	$4\frac{29}{32}$
$\frac{7}{8}$	$18\frac{27}{32}$	$\frac{7}{8}$	$9\frac{3}{4}$	$\frac{7}{8}$	$6\frac{29}{32}$	$\frac{7}{8}$	$5\frac{5}{8}$	$\frac{7}{8}$	$5\frac{1}{32}$
5	$19\frac{5}{16}$	5	10	5	$7\frac{1}{16}$	5	$5\frac{25}{32}$	5	$5\frac{3}{16}$
$\frac{1}{8}$	$19\frac{13}{16}$	$\frac{1}{8}$	$10\frac{1}{4}$	$\frac{1}{8}$	$7\frac{1}{4}$	$\frac{1}{8}$	$5\frac{25}{16}$	$\frac{1}{8}$	$5\frac{5}{16}$
$\frac{1}{4}$	$20\frac{9}{32}$	$\frac{1}{4}$	$10\frac{1}{2}$	$\frac{1}{4}$	$7\frac{7}{16}$	$\frac{1}{4}$	$6\frac{1}{16}$	$\frac{1}{4}$	$5\frac{7}{16}$
$\frac{3}{8}$	$20\frac{25}{32}$	$\frac{3}{8}$	$10\frac{3}{4}$	$\frac{3}{8}$	$7\frac{19}{32}$	$\frac{3}{8}$	$6\frac{7}{32}$	$\frac{3}{8}$	$5\frac{9}{16}$
$\frac{1}{2}$	$21\frac{1}{4}$	$\frac{1}{2}$	11	$\frac{1}{2}$	$7\frac{25}{32}$	$\frac{1}{2}$	$6\frac{11}{32}$	$\frac{1}{2}$	$5\frac{11}{16}$
$\frac{5}{8}$	$21\frac{3}{4}$	$\frac{5}{8}$	$11\frac{1}{4}$	$\frac{5}{8}$	$7\frac{31}{32}$	$\frac{5}{8}$	$6\frac{1}{2}$	$\frac{5}{8}$	$5\frac{27}{32}$
$\frac{3}{4}$	$22\frac{7}{32}$	$\frac{3}{4}$	$11\frac{1}{2}$	$\frac{3}{4}$	$8\frac{1}{8}$	$\frac{3}{4}$	$6\frac{5}{8}$	$\frac{3}{4}$	$5\frac{31}{32}$
$\frac{7}{8}$	$22\frac{11}{16}$	$\frac{7}{8}$	$11\frac{3}{4}$	$\frac{7}{8}$	$8\frac{5}{16}$	$\frac{7}{8}$	$6\frac{25}{32}$	$\frac{7}{8}$	$6\frac{3}{32}$
6	$23\frac{3}{16}$	6	12	6	$8\frac{15}{32}$	6	$6\frac{15}{16}$	6	$6\frac{7}{32}$

For bending hard or soft tubing — Copper, Aluminum, Brass, Steel and Stainless Steel

1. To place tube in bender, pull out slip joint and move handles of bender so that they are at right angles. Raise the clip and place the tube in the space between the handle slide block and the bending form.

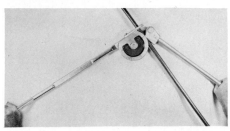

2. Drop clip over tube and push in slip joint. Note that zero mark on bending form will be even with front of handle slide block.

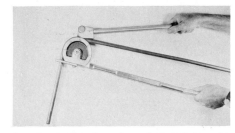

3. Proceed to bend desired angle as indicated by calibrations on bending form. Bends to any angle up to 180° can be made with the Imperial Tube Bender.

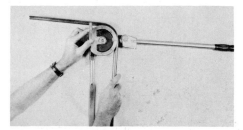

4. To remove the bent tube from bender, pull out slip joint and raise the clip. Tube can then readily be removed.

NOTE: Occasionally placing a drop of oil on the pin and bending shoe will assure a smoother working tool.

GUIDE FOR MAKING DIMENSIONAL BENDS

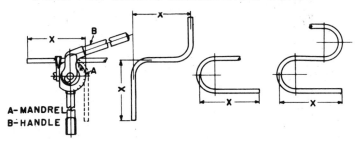

A—MANDREL
B—HANDLE

BENDS FROM HOOK SIDE

The illustration above is for bends measured from the hook-side of bender. For 7/16″ and 1/2″ benders, line up mark on tube which is to be the center of the bend with the edge of the Forming Wheel. On 1/8″ to 3/8″ benders, use either the edge of the Forming Wheel or the mark on the lever designed for that purpose.

TO OBTAIN "X" DIMENSION TUBE SHOULD BE PLACED IN BENDER AS ILLUSTRATED IN FIGURES 1 AND 2

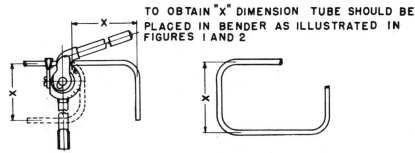

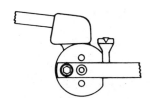

The three position start feature furnished on benders 3/8″, 7/16″ and 1/2″ sizes only.

BENDS FROM LEVER SIDE

For 7/16″ and 1/2″ benders — on bends measured from the lever-side of bender — the tube mark must be lined up with the reference mark on the Forming Wheel as illustrated. For 1/8″ to 3/8″ benders, use the reference mark stamped on lever.

FIGURE 4-17 Operation of lever-type benders. (Courtesy of Imperial-Eastman.)

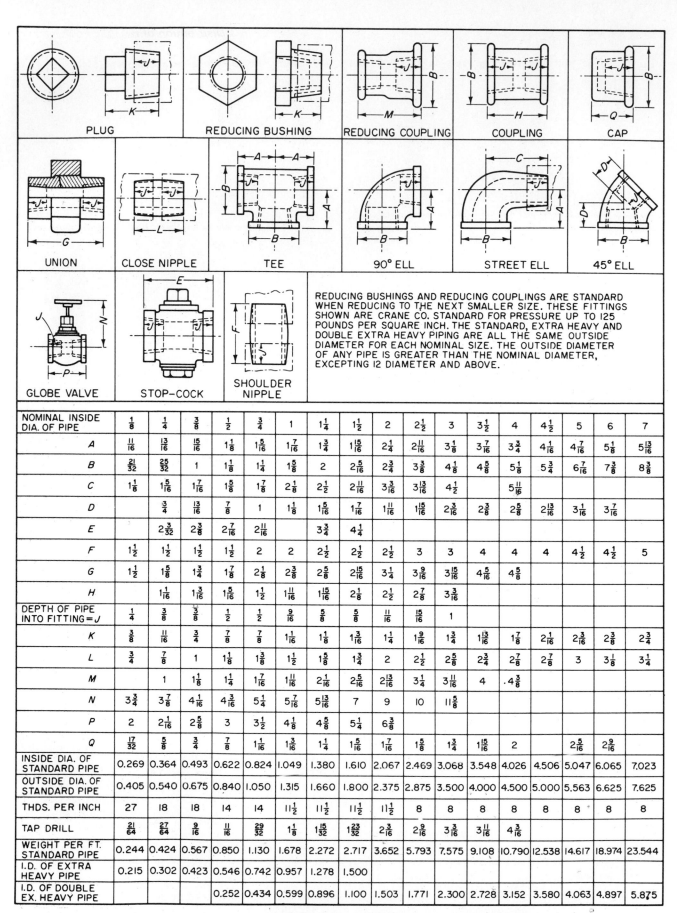

REDUCING BUSHINGS AND REDUCING COUPLINGS ARE STANDARD WHEN REDUCING TO THE NEXT SMALLER SIZE. THESE FITTINGS SHOWN ARE CRANE CO. STANDARD FOR PRESSURE UP TO 125 POUNDS PER SQUARE INCH. THE STANDARD, EXTRA HEAVY AND DOUBLE EXTRA HEAVY PIPING ARE ALL THE SAME OUTSIDE DIAMETER FOR EACH NOMINAL SIZE. THE OUTSIDE DIAMETER OF ANY PIPE IS GREATER THAN THE NOMINAL DIAMETER, EXCEPTING 12 DIAMETER AND ABOVE.

NOMINAL INSIDE DIA. OF PIPE	$\frac{1}{8}$	$\frac{1}{4}$	$\frac{3}{8}$	$\frac{1}{2}$	$\frac{3}{4}$	1	$1\frac{1}{4}$	$1\frac{1}{2}$	2	$2\frac{1}{2}$	3	$3\frac{1}{2}$	4	$4\frac{1}{2}$	5	6	7
A	$\frac{11}{16}$	$\frac{13}{16}$	$\frac{15}{16}$	$1\frac{1}{8}$	$1\frac{5}{16}$	$1\frac{7}{16}$	$1\frac{3}{4}$	$1\frac{15}{16}$	$2\frac{1}{4}$	$2\frac{11}{16}$	$3\frac{1}{8}$	$3\frac{7}{16}$	$3\frac{3}{4}$	$4\frac{1}{16}$	$4\frac{7}{16}$	$5\frac{1}{8}$	$5\frac{13}{16}$
B	$\frac{21}{32}$	$\frac{25}{32}$	1	$1\frac{1}{8}$	$1\frac{1}{4}$	$1\frac{5}{8}$	2	$2\frac{5}{16}$	$2\frac{3}{4}$	$3\frac{3}{8}$	$4\frac{1}{8}$	$4\frac{5}{8}$	$5\frac{1}{8}$	$5\frac{3}{4}$	$6\frac{7}{16}$	$7\frac{3}{8}$	$8\frac{3}{8}$
C	$1\frac{1}{8}$	$1\frac{5}{16}$	$1\frac{7}{16}$	$1\frac{5}{8}$	$1\frac{7}{8}$	$2\frac{1}{8}$	$2\frac{1}{2}$	$2\frac{11}{16}$	$3\frac{3}{16}$	$3\frac{13}{16}$	$4\frac{1}{2}$		$5\frac{11}{16}$				
D		$\frac{3}{4}$	$\frac{13}{16}$	$\frac{7}{8}$	1	$1\frac{1}{8}$	$1\frac{5}{16}$	$1\frac{7}{16}$	$1\frac{11}{16}$	$1\frac{15}{16}$	$2\frac{1}{16}$	$2\frac{3}{8}$	$2\frac{5}{8}$	$2\frac{13}{16}$	$3\frac{1}{16}$	$3\frac{7}{16}$	
E		$2\frac{3}{32}$	$2\frac{3}{8}$	$2\frac{7}{16}$	$2\frac{11}{16}$		$3\frac{3}{4}$	$4\frac{1}{4}$									
F	$1\frac{1}{2}$	$1\frac{1}{2}$	$1\frac{1}{2}$	$1\frac{1}{2}$	2	2	$2\frac{1}{2}$	$2\frac{1}{2}$	$2\frac{1}{2}$	3	3	4	4	4	$4\frac{1}{2}$	$4\frac{1}{2}$	5
G	$1\frac{1}{2}$	$1\frac{5}{8}$	$1\frac{3}{4}$	$1\frac{7}{8}$	$2\frac{1}{8}$	$2\frac{3}{8}$	$2\frac{5}{8}$	$2\frac{15}{16}$	$3\frac{1}{4}$	$3\frac{9}{16}$	$3\frac{15}{16}$	$4\frac{5}{16}$	$4\frac{5}{8}$				
H		$1\frac{1}{16}$	$1\frac{3}{16}$	$1\frac{5}{16}$	$1\frac{1}{2}$	$1\frac{11}{16}$	$1\frac{15}{16}$	$2\frac{1}{8}$	$2\frac{1}{2}$	$2\frac{7}{8}$	$3\frac{3}{8}$						
DEPTH OF PIPE INTO FITTING = J	$\frac{1}{4}$	$\frac{3}{8}$	$\frac{3}{8}$	$\frac{1}{2}$	$\frac{1}{2}$	$\frac{9}{16}$	$\frac{5}{8}$	$\frac{5}{8}$	$\frac{11}{16}$	$\frac{15}{16}$	1						
K	$\frac{3}{8}$	$\frac{11}{16}$	$\frac{3}{4}$	$\frac{7}{8}$	$\frac{7}{8}$	$1\frac{1}{16}$	$1\frac{1}{8}$	$1\frac{3}{16}$	$1\frac{1}{4}$	$1\frac{9}{16}$	$1\frac{3}{4}$	$1\frac{13}{16}$	$1\frac{7}{8}$	$2\frac{1}{16}$	$2\frac{3}{16}$	$2\frac{3}{8}$	$2\frac{3}{4}$
L	$\frac{3}{4}$	$\frac{7}{8}$	1	$1\frac{1}{8}$	$1\frac{3}{8}$	$1\frac{1}{2}$	$1\frac{5}{8}$	$1\frac{3}{4}$	2	$2\frac{1}{2}$	$2\frac{5}{8}$	$2\frac{3}{4}$	$2\frac{7}{8}$	$2\frac{7}{8}$	3	$3\frac{1}{8}$	$3\frac{1}{4}$
M		1	$1\frac{1}{8}$	$1\frac{1}{4}$	$1\frac{7}{16}$	$1\frac{11}{16}$	$2\frac{1}{16}$	$2\frac{5}{16}$	$2\frac{13}{16}$	$3\frac{1}{4}$	$3\frac{11}{16}$	4	$.4\frac{3}{8}$				
N	$3\frac{3}{4}$	$3\frac{7}{8}$	$4\frac{1}{16}$	$4\frac{3}{16}$	$5\frac{1}{4}$	$5\frac{7}{16}$	$5\frac{13}{16}$	7	9	10	$11\frac{5}{8}$						
P	2	$2\frac{1}{16}$	$2\frac{5}{8}$	3	$3\frac{1}{2}$	$4\frac{1}{8}$	$4\frac{5}{8}$	$5\frac{1}{4}$	$6\frac{3}{8}$								
Q	$\frac{17}{32}$	$\frac{5}{8}$	$\frac{3}{4}$	$\frac{7}{8}$	$1\frac{1}{16}$	$1\frac{3}{16}$	$1\frac{1}{4}$	$1\frac{5}{16}$	$1\frac{7}{16}$	$1\frac{5}{8}$	$1\frac{3}{4}$	$1\frac{15}{16}$	2		$2\frac{5}{16}$	$2\frac{9}{16}$	
INSIDE DIA. OF STANDARD PIPE	0.269	0.364	0.493	0.622	0.824	1.049	1.380	1.610	2.067	2.469	3.068	3.548	4.026	4.506	5.047	6.065	7.023
OUTSIDE DIA. OF STANDARD PIPE	0.405	0.540	0.675	0.840	1.050	1.315	1.660	1.800	2.375	2.875	3.500	4.000	4.500	5.000	5.563	6.625	7.625
THDS. PER INCH	27	18	18	14	14	$11\frac{1}{2}$	$11\frac{1}{2}$	$11\frac{1}{2}$	$11\frac{1}{2}$	8	8	8	8	8	8	8	8
TAP DRILL	$\frac{21}{64}$	$\frac{27}{64}$	$\frac{9}{16}$	$\frac{11}{16}$	$\frac{29}{32}$	$1\frac{1}{8}$	$1\frac{15}{32}$	$1\frac{23}{32}$	$2\frac{3}{16}$	$2\frac{9}{16}$	$3\frac{3}{16}$	$3\frac{11}{16}$	$4\frac{3}{16}$				
WEIGHT PER FT. STANDARD PIPE	0.244	0.424	0.567	0.850	1.130	1.678	2.272	2.717	3.652	5.793	7.575	9.108	10.790	12.538	14.617	18.974	23.544
I.D. OF EXTRA HEAVY PIPE	0.215	0.302	0.423	0.546	0.742	0.957	1.278	1.500									
I.D. OF DOUBLE EX. HEAVY PIPE				0.252	0.434	0.599	0.896	1.100	1.503	1.771	2.300	2.728	3.152	3.580	4.063	4.897	5.875

FIGURE 4-18 Pipe data.

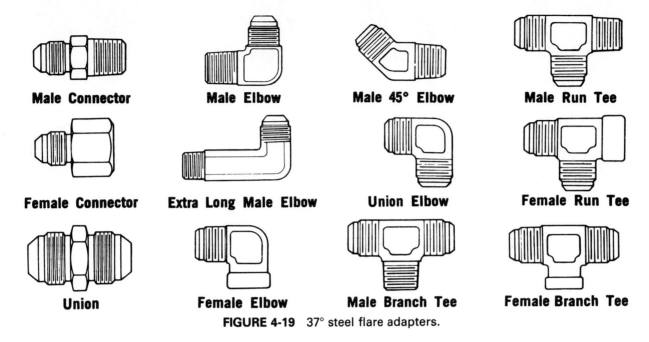

Male Connector **Male Elbow** **Male 45° Elbow** **Male Run Tee**

Female Connector **Extra Long Male Elbow** **Union Elbow** **Female Run Tee**

Union **Female Elbow** **Male Branch Tee** **Female Branch Tee**

FIGURE 4-19 37° steel flare adapters.

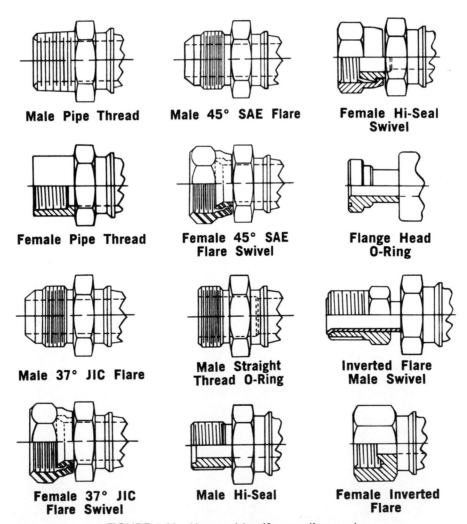

Male Pipe Thread **Male 45° SAE Flare** **Female Hi-Seal Swivel**

Female Pipe Thread **Female 45° SAE Flare Swivel** **Flange Head O-Ring**

Male 37° JIC Flare **Male Straight Thread O-Ring** **Inverted Flare Male Swivel**

Female 37° JIC Flare Swivel **Male Hi-Seal** **Female Inverted Flare**

FIGURE 4-20 How to identify coupling ends.

QUESTIONS

1. How can the fabrication of a system minimize friction and reduce power loss?

2. What effect does hydraulic shock have on fluid conductors?

3. What variables determine the wall thickness and safety factor of a conductor for any given operating pressure?

4. What is a hydrauliscope, and what is its use in connection with a hydraulic system?

5. Name several considerations for safeguarding a system from overpressure conditions.

6. What causes pressure loss through conductors and connectors?

7. What flow velocity is generally recommended for the discharge side of a pump?

8. What is the recommended velocity for the inlet side of the pump?

9. What effect does vibration have on a hydraulic system, and how is it controlled?

10. What size inlet line would you use for a 12-gpm pump?

11. What size discharge line would you use for a 12-gpm pump?

5
Conditioning Power Fluids

Reservoirs, heat exchangers, strainers, filters, and magnets are all important components of a fluid power system to assure smooth and trouble-free operation of machines and equipment. Proper conditioning of the hydraulic fluid is probably one of the most neglected considerations in the design of hydraulic circuits. Consequently, 40 to 60% of all the trouble in hydraulic circuits results from improper care of the hydraulic fluid.

Conditioning the hydraulic fluid means to provide ample storage for the fluid in a system, to maintain its proper operating temperature, and to keep it clear and free of contamination.

Different fields of applying fluid power will require varying degrees of fluid conditioning. For instance, such applications as numerically controlled machines, high-performance spacecraft, and nuclear-powered submarines require positive control of the fluid's cleanliness and physical conditions at all times. Hydraulic systems of this nature use servo valves and other precision components that have extremely close tolerances between moving parts. Because of these close tolerances, a very small particle of dirt could be detrimental to the function of a control valve and hence cause a serious mishap. A sharp rise in temperature could also cause various parts to expand and prevent the proper function essential for the control of system actuators. The hydraulic system on a space capsule, as shown in Fig. 5-1, is responsible for guiding the craft through space and back to earth safely.

FIGURE 5-1 Apollo command module. (Courtesy of NASA and North American Aviation, Inc. Reprinted from F. Ordway, J. Gardner, M. Sharpe, and Wakeford, *Applied Astronautics: An Introduction to Space Flight*, Prentice-Hall, Inc.)

Extreme measures must be taken in these special areas of fluid power application to keep the power fluids clean and stable and at the proper temperature. Many times, costly equipment and human lives depend on the hydraulic system and its ability to function properly.

Hydraulic systems used on mobile equipment are rugged and less sophisticated, with components that can withstand high shock loads under intermittent duty operation. The components such as pumps, motors, and control valves are built to be more compatible with dirt, but there are still limits to the amount of contamination that can be tolerated. Mobile hydraulic circuits often use a pressure cap on the reservoir to seal the fluid from the dusty or dirty atmospheric conditions under which the equipment must work. Scrapers or boots are used on cylinder rods to prevent dirt from getting into the hydraulic fluid as the cylinders operate (see Fig. 5-2).

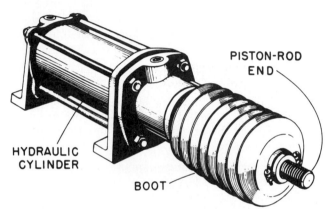

FIGURE 5-2 Booted cylinder. (Courtesy of Mobil Oil Corporation.)

Space limitations on machinery may dictate the use of smaller reservoirs, which means that temperature control of the working fluid may be critical. Heavy dirt and mud caked on the outside of the reservoir and on the hydraulic hoses and tubing could prevent heat transfer to the surrounding atmosphere; therefore, these areas should be kept clean.

The machine tool industry presents a slightly different problem, because hydraulic machinery must operate continuously. The cleanliness, stability, and temperature of the fluid must be properly maintained for continuity of production and the quality control of machined parts. Long life expectancy of equipment and good maintenance are important factors in preventing down time and loss of production (Fig. 5-3).

It is also easier to control the proper condition of the operating fluid in machine tools because of the accessibility of cooling water and the availability of preventive maintenance. Reservoirs, coolers, filters, and magnets can be maintained and serviced regularly by skilled personnel. This means that machine designers and original-equipment manufacturers (OEM) should provide or recommend the proper conditioning equipment for each installation. Generally, in this type of application, there is little excuse for improper installation of equipment or lack of a good preventive maintenance program to keep the hydraulic systems trouble-free and operating smoothly.

Regardless of the field of application, there are certain important factors which must be considered when conditioning equipment is selected for any hydraulic system.

THE HYDRAULIC RESERVOIR

Reservoir Capacity. The hydraulic reservoir should contain enough fluid that its working level is always maintained during the system's operation. It should also have additional capacity to hold all the fluid in the system during shutdowns. The reservoir capacity is generally between two and three times the capacity of the hydraulic pump. In other words, a pump delivering 10 gpm would need a reservoir with a capacity of about 30 gallons. If a hydraulic system has a great number of cylinders operating simultanously, more careful consideration should be given to the proper reservoir size by calculating the entire system's volume; for example, an agricultural tractor with fluid power to operate auxiliary equipment, such as a loader, ripper, snow plow, etc. It should also be understood that systems blowing high-pressure fluid over the relief valve cause rapid heat buildup, and a proper amount of fluid should be maintained and circulated within the reservoir to help dissipate its heat. The side wall area of the reservoir must also be adequate to help provide effective transfer of heat through the walls to the surrounding atmosphere.

Actually, a detailed analysis of the heat transfer between the fluid, reservoir, and atmosphere is very complicated and often inaccurate because of changes in convection rate due to local atmospheric conditions. The complications also involve making allowances for radiation, conduction, and convection of the various materials, such as color and thickness of paint, thickness of metal wall, position of surfaces such as surfaces of sides, top, bottom, or ends. For instance, the bottom surface may be a short distance from the floor and have uncertain ventilation, making it extremely difficult to calculate.

Experience has proved, however, that within moderate ambient temperatures, reservoirs generally have the capacity to maintain the proper fluid operating temperatures.

Other features of a well-designed reservoir, as shown in Fig. 5-4, may include the following:

Baffle Plate. The baffle plate separates the system return fluid from that going to the pump inlet. This aids in cooling by promoting the flow of fluid along the side walls, and also helps to separate water, entrained air, and dirt from the hydraulic fluid.

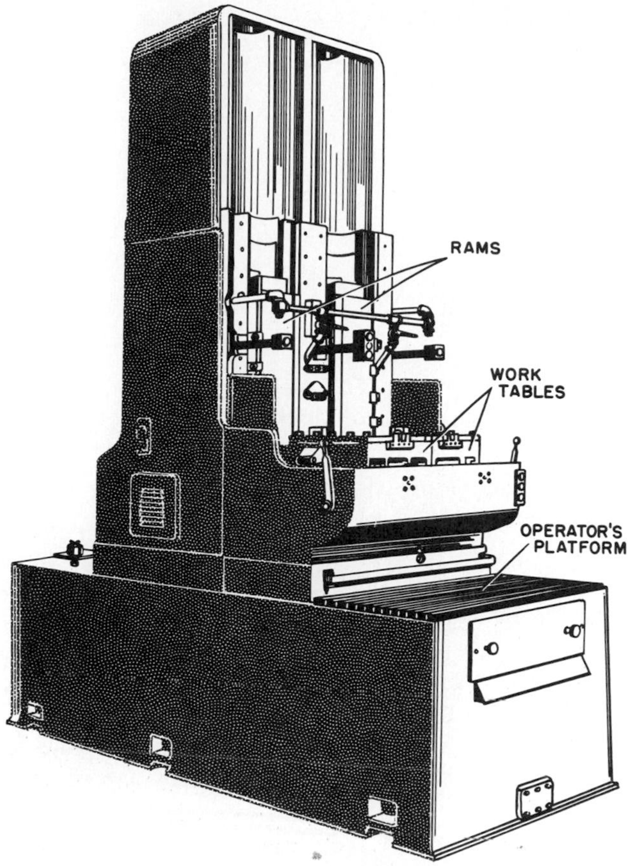

FIGURE 5-3 (a) Vertical duplex hydro-broach; (b) circuit.

Vertical Duplex Hydro-Broach.   Broaching is roughly comparable to drawing a milled file of suitable shape over the surface to be machined. Each succeeding tooth takes off a small part of the metal to be removed, but, usually, a broach is made long enough and with enough teeth so that the work is finished in one pass. Bot internal surfaces (holes of all shapes) and external surfaces (one or several at a time) can be broached. Since many broach teeth are cutting at the same time, greater force is required than when comparable work is done on a milling machine, shaper, or planer. Hydraulic actuation easily provides this force as well as ease of control and adjustment of speed. The cost of broaches and work-holding fixtures limits this method to mass production where a large number of identical parts are machined. The duplex broach shown has two rams, two work tables, etc. At the end of a cutting stroke of one ram, the work table on that side moves away from the broach so that the operator can unload and reload the work pieces safely. During this time, the ram makes it return stroke and the second ram is on its cutting stroke. The ram speeds are adjusted to suit the work and the operator's speed. Cutting is then virtually continuous, with rams moving up and down and work table moving in and out in synchronism. The main hydraulic circuit for this machine is shown below with control valves in position for the downward work stroke of the right ram. High-pressure oil is supplied by the constant-volume-type man pump through start-stop-reverse, ram-direction, and right-ram valves to the rod end of the right-ram cylinder. Oil under back pressure is forced from the blank end of this cylinder throught the ram valves to the blank end of the left-ram cylinder, driving the left ram upward on it return stroke. Oil flows from the rod end of the left-ram cylinder under exhaust pressure to the resevoir. At this time, the work tables are held stationary. At the end of the work stroke, the ram and table control valves are shifted hydraulically by means of a low-pressure pilot circuit (not shown) that has its own pump. The work tables are reciprocated, and the left ram begins it work stroke. The table reciprocating linkages are driven from cranks on a shaft that is rotated by means of a rack on the table-piston rod. Bearings and ways are automatically lubricated with oil under pressure.

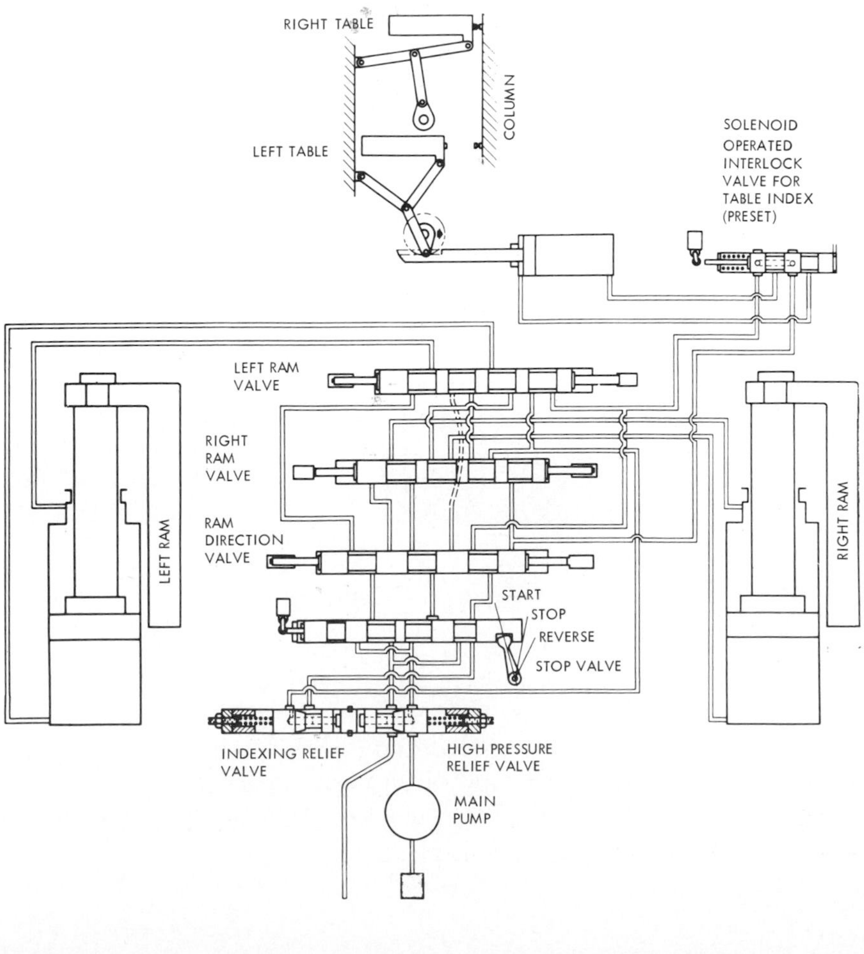

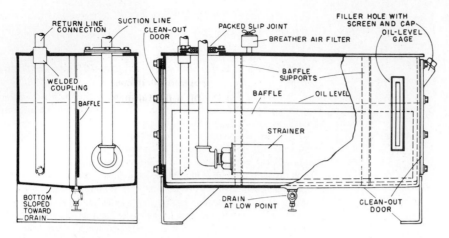

FIGURE 5-4 Reservoir. (Courtesy of Mobil Oil Corporation.)

Bottom Drain. A drain connection on the dished bottom of the reservoir provides a means of draining water and sediment from the system. This is also important when fluid in the system is being changed.

Sight Glass. The sight glass provides a visual check on the fluid level. Proper fluid operating levels must be maintained.

Filling Cap and Hole. A capped hole fitted with a fine mesh screen prevents accidental contamination when the reservoir is being replenished to proper operating level. Makeup fluid should be prefiltered and transferred by means of clean equipment, such as containers, hose, or funnel.

Breather Air Filter. The air filter must be of adequate size and equipped with a filter element suitable to the local atmospheric conditions under which it must work. Since the fluid level in the reservoir must rise and fall as the cylinders stroke, air comes and goes through the breather. The filter element must be kept clean to allow atmospheric pressure to be in contact with the liquid at all times.

Clean-out Cover or Door. A large opening to facilitate cleaning, change of strainer, or inspection is a desirable feature. Also, some reservoirs have a protective coating which may have to be removed if one is changing from one type of hydraulic fluid to another.

Return Line. The return line should be cut on an angle before it is fitted into the reservoir. This prevents any restriction on the returning fluid by jamming the end of the return pipe against the bottom of the reservoir. The size of the return line should be large enough to prevent excessive back pressures in the system (usually less than 30 psi).

Drain Line. An extra drain line must be fitted into the reservoir when back pressures in the return line are too great. The drain line would be used exclusively to allow the free drain of any components fitted with external drain connections.

Pump Inlet Line. The pump inlet line should be large enough to assure satisfaction of the inlet requirements of the hydraulic pump to prevent cavitation. A slip joint or other means should be employed to facilitate the complete removal of the inlet line and strainer to permit regular inspection or cleaning of the strainer element. Figure 5-5 shows one method used to facilitate servicing inlet line strainers.

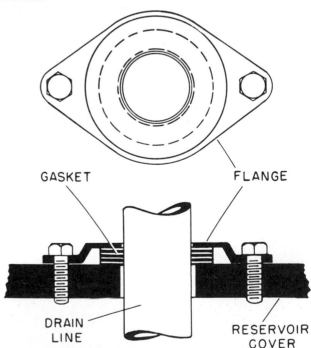

FIGURE 5-5 Flange connection for removing pump inlet and strainer. (Courtesy of Mobil Oil Corporation.)

Drive Mounting Base. Many installations use the top of the reservoir to mount an additional steel plate to support the drive motor and pump. This is an excellent way to keep the pump mounted closer to the fluid being pumped, and to help the inlet characteristics of the pump.

Casters. Another convenient feature to facilitate the portability of a complete hydraulic power source is to mount the reservoir, drive, pump, and sometimes, controls as a complete package on casters. A spare power unit of this design is used many times as a standby to facilitate quick change of power in cases of emergency or for controlled maintenance.

The design of any reservoir is generally dictated by the type of application for which the system is intended. For example, in a closed-circuit system (see Fig. 5-6) the fluid transmits power from either side of the pump to change the direction of the fluid motor. Because the same fluid is being circulated constantly, it is necessary only to make up the fluid losses which occur because of clearance spaces between integral moving parts.

Systems of this design generally replenish fluid losses through the use of a second pump, which takes its fluid from a reservoir. The replenishing system should be sized in accordance with the efficiencies of pump and motor drive units. The expected losses of fluid during operation may be computed through actual testing, or may be obtained directly from the manufacturers of the equipment. The size of the

reservoir to be used depends upon the amount of fluid makeup needed to keep the system fully primed at all times.

Proper selection of a reservoir for any hydraulic system depends largely on the field of application and the type of duty under which the system must operate. In some cases, a given length and size of pipe may be very adequate for a reservoir, whereas in other cases much more elaborate equipment is needed.

HEAT EXCHANGERS

Coolers or heaters are called heat exchangers and are used to control fluid operating temperatures in a hydraulic system. When high or low ambient temperatures are encountered, such as when machinery is operating close to a hot furnace or when mobile equipment is operating in subzero weather, heat exchangers are used to control and maintain proper operating temperatures of the system's fluid. The steady temperature reached by the fluid in a hydraulic system depends on both the *amount of heat generated* and the *heat-dissipating ability of the system*.

Heating units employ hot water, steam, or electricity and are generally restricted to use for startups or for fluid storage and transfer tanks. Electric heaters are generally used for hydraulic equipment or mobile machinery operating in cold temperatures.

Critical and complex machinery may require conditioning of the fluid to proper operating temperatures before startups are made, which would require positive control of both heating and cooling. Generally, the natural heat-generating characteristics of a system after warmups are sufficient or greater than required. Therefore, the greatest concern in designing hydraulic systems is to provide adequate cooling.

Heat is generated naturally in a hydraulic system because it is impossible to maintain 100% efficiency of operation. This is true with any power-transmitting system using fluid, electrical, or mechanical drives.

If heat energy is allowed to accumulate in a hydraulic system, high temperatures result with possible damage to seals, fluid, and moving parts. High temperature also affects the viscosity of the fluid, which may change the operating performance characteristics of a machine or piece of equipment. Desired temperature ranges are usually between 120 and 140°F, and if the natural processes of heat transfer do not maintain the proper temperature range, some method of cooling must be used. Figure 5-7 shows the installation of a heat exchanger on a hydraulic system.

The amount of cooling necessary to maintain the proper temperature in a hydraulic system can be estimated in a number of ways. For example, the heat generated is equal to the total input power less the actual mechanical work done based on a definite time cycle of operation. Thus, the more efficient hy-

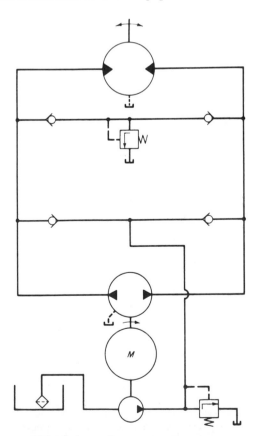

FIGURE 5-6 Closed-circuit system.

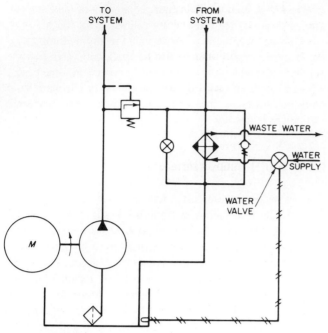

FIGURE 5-7 Hydraulic application showing heat exchanger.

draulic systems require less cooling. It is also an important fact that the kinetic energy in a fluid under pressure is converted into heat when the fluid is passed over a relief valve or undergoes pressure drop through a component.

METHOD OF ESTIMATING FLUID TEMPERATURE RISE IN A HYDRAULIC SYSTEM

EXAMPLE 1

When a given quantity of fluid is forced through a restriction, power is required. The power or energy required is converted *into heat* and results in a rise of fluid temperature. For example,

$$\text{Flow rate} = 10 \text{ gpm}$$

$$\text{Pressure drop} = 100 \text{ psi}$$

The hydraulic horsepower involved is

$$\text{hp} = \text{gpm} \times \text{psi (drop)} \times 0.000583$$

$$= 10 \times 100 \times 0.000583 = 0.583$$

The conversion factor is

$$2544 \text{ Btu/hr} = 1 \text{ hp}$$

One Btu (British thermal unit) is the amount of heat required to raise the temperature of 1 pound of water 1°F. Since our hp formula includes flow in gallons per minute, we must reduce the conversion factor to Btu/min.

$$\frac{2544}{60} = 42.4 \text{ Btu/min (1 hp)}$$

The energy converted into heat is

$$42.4 \text{ Btu/min} \times 0.583 \text{ hp} = 24.7 \text{ Btu/min}$$

The specific heat of oil is approximately 0.42 Btu/lb-°F. The specific gravity of oil is approximately 0.85. The energy converted into heat (24.7 Btu/min) will raise the temperature of each gallon of oil as follows:

$$\text{Temp. (°F)} = \frac{\text{Btu/min}}{\text{sp. heat of oil} \times \text{quantity of oil (lb/min)}}$$

$$\text{Btu/min} = 24.7$$

$$\text{Specific heat of oil} = 0.42 \text{ Btu/lb-°F}$$

$$\text{Quantity of oil flowing in lb/min} = 71 \text{ lb/min}$$

$$1 \text{ gallon of oil} = 231 \text{ in.}^3$$

$$1 \text{ ft}^3 = 1728 \text{ in.}^3$$

$$\text{Weight of oil} = 62.4 \times 0.85 = 53 \text{ lb/ft}^3$$

Weight of 1 gallon of oil

$$= \frac{231}{1728} \times 62.4 \times 0.85 = 7.1 \text{ lb}$$

$$\text{Temp. (°F)} = \frac{24.7}{0.42 \times 71} = 0.84°F$$

EXAMPLE 2

A hydraulic pump with a capacity of 20 gpm at 3000 psi discharges 50% of its volume over the relief valve during the feed portion of the machining cycle. If the feed stroke takes 3 minutes, how much heat is generated in Btu?

Solution:

$$\text{Hydraulic horsepower} = \frac{\text{gpm} \times \text{psi}}{1714}$$

$$= \frac{20 \times 0.50 \times 3000 \text{ psi}}{1714}$$

$$= \frac{30,000}{1714}$$

$$= 17$$

Since there are 42.4 Btu/min in 1 hp, Total heat gained (Btu)

$$= \frac{42.4 \text{ Btu/min} \times 17 \text{ hp} \times 3 \text{ min}}{1 \text{ hp}}$$

$$= 2162.4 \text{ Btu}$$

The result in heat gained due to the high-pressure fluid's being discharged over the relief valve shows the importance of selecting the proper cooling equipment based on an analysis of the system's operations (see Fig. 5-8).

High fluid velocities, high ambient temperature, mechanical friction, and fluid friction are all factors of heat generation. Experience has shown that heat generation due to these factors is approximately 20% of the input horsepower (prime mover).

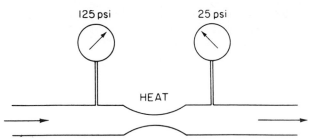

FIGURE 5-8 Fluid temperature rise in a hydraulic line.

EXAMPLE 3

A hydraulic system has a pump which is driven by a 20-hp electric motor. What is the expected heat generation?

SOLUTION:

$$1 \text{ Btu} = 778 \text{ ft-lb}$$

$$\text{Heat generated} = (0.20)(20 \text{ hp}) = 4 \text{ hp}$$

$$4 \text{ hp} \left(\frac{33,000 \text{ ft-lb/min}}{1 \text{ hp}} \right) = 132,000 \text{ ft-lb/min}$$

$$132,000 \text{ ft-lb/min} \left(\frac{1 \text{ Btu}}{778 \text{ ft-lb}} \right)$$

$$= 169.4 \text{ Btu/min}$$

Therefore, the system must dissipate 169.4 Btu/min. If the system itself cannot get rid of this much heat, a cooler must be added to control the proper system temperature.

The most popular types of heat exchangers used with hydraulic systems use air or water as the cooling medium. The law of applying heat exchangers states that the heat given off by the hot object must equal the heat gain by the cold object. Generally speaking, if all other factors remain constant, heat transferred in any given heat exchanger will be directly proportional to the temperature difference between the hydraulic fluid and the cooling media.

Air coolers (see Fig. 5-9) are used where water may be too expensive or not readily available. Air-to-oil coolers are generally used on mobile-type equipment. These coolers consist of a core section of oval or round tubes and plate fins, or individual round finned tubes. The oval core with plate fins has a compact surface which provides good surface efficiency, low air resistance, and minimum power-to-heat removal ratio.

The round core with plate fins is generally used for fluids having a high rate of heat transfer. It can be used also for systems with high operating pressures. This design operates more efficiently with forced air circulation when used for cooling viscous fluids. The individual round fin tube core section is comparable with the round core design, except that individual tubes can be removed from the core section. These coolers are used for applications where the hydraulic fluid enters the cooler at a minimum of 130°F with an ambient air temperature of 100°F or more.

Water-cooled heat exchangers of the shell and tube design are generally used for hydraulic systems when ample water supply is available. A heat exchanger of this type consists of an assembled bundle of tubes inserted into a shell. The tubes are baffled to direct the hydraulic fluid through the shell side of the unit at right angles to the tube bundle. The cooling medium generally flows through the tubes.

There are a number of design variations; however, the straight tube design with fixed tube bundle is popular for

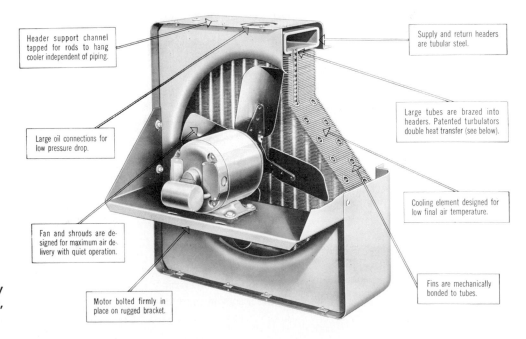

Header support channel tapped for rods to hang cooler independent of piping.

Supply and return headers are tubular steel.

Large oil connections for low pressure drop.

Large tubes are brazed into headers. Patented turbulators double heat transfer (see below).

Fan and shrouds are designed for maximum air delivery with quiet operation.

Cooling element designed for low final air temperature.

Motor bolted firmly in place on rugged bracket.

Fins are mechanically bonded to tubes.

FIGURE 5-9 Air cooler. (Courtesy of Vickers, Incorporated, Troy, Mich.)

use with hydraulic systems. The units are available with various baffle arrangements to create single- or multiple-pass exchangers. The multiple-pass designs use less water and can be used more efficiently and at less cost when colder circulating water is available. Tubes are accessible from either end for cleaning.

The main factors for selecting the proper heat exchangers are as follows:

1. Determine the actual heat (Btu/min) generated during system operations on the basis of a 1-minute time cycle (total input horsepower minus mechanical work output).

2. Select the hydraulic fluid with careful consideration of type of fluid, viscosity, density, specific heat, flow rate, and inlet temperature.

3. For the cooling fluid, determine:
   a. Flow and inlet temperature for water-cooled exchangers.
   b. Ambient temperature for air-cooled units. When water for cooling is readily available the type of cooling system shown in Fig. 5-10 is used to maintain proper fluid temperature.

FIGURE 5-10 Water-type water cooler. (Courtesy of Young Radiator.)

Heat exchangers installed on the low-pressure side of the hydraulic circuit eliminate the need for high-pressure units. When the pressure differential is exceedingly high between the hydraulic fluid and the coolant fluid, it enhances the possibility of water leakage into the hydraulic system. Heat exchangers should be protected against overpressures as well as shock loads. Relief valves properly placed help to provide protection for heat exchangers.

A separate low-pressure circuit from the reservoir through the heat exchanger and back to the reservoir permits the hydraulic fluid to circulate without being affected by flow conditions found in the main circuit.

Excessive pressure drops should be avoided when heat exchangers are selected. They should be inspected periodically and cleaned at regular intervals in order to maintain their efficiency.

SOURCES OF CONTAMINATION IN A HYDRAULIC SYSTEM

Modern hydraulic power and control systems must be reliable and dependable, and must provide greater accuracies than ever before. This means that each component used in a hydraulic system must be precisely machined.

The biggest enemy of a precision machine is contamination. Contamination is any substance in a fluid power system which is detrimental to its proper function. Contamination may be in the form of a liquid, gas, or a solid.

The source of contaminants may be any of the following:

1. Assembly or machine fabrication
2. The system's operational factors
3. Careless maintenance and service practices

The contaminants found to result from component assembly or machine fabrication include metal chips, parts of pipe threads, tubing burrs, pipe dope, shreds of plastic tape, lint, bits of sealing material, scale, welding beads, bits of hose, plain dirt, and sludge. New hydraulic machinery may be seriously damaged during startups unless every precaution is observed to free system components from such contaminants.

RECOMMENDED PROCEDURES FOR USERS

1. Uncrate equipment when it is received and carefully inspect for any damage that may have occurred during shipment.

2. Carefully store in a clean, dry place, making sure that ports or openings are sealed tight, until ready for use.

3. Use accepted practices of good workmanship during fabrication or setting up equipment for use, and use clean and compatible materials.

4. Flush system thoroughly (put in blanks for any servo valves during flushing procedures) with a good grade of recommended flushing fluid; then drain completely.

5. Flush system with the hydraulic fluid (filtered) to be used, and then drain completely.

6. Inspect and clean or replace all filters in the system.

7. Fill system with the filtered hydraulic fluid to be used for the machine operation.

This may seem a complex procedure simply to put a hydraulic system into operation, but, compared in cost to expensive downtime and loss of production, it is well worth the effort.

During the normal operation of hydraulic equipment, there are many sources of contamination that may cause serious trouble unless proper maintenance practices are used. Contamination due to operation may be caused by water or moisture getting into the system, entrained gases, scale caused by rust, bits of sealing material, metal slivers, particles of the metal due to wear, and sludge or varnishes due to oxidation of the fluid.

Installation of adequate filtration and magnets at vulnerable points throughout the system and a good preventive maintenance program are the best safeguards against these types of contamination. It is also important that the right fluid be used and that its temperature be properly maintained.

Improper maintenance and service practices can also add harmful contamination to the hydraulic system. One of the main sources of contamination is the use of dirty equipment, such as buckets or funnels, to add fluid to a hydraulic system. It has also been found that hydraulic components during assembly or disassembly have been placed on a dirty workbench or wiped off with a dirty rag.

Clean equipment and clean workstations are essential when servicing or maintaining hydraulic equipment. When it is necessary to do field repair or work in contaminated areas, spread an old sheet or wrapping paper on top of the work surface. Wash all parts and dip in clean hydraulic fluid just before assembly.

Good maintenance and service practices can prevent contamination and expensive downtime.

STRAINERS AND FILTERS

The importance of keeping the hydraulic fluid clean and free of harmful contamination cannot be overemphasized. Strainers and filters are used to remove contaminating particles from the hydraulic system. The terms *strainer* and *filter* are often used interchangeably because they have a common function. The industry has attempted to distinguish between the two types of units according to a basic difference in their construction. Thus a strainer may be defined as "a device for the removal of solids from a fluid wherein the resistance to motion of such solids is in a straight line." This filter may be defined as "a device for the removal of solids from a fluid wherein the resistance to motion of such solids is in a tortuous path."

Strainers are generally designed of a fine mesh wire screen or a screening element made of specially processed wire of varying thickness wrapped around a metal frame (see Fig. 5-11). Strainers do not provide the fine filtrating action as do filters; thus they offer less resistance to flow. For this reason, strainers may be used in a pump inlet line to protect the pump from the larger sizes of detrimental particles. Differently constructed hydraulic pumps require different inlet conditions. It is important that the capacity of the inlet strainer, including case and fittings, be sufficient to keep pressure drop at a minimum. The finer the screen mesh, the greater the pressure loss. Pressure drop also increases during operating as dirt clogs the small screen openings. If the pressure drop reaches a critical limit for a given pump design, cavitation results and the pump life is shortened.

There are a number of arrangements for using strainers in a pump inlet line. If single strainer units are insufficient for the demands of a given pump, two or more can be used, as shown in Fig. 5-12. To help facilitate regular cleaning, strainers need to be installed only hand tight, provided that the fittings are constantly submerged.

A 60-mesh screen is generally recommended where actual pressure drop is unknown. However, if the pressure drop at the lowest operating temperature of the fluid used is known, screens as fine as 120 mesh can be used, if of ample capacity.

Filtration is fast becoming an art in itself, and the modern tendency is to measure detrimental particle sizes in microns. A micron equals one-millionth of a meter, or 0.000039 in. The limit of visibility with the unaided eye is about 40 microns [or micrometers (μm)]; thus many detrimental particles in a hydraulic system are not normally visible (see Fig. 5-13).

Modern high-flow, high-pressure hydraulic systems may require a combination of strainers and filters for protection against detrimental contaminants. The pump is generally the most expensive component in the system and should have the protection of an inlet strainer. It is also a fact that filters applied on the discharge side of the pump or in the return line to the reservoir can be of real significance in preventing faulty operation and loss of production.

The following information is important in order to select the proper filter for any application:

1. The size and physical nature of the detrimental contaminates to be removed

2. The viscosity, density, operating temperature, and

**FLOW RATES (gpm) FOR HYDRAULIC FLUIDS OF COMMON VISCOSITIES
FOR 100 MESH AND PRESSURE DROPS WHEN STRAINER IS CLEAN**

Strainer sizes	Approximately 6 sq in.			Approximately 14 sq in.			Approximately 32 sq in.			Approximately 45 sq in.		
psi drop	5	10	15	5	10	15	5	10	15	5	10	15
50 SUS	4.9	7.0	8.5	9.6	13.5	16.6	26	37	45	93	132	161
100 SUS	4.1	5.9	7.2	8.0	11.3	13.9	22	31	38	78	111	136
200 SUS	3.5	4.9	6.0	6.7	9.5	11.6	18	26	32	65	92	113
300 SUS	3.1	4.4	5.4	6.0	8.5	10.4	16	23	28	56	80	98

FIGURE 5-11   Strainer. (Courtesy of Schroeder Brothers.)

corrosive properties of the hydraulic fluid being used

3. The materials used for fabricating the system before the filter and after the filter; seals, fittings, tubing, hose, instruments, and other parts should be analyzed

4. The operating characteristics of the system, such as shock load, fluid velocity, and direction

5. Pressures and pressure drops that the filter media must encounter during the system's operation

6. The component in the system that requires the closest tolerances and hence the least contamination

Filters provide much finer filtration than strainers and may be applied at several different places in the circuit. Where the filter is placed depends a great deal on the application and the design of the hydraulic equipment. Each application must be analyzed to determine the degree of filtration needed, the vulnerable points in the circuit, and the type of filtering media that should be used. Figures 5-14 through 5-17 show examples of where filters may be placed in a hydraulic system.

Filters are rated in terms of "nominal" and "absolute" filtration. Both terms are expressed in microns. The term *nominal filtration* indicates the ability to remove about 98% of the solid particles from the fluid which are equal to or larger than the micron rating of the filter.

The *absolute rating* indicates the size of the filter element pores through which the fluid must flow, and gives the largest possible size solid that can pass through the element. Many laboratory-controlled tests have indicated that irregularly shaped particles and parts of O-rings and so on, that may be elastic in nature have no true absolute rating. As a filter element removes contaminating particles from the fluid stream, they build up on the media or element surface. This reduces the size of the pores through which

FIGURE 5-12   Multiple strainer arrangements. (Courtesy of Vickers, Incorporated, Troy, Mich.)

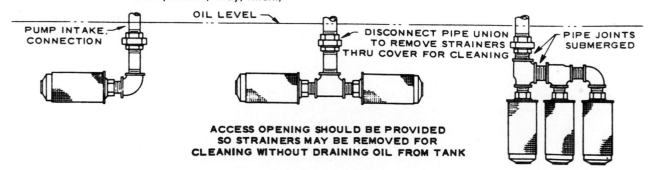

the fluid must flow. Thus a filter rated 60 μm absolute may have a nominal rating of 35 μm because 98% of the particles 35 μm or larger are removed. Some of the basic filter designs are illustrated in Fig. 5-18.

Filter accessories include devices that assist in maintenance, such as indicators (see Fig. 5-19) that operate on the pressure differential of the filter to show that a filter element needs to be cleaned or replaced. Many filters are equipped with relief valves (see Fig. 5-20) which are set to open when the differential pressure reaches a predetermined value. Also, pressure gauges on either side of a filter are sometimes used to show the pressure differential across the filter. When the gauges reach a certain pressure difference above that which is normal when the filter is clean, it indicates a dirty filter. Manual shutoff and bypass valves are also used to isolate the filter during service (see Fig. 5-21).

Portable units, such as that shown in Fig. 5-22, aid during service and repair or when a system is accidentally contaminated to filter large quantities of fluid in order to

STRAINER MESH SIZES AND SIZE OF SOLIDS REMOVED WHEN STRAINER IS CLEAN

Mesh	Wire	Size in.	Microns
8	0.028	0.097	2500
30	0.012	0.021	540
50	0.009	0.011	282
60	0.0075	0.0092	236
70	0.0065	0.008	205
80	0.0055	0.007	180
90	0.005	0.0061	156
100	0.0045	0.0055	141
150	0.0026	0.0041	105
200	0.0021	0.0029	75
250	0.0016	0.0024	60

(One micron equals 0.000039 in.)
Recommended strainer mesh

700 SUS Hyd. Oil	30 mesh
350 SUS Hyd. Oil	60 mesh
200 SUS Hyd. Oil	100 mesh

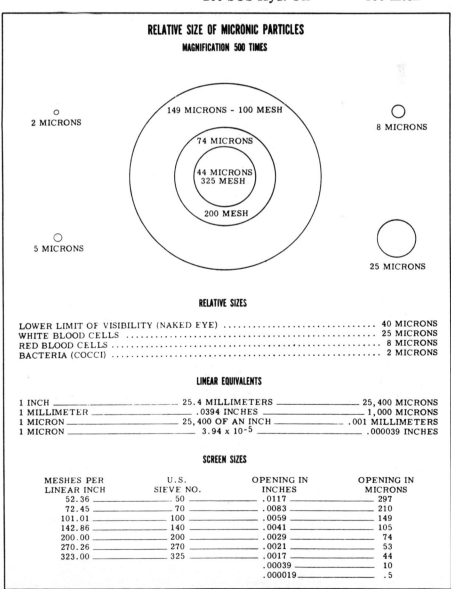

RELATIVE SIZE OF MICRONIC PARTICLES
MAGNIFICATION 500 TIMES

2 MICRONS

149 MICRONS - 100 MESH

74 MICRONS

44 MICRONS
325 MESH

200 MESH

8 MICRONS

5 MICRONS

25 MICRONS

RELATIVE SIZES

LOWER LIMIT OF VISIBILITY (NAKED EYE) .... 40 MICRONS
WHITE BLOOD CELLS .... 25 MICRONS
RED BLOOD CELLS .... 8 MICRONS
BACTERIA (COCCI) .... 2 MICRONS

LINEAR EQUIVALENTS

1 INCH	25.4 MILLIMETERS	25,400 MICRONS
1 MILLIMETER	.0394 INCHES	1,000 MICRONS
1 MICRON	25,400 OF AN INCH	.001 MILLIMETERS
1 MICRON	3.94×10^{-5}	.000039 INCHES

SCREEN SIZES

MESHES PER LINEAR INCH	U.S. SIEVE NO.	OPENING IN INCHES	OPENING IN MICRONS
52.36	50	.0117	297
72.45	70	.0083	210
101.01	100	.0059	149
142.86	140	.0041	105
200.00	200	.0029	74
270.26	270	.0021	53
323.00	325	.0017	44
		.00039	10
		.000019	.5

FIGURE 5-13 Mesh strainers (magnified 500 times). (Courtesy of Vickers, Incorporated, Troy, Mich.)

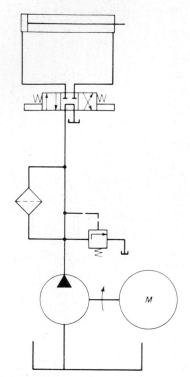

FIGURE 5-14 Bypass filter; pressure side.

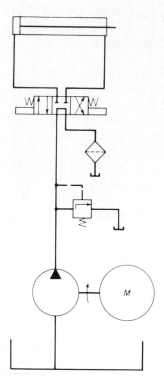

FIGURE 5-15 Full flow in return to tank line.

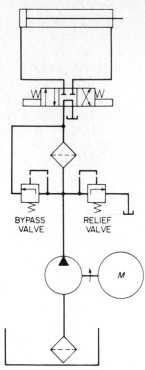

FIGURE 5-16 Suction line and pressure line filtration.

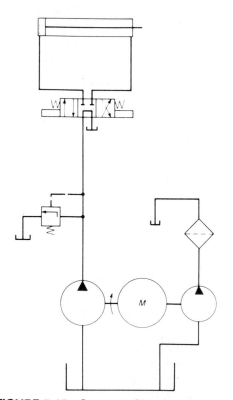

FIGURE 5-17 Separate filtration system.

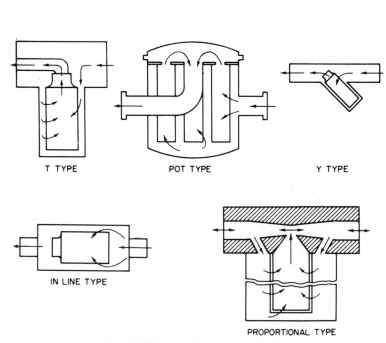

FIGURE 5-18 Basic filter designs.

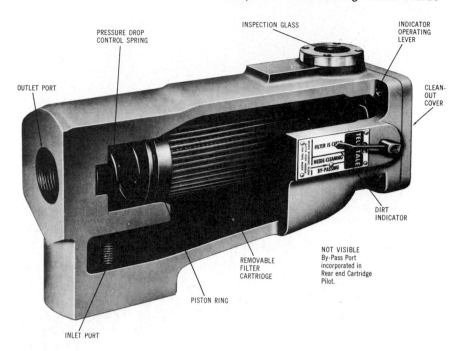

INSPECTION GLASS

INDICATOR
OPERATING
LEVER

PRESSURE DROP
CONTROL SPRING

CLEAN-
OUT
COVER

OUTLET PORT

FILTER IS CLEAN

NEEDS CLEANING

BY-PASSING

TELL-TALE

DIRT
INDICATOR

NOT VISIBLE
By-Pass Port
incorporated in
Rear end Cartridge
Pilot.

REMOVABLE
FILTER
CARTRIDGE

PISTON RING

INLET PORT

FIGURE 5-19 Tell-tale filter.

FIGURE 5-20 (a) Full-flow filter with relief valve. (Courtesy of Vickers, Incorporated, Troy, Mich.) (b) High-pressure filter. (Courtesy of Schroeder Brothers.)

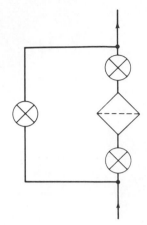

FIGURE 5-21 Manual bypass.

FIGURE 5-22 Portable Filter. This filter has its own pump and hose connections. Oil is drawn from a reservoir—from the low point, if possible—is pumped through the filter, and is returned to the clean oil side of the reservoir. The filter has a replaceable filter cartridge, a relief valve, and a pressure gauge. The pressure reading increases as the cartridge becomes loaded with dirt. When a specified pressure is reached, the cartridge should be changed. A portable filter is usually connected to a reservoir for 24 hours at a time. (Courtesy of William W. Nugent & Co.)

get the equipment back in service. The portable unit is also excellent in reclaiming hydraulic fluid.

The materials selected for the construction of filter housings depend primarily on the intended service. Aircraft and missile usage require materials of high strength and low weight, such as stainless steels and certain aluminum alloys.

Aluminum is generally restricted to medium-temperature ranges where noncorrosive conditions are essential. For marine, industrial, and some military applications, carbon steels and copper–nickel alloys are used. Because of the trend in all fields of usage to use full-flow filtration, the body must be rated for its intended operating pressure.

There are several basic types of filter media, such as paper, sintered metal powder, woven wire cloth, and certain kinds of ceramic or plastic. In the selection of a filter media, it is important to consider the type of fluid, chemical compatibility, temperature, and the ability to withstand high flow rates.

Paper filter elements are made of special cellulosic materials which are impregnated with a resin to provide greater strength. The paper media may be used as an extended area filter where particles removed depend on the amount of surface exposed and the porosity of the filter cake that is formed. Two popular filters of the type are made by folding sheets of resin-impregnated paper into accordion-like pleats or into "concertina folds" and the forming into the filter housings. Other types are made by stacking cellulose discs of various designs with alternate layers of cellulose spacers. The fluid is filtered laterally, vertically, and between the discs in an edge-type manner. The design generally determines the efficiency of the filter. Recently it was discovered that by adding glass fiber greater strength can be obtained. Cellulose filter elements can be made with accurate control of pore sizes between 2 and 25 μm. They can be used for temperatures as high as 275°F. The large amount of surface area helps to increase the actual operation time between servicing. These filter elements are available for nominal ratings of 0.5 to 100 μm. Absolute ratings range from 5 μm and up.

In general, the sintered metal powder media, including bronze and stainless steel, provide depth filtration with nominal ratings of 2 to 65 μm. The absolute ratings range from 13 to 100 μm. Media migration or self-contamination may be a problem unless proper manufacturing methods are used.

Ceramic or plastic media are also used, but higher pressure drops may be expected as well as less resistance to mechanical and thermal shock.

Woven wire cloth elements are usually made of stainless steel and provide nominal filtration ratings of 2 to 100 μm with absolute ratings from 12 to 200 μm. This type of filter media provides resistance to corrosion and fatigue. It is free from migration and resists penetration by sharp particles.

Filtration is accomplished by essentially two classes of filters, surface filters and depth filters. The surface filter (see Fig. 5-23) accomplishes all its filtering action at the surface of the media. The pores are all very nearly of uniform size, and the action approaches the absolute. This class of filter is easy to clean and resists migration, but it has a comparatively low dirt-holding capacity.

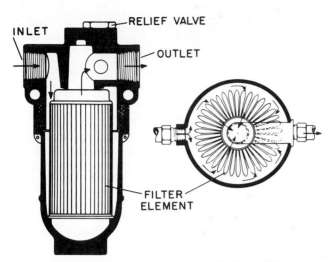

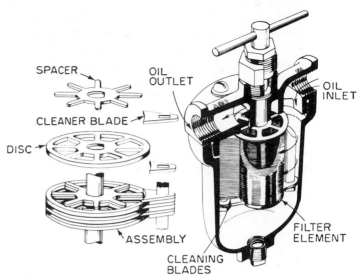

FIGURE 5-23 Surface-Type Full-Flow Filter. This model is suitable for line pressures up to 1500 psi. The replaceable filter element is made of resin-impregnated paper having pores of up to 5-micron size. The paper is pleated and formed into a cylinder as shown. Hydraulic fluid passes through the paper leaving behind dirt and other solids in a layer on the outer surface. In service, particles down to about one-micron (about 0.00004-in.) size are held back. If pressure drop across the filter becomes too great, the relief valve opens and by-passes unfiltered fluid. (Courtesy of Socony Mobil Oil Corporation.)

FIGURE 5-24 Depth-Type Size Filter. Dirty oil surrounds the renewable filter cartridges and flows radially inward through a considerable depth of filter material such as cellulose. Clean oil flows downward and out at the left. These filters are capable of very fine filtration. (Courtesy of Socony Mobil Oil Co.)

FIGURE 5-25 Edge-Type Size Filter. The filter element consists of a stack of wheel-shaped metal discs, each separated from the next by a thin metal spacer. The fineness of filtration is determined by the thickness of these spacers. The stack is closed at the bottom and opened to the filter outlet at the top. Dirty inlet oil flows into the space around the element and is forced through the small openings between the discs. All solids larger than the openings are held back. Clean oil flows upward through the stack to the outlet. Stationary cleaner blades extend into the space between the discs (see assembly sketch), and the filter element can be rotated against these for cleaning, using the handle shown. Solids fall to the bottom of the housing and should be removed at regular intervals. (Courtesy of Socony Mobil Oil Corporation.)

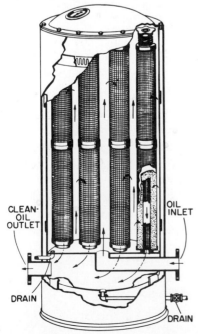

The depth filter (see Fig. 5-24) operates throughout the volume of the filter material by presenting many tortuous passages through which the fluid must flow. The pores or passages are not of uniform size, and the entrapment of particles depends on the depth and nature of the various passages. The action of this type of filter is primarily statistical because of the different sizes of the passages and the effect of fluid velocity, which could dislodge trapped particles if rapidly increased. Depth filters operate successfully at lower flow rates and at relatively low pressure drops. The types of filters used to eliminate contaminants from a hydraulic system are shown in Figure 5-25.

The use of magnets has been found to be advantageous for removing iron and steel particles from the hydraulic fluid. Magnetic plugs are installed at strategic locations in the reservoir to prevent the particles of steel and iron from getting back up into the system.

Magnetic rings and pencils are also assembled on pump

inlet strainers to attract and hold any steel or iron particles trying to get through the wire mesh screen. Magnets help to prevent the sliver-type particles such as broken-off threads from penetrating through the strainer and into the system. Magnets can be installed easily at any time and may save valuable equipment from being damaged.

BENEFITS FROM PROPER CONDITIONING OF THE HYDRAULIC FLUID

Using the correct hydraulic fluid, selected especially for the equipment used as well as for the operating conditions involved, is most important. Proper conditioning of the fluid selected must be maintained continually to keep the fluid clean and at proper operating temperature if the hydraulic system is to function properly and give long life. Proper conditioning provides the following benefits:

1. *Reduction of power losses.* The use of properly sized reservoirs and adequate coolers reduces system leakage by controlling the viscosity of the fluid with temperature control. This also helps to minimize friction by maintaining adequate lubrication between moving parts in pumps, fluid motors, and cylinders. When these properties are maintained, power losses are minimized.

2. *More continuous production.* When the hydraulic fluid is properly conditioned, less pressure drop occurs, which maintains good control, fast response, and accurate timing of machine operations. Strainers, filters, and magnets play an important part in keeping the system clean and preventing costly interruptions in production. The cost of installing a filter or magnet in a hydraulic system is incidental, compared to the loss of a machine and the time involved to make costly repairs.

3. *Reduction of cost of maintenance.* The life expectancy of hydraulic equipment under normal operating conditions can be many years. Hydraulic equipment has been known to operate 15 to 20 years or more without trouble. The main reason for this kind of service life is the fact that the hydraulic fluid was kept clean and properly conditioned. Detrimental contamination and high temperatures can create excessive maintenance costs, if not controlled. When system fluids are properly conditioned, it requires a minimum amount of maintenance to keep equipment in good operating condition.

4. *Reduction of the cost of hydraulic fluid.* Good tight connections are, of course, needed to prevent external leakage. Temperature control is also important, because hot oil leaks more readily than oil at its proper operating temperature. When hydraulic fluid is overheated, it oxidizes rapidly and loses its important physical properties. This causes a kind of chain reaction which quickly contaminates the entire system, which necessitates a changing of the system's fluid. A good hydraulic fluid does not wear out or lose its desirable properties unless it is overheated or contaminated with dirt. Good control of the hydraulic fluid with proper conditioning equipment increases the life of the fluid and the system's components.

Note: Moisture in hydraulic oil can have an adverse effect on some critical hydraulic systems. When this is a factor, driers and/or dessicants (such as activated charcoal or aluminum oxides, etc.) may be used with the filters to maintain the needed "dry oil."

QUESTIONS

1. What is meant by *conditioning* hydraulic fluids?
2. Name the important features found on a well-designed hydraulic reservoir.
3. What is the general rule of thumb for sizing a reservoir's capacity?
4. What is the function of a heat exchanger when used with the hydraulic system?
5. What effect would high temperatures have on the operating characteristics of a hydraulic system?
6. How many Btu per hour would be gained by a system blowing 7.5 gpm over the relief valve at 1500 psi every 18 minutes out of each hour of operation?
7. What are the main factors in selecting the proper heat exchangers for a hydraulic system?
8. How does contamination get into a hydraulic system?
9. What are the contaminants that are generally found in a hydraulic system?
10. How is contamination controlled in a hydraulic system?
11. How are metal chips eliminated from a hydraulic system?
12. In general, what is the difference between a strainer and a filter?
13. How are the sizes of detrimental particles determined?
14. What are some of the important factors to consider in the selection of filtration devices?

15. Explain the difference between a high-pressure full-flow filter and a low-pressure full-flow filter.

16. Why are indicators used on filters, and on what principle do they operate?

17. Name some of the basic types of filter media.

18. Describe the difference between a surface filter and a depth filter.

19. What are some of the benefits from properly conditioned hydraulic systems?

6

The Source of Hydraulic Power

Transmitting hydraulic power requires a source of input energy, one or more pumps to convert this mechanical energy into hydraulic energy, controls, and actuating cylinders or fluid motors that utilize the fluid power to provide linear or rotary motion (see Fig. 6-1).

The input energy source may be an electric motor, internal combustion engine, turbine drive, power-takeoff arrangement, or other mechanical device that can impart force and motion to operate the pump. The pump, in turn, must supply hydraulic fluid to the system. A hydraulic pump is defined as a device that converts mechanical force and motion into fluid power. It is the heart of the system and generally one of the more costly components used in the transmission of hydraulic power.

Where electric power is readily available, such as in most industrial plants, electric motors are used to drive the hydraulic pumps. Power units, as shown in Fig. 6-2, are especially popular in the machine tool field of application. Electric motors are readily available in speeds (rpm) which are compatible with most hydraulic pumps.

Hydraulic power systems found on farm tractors, road machinery, earth-moving equipment, utility vehicles, and various other devices using internal combustion engines use power-takeoff arrangements to operate the hydraulic pump (see Fig. 6-3).

When power-takeoff arrangements are not available on machines, auxiliary internal combustion engines are generally used to supply the input energy for the hydraulic system. The operating speed (rpm) of the engine is selected to be compatible with that of the recommended speed (rpm) of the hydraulic pump.

Figure 6-4 shows a method of directly coupling the hydraulic pump to the main drive. Engine speed in rpm must be compatible with that of the pump's rated speed.

Aircraft and other equipment using high-speed prime movers, such as turbines or jet engines, require specially designed high-speed pumps, or employ the use of speed-reduction devices (see Fig. 6-5). Speed reducers coupled to the main drive change the economical operating speed of the prime mover to an economical operating speed of the hydraulic pump.

Belt and chain drives are also used to transfer input force and motion from a prime drive for the operation of a hydraulic pump. Operating speeds (rpm) are adjusting by selecting the proper pulley diameters or sprocket tooth ratios (see Fig. 6-6a–c).

Speed reduction by means of gears can be expressed as the ratio of the number of gear teeth on one gear to the number of teeth on the other. When N represents the number of teeth on the large gear and n the number on the smaller gear, the speed reduction is given by

$$\text{Speed reduction} = \text{gear ratio} = \frac{N}{n}$$

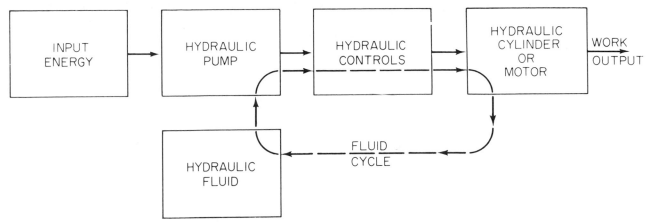

FIGURE 6-1　Elements of hydraulic power.

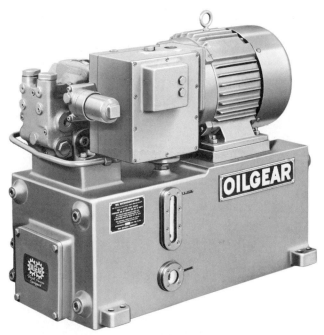

FIGURE 6-2　Power unit showing electric motor and hydraulic pump. (Courtesy of The Oilgear Company, Milwaukee, Wis.)

FIGURE 6-3　Tractor. (Courtesy of J. I. Case.)

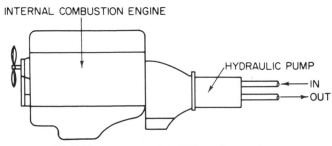

FIGURE 6-4　Direct-coupled engine and pump.

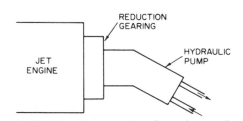

FIGURE 6-5　Power taken from jet engine.

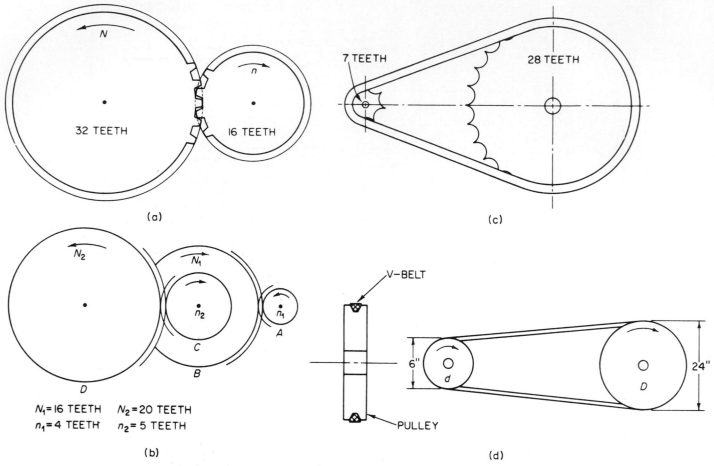

FIGURE 6-6 (a) Spur gear; (b) gear train; (c) sprocket and chain; (d) pulley and V belt.

EXAMPLE

$$\text{Gear ratio} = \frac{N}{n}$$

$$= \frac{32}{16}$$

$$= 2 : 1$$

EXAMPLE

Consider *A* and *B* as a system, and *C* and *D* as a system (speed of *B* equals speed of *C*). This results as follows:

$$\text{Speed reduction} = \text{gear ratio} = \frac{N_1}{n_1} \times \frac{N_2}{n_2}$$

$$N_1 = 16 \text{ teeth}$$

$$n_1 = 4 \text{ teeth}$$

$$N_2 = 20 \text{ teeth}$$

$$n_2 = 5 \text{ teeth}$$

$$\text{Speed reduction} = \text{gear ratio} = \frac{16}{4} \times \frac{20}{5}$$

$$= \frac{320}{20}$$

$$= 16 : 1$$

$$\text{Speed ratio} = \frac{N_1}{n_2}$$

$$= \frac{28}{7} = \frac{4}{1}$$

Pulleys and belts are also used for connecting hydraulic pumps to the main drive. Generally, the V belt is employed to provide adequate strength and area of contact for the friction required to prevent slippage (see Fig. 6-6d).

In general, pulley and belt drive speeds can be expressed in terms of pulley diameters.

$$\text{Speed ratio} = \frac{D}{d}$$

$$= \frac{24}{6} = 4 : 1$$

In most applications of hydraulic power, the pump is connected directly to the input drive shaft by a flexible coupling. A number of different designs are available for a wide choice of power ratings and operating speeds and for a given allowable misalignment.

Important safety rules for the installation of guards covering rotating machinery should be strictly observed, regardless of the type of power drive used.

In general, pump manufacturers limit the amount of side loading when pumps are driven by belts, chains, and other devices. It is more efficient, safer, and less wearing on the equipment when direct coupling arrangements can be used (Fig. 6-7).

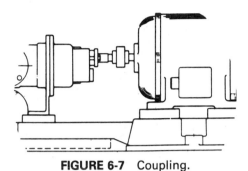

FIGURE 6-7 Coupling.

INPUT HORSEPOWER

The input horsepower for hydraulic power systems can be determined by the following equation:

$$hp = gpm \times psi \times 0.000583$$

EXAMPLE

What horsepower electric motor would be needed to drive a hydraulic pump whose actual delivery is 20 gpm at 1500 psi?

Solution:

$$hp = 20 \text{ gpm} \times 1500 \text{ psi} \times 0.000583$$

$$= 17.4$$

It would not be practical to buy an electric motor that has to be specifically built, so the logical choice would be a standard 20-hp electric motor. It is good practice to select a large enough electric motor to handle emergency overloads. Some industrial plants have a standard practice of selecting the next largest size motor automatically to prevent any future problems.

Output power of a hydraulic pump is expressed as a function of flow volume and pressure. Some engineers use the simple rule that one hydraulic hp equals the delivery of one gallon per minute at 1500 psi. The previous example problem indicates that this is close for a quick estimate of power requirements.

Another version of the hydraulic horsepower formula is shown in the following:

$$\text{Hydraulic hp} = \frac{\text{gpm} \times \text{psi}}{1714}$$

EXAMPLE

A farm tractor is equipped with a 40-hp internal combustion engine. What portion of this hp will be needed to operate a front-end hydraulic loader if the pump delivers 12 gpm at 2000 psi?

Solution:

$$hp = \frac{\text{gpm} \times \text{psi}}{1714}$$

$$= \frac{12 \times 2000}{1714}$$

$$= 14$$

The input horsepower requirements of any hydraulic system depend mainly on the maximum needs at any one time during a cycle of the equipment operation. The operating pressure at any given moment in a hydraulic system depends primarily on the resistance due to the work load. The number of gallons per minute being delivered to a hydraulic cylinder or motor at any given moment determines how fast the work is being done. Therefore, the input power requirements must be adequate to handle the maximum pressure and volume flow at any given moment in a hydraulic system.

In comparison with mechanical systems, one of the main advantages of using hydraulic power is the fact that a well-designed system provides the correct amount of operating horsepower at all times.

PUMP CLASSIFICATION

There are two general classifications of pumps which are being used for the transmission of hydraulic power:

1. Nonpositive displacement pumps
2. Positive displacement pumps

Nonpositive displacement pumps are primarily velocity-type units which have a great deal of clearance between the rotating and stationary parts of the pump. These pumps are generally used for low-pressure, high-volume flow applications. Because of the large clearance space, these pumps are not self-priming. In other words, the pumping action has too much clearance space to seal against atmospheric pressure. The displacement between inlet and outlet is not positive. Therefore, the volume of fluid delivered by the pump depends on the speed at which the pump is operated and the resistance at the discharge side. As the resistance builds up at the discharge side, fluid slips back into the clearance spaces, or, in other words, follows the path of least resistance. When the resistance gets to a certain value, no fluid at all will be delivered to the system and the volumetric efficiency of the pump will drop to zero for a given speed. Increased speed will increase pressure. Figure 6-8 shows an impeller-type nonpositive displacement pump which is generally used for transferring large quantities of liquid at a low pressure.

Figure 6-9 shows the axial flow design of a nonpositive displacement pump design which handles up to thousands of gallons of liquid per hour. Figure 6-10 shows a centrifugal pump, which has the highest potential among nonpositive displacement-type pumps.

The main uses for nonpositive displacement pumps are for transferring fluids, coolant supply, and to supercharge the inlet of other pumps. When centrifugal pumps are multistage units (see Fig. 6-11) they can be used for higher operating pressures. They operate on the principle that fluid taken in at the inlet of the first stage is delivered to the inlet of the second stage and so on through each successive stage of the pump. In this manner the potential of the entire pumping unit is substantially increased. In other words, it is a number of centrifugal pumps hooked up in a series. These units are used for large hydraulic systems using

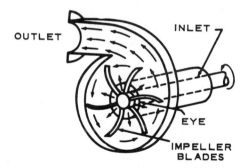

FIGURE 6-8   Impeller unit. (Courtesy of Vickers, Incorporated, Troy, Mich.)

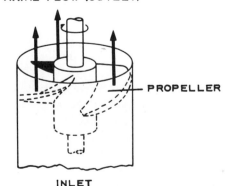

FIGURE 6-9   Axial flow unit. (Courtesy of Vickers, Incorporated, Troy, Mich.)

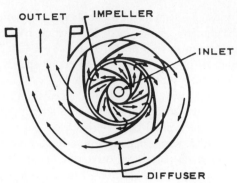

FIGURE 6-10   Centrifugal pump: single stage. (Courtesy of Vickers, Incorporated, Troy, Mich.)

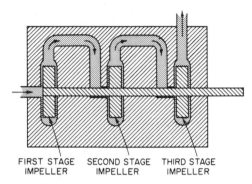

FIGURE 6-11   Centrifugal pump: multistage.

water, such as forge presses where hot molten metal presents a fire hazard.

Positive displacement pumps are generally used for hydraulic power. A positive displacement pump has a very close clearance between rotating and stationary parts. As the classification implies, pumps of this design provide a specific amount of fluid to the hydraulic system for each revolution that they are turned.

The volumetric efficiency of any pump is the quantity of fluid delivered from the outlet port, compared to the theoretical displacement of the pump.

$$\text{Vol. eff.} = \frac{\text{actual output (gpm)}}{\text{theoretical output (gpm)}} \times 100$$

Theoretical displacement depends primarily on the geometry of the pump by design (cubic inches of fluid displacement per revolution) and the operating speed (rpm).

Theoretical output (gpm)

$$= \frac{\text{displacement (in.}^3\text{)/rev.} \times \text{rpm}}{231 \text{ in.}^3}$$

The actual delivery of the pump depends on how well the pump is made and is computed by actual laboratory testing.

Overall efficiency is expressed as follows:

$$\text{Pump overall eff. } (\%) = \frac{\text{output hyd. hp}}{\text{input hp to pump}} \times 100$$

Mechanical efficiency as well as volumetric efficiency have an important bearing on overall efficiency. Mechanical efficiency is the ratio of the calculated power required to drive a hydraulic pump compared to the actual input power.

$$\% \text{ mech. eff. } = \frac{\begin{array}{c}\text{theoretical torque required}\\\text{to drive the pump (lb-in.)}\end{array}}{\begin{array}{c}\text{actual torque supplied to}\\\text{pump (lb-in.)}\end{array}} \times 100$$

Theoretical input torque is

$$\text{Torque (lb-in.) } = \frac{\text{displacement (in.}^3)/\text{rev.} \times \text{psi}}{2\pi}$$

Actual input torque is calculated by laboratory testing:

$$\text{Actual torque (lb-in.) } = \frac{\text{brake hp} \times 63,025}{\text{rpm}}$$

Brake horsepower can be calculated as follows:

$$\text{Brake hp } = \frac{\text{hyd. output hp} \times 100}{\text{pump vol. efficiency}}$$

EXAMPLE 1

What is the volumetric efficiency of a pump that displaces 0.7 in.^{3} of fluid per revolution if the actual delivery is 4.5 gpm at 1800 rpm?

Solution:

$$\text{Vol. eff. } = \frac{\text{actual output (gpm)}}{\text{theoretical output (gpm)}} \times 100$$

Actual output given = 4.5 gpm

$$\text{Theoretical output (gpm) } = \frac{\text{dis. (in.}^3)/\text{rev.} \times \text{rpm}}{231}$$

$$= \frac{0.7 \times 1800}{231}$$

$$= \frac{1260}{231}$$

$$= 5.4 \text{ gpm}$$

so

$$\text{Vol. eff. } = \frac{4.5}{5.4} \times 100$$

$$= 83\%$$

EXAMPLE 2

What is the overall efficiency of a hydraulic pump that is driving a fluid motor at 2000 rpm with a torque of 550 lb-in.? The pump is driven by a 20-hp electric motor.

Solution:

$$\text{Pump overall eff. } = \frac{\text{output hyd. hp}}{\text{input hp}} \times 100$$

$$\begin{array}{c}\text{Fluid motor hp}\\(\text{output hp})\end{array} = \frac{\text{torque (lb-in.)} \times \text{rpm}}{63,025}$$

$$= \frac{550 \times 2000}{63,025}$$

$$= 16$$

The input hp is given as 20 hp. Substituting, we obtain

$$\text{Eff. } = \frac{16 \text{ hp}}{20 \text{ hp}} \times 100$$

$$\text{Overall eff. } = 80\%$$

EXAMPLE 3

What is the mechanical efficiency of a hydraulic pump with a displacement of 0.7 in.^{3}/rev. operating at 2000 psi? The pump is operating at 1800 rpm and the brake horsepower is 7.2.

Solution:

$$\% \text{ mech. eff. } = \frac{\begin{array}{c}\text{theoretical torque required}\\\text{to drive pump (lb-in.)}\end{array}}{\begin{array}{c}\text{actual torque supplied to}\\\text{pump}\end{array}} \times 1000$$

$$\text{Theoretical torque required } = \frac{\text{dis. (in.}^3)/\text{rev.} \times \text{psi}}{2\pi}$$

$$= \frac{0.7 \text{ in.}^3 \times 2000 \text{ psi}}{6.28}$$

$$= 223 \text{ lb-in.}$$

$$\text{Actual torque (lb-in.) } = \frac{\text{bhp} \times 63,025}{\text{rpm}}$$

$$= \frac{7.2 \times 63,025}{1800 \text{ rpm}}$$

$$= 246 \text{ lb-in.}$$

$$\text{Mech. eff. } = \frac{223 \text{ lb-in.}}{246 \text{ lb-in.}} \times 100$$

$$= 90\%$$

EXAMPLE 4

What is the brake horsepower of a hydraulic pump delivering 36 hyd. hp if the pump has an efficiency of 87%?

Solution:

$$\text{Brake} = \frac{\text{hyd. output hp}}{\text{pump eff.}} \times 100$$

$$= \frac{36}{87} \times 100$$

$$= 41.3$$

PUMPING THEORY

When a positive displacement pump is operated, trapped fluid is displaced from the inlet to the outlet because of the close tolerance between mating parts. In Fig. 6-12, as the gears turn, fluid trapped between the gear teeth and the housing is carried from the inlet side to the discharge side of the pump. When the pump is first started, the fluid displaced is air, and as the air is taken from the inlet pipe a vacuum is created. When the pressure within the inlet line to the pump becomes less than the atmospheric pressure, fluid is forced by the atmospheric pressure through the strainer and up into the pump inlet. Thus it becomes obvious that fluid supplied to the pump is dependent on atmospheric pressure to raise it up into the pump inlet.

Since atmospheric pressure is 14.7 psi at sea level, it is important to control other variables, such as pipe inlet size and length of inlet pipe, within the working range of atmospheric pressure. The recommended fluid velocity for good inlet conditions of the pump should be maintained between 2 and 5 ft/sec.

$$\text{Fluid velocity (ft/sec)} = \frac{\text{gpm} \times 0.3208}{\text{inlet pipe area}}$$

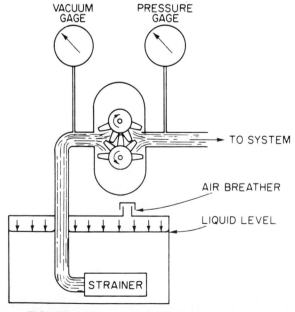

FIGURE 6-12   Positive-displacement pump.

EXAMPLE

What area of pipe is needed for the inlet of a 20-gpm pump? (The value for velocity must be selected between 2 and 5 ft/sec.)

Solution: Substituting, we have

$$\text{Velocity (3 ft/sec)} = \frac{20 \text{ gpm} \times 0.3208}{\text{area of pipe (in.}^2)}$$

$$\text{Area of pipe (in.}^2) = \frac{20 \text{ gpm} \times 0.3208}{3 \text{ ft/sec}}$$

$$= 2.208 \text{ in.}^2$$

To find the diameter from the area:

$$\text{Diameter (in.)} = \sqrt{\frac{\text{Area (in.}^2)}{0.7854}}$$

$$= \sqrt{\frac{2.208}{0.7854}}$$

$$= \sqrt{2.8}$$

$$\text{Diameter}^2 = 2.8$$

$$\text{Diameter} = 1.7 \text{ in.}$$

The nearest nominal standard pipe size would be $1\frac{1}{2}$-in.-diameter pipe with an area of 2.03 in.² This would increase the velocity rate a little, but it would still remain close to 3 ft/sec.

Pump inlet conditions are also affected by the vicosity, specific gravity, fluid temperature, and the type of fluid being used in the system. It is essential that all these variables be carefully considered to keep inlet conditions of the pump under control for efficient operation. When control of proper conditions is impossible because of the nature of the application or environmental factors, the pump inlet should be pressurized.

The pump inlet can be pressurized either through the use of a liquid or a gas. Pressurization of the hydraulic pump inlet may be necessary for the following reasons:

1. When the hydraulic application occurs in high-altitude areas, where atmospheric pressure is less than normal.
2. If the pump is driven at high rpm and atmospheric pressure cannot act quickly enough to fulfill the inlet requirements.
3. When the specific gravity of the hydraulic fluid is greater than that of water.
4. To provide greater machinery stability for smoother finishes, especially if the work loads are erratic by nature.
5. To allow hydraulic fluids to operate at high temperatures without vaporization.

6. To assure safety of operation for high-performance machinery, or to protect human lives.

If the hydraulic fluid itself is used to pressurize the pump inlet, a second pump is generally used. This second pump may be an independent system or can be attached to the same input shaft of the main pump. The inlet supply pressure to the main pump varies with the application but is usually maintained below 100 psi because of the effect on shaft seals or other low-pressure parts. Figure 6-13 illustrates the use of an auxiliary pump to supply the main pump when atmospheric pressure is incapable of forcing an adequate supply of hydraulic fluid to the pump.

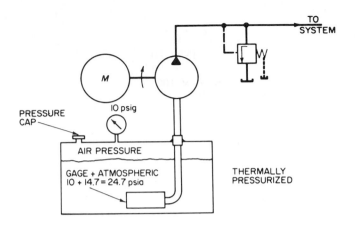

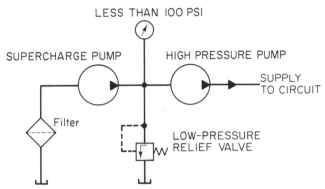

FIGURE 6-13 Pressurized high-pressure pump inlet using a supercharge pump.

When a gas is used to pressurize the pump inlet, the reservoir is completely enclosed. The space above the liquid in the tank is maintained under a pressure greater than that of the atmosphere. This may be done by pumping the gas into the space or by relying on the natural thermal expansion of the trapped gas as the hydraulic liquid heats up as a result of the hydraulic system's operations. Pressure caps similar to the automobile radiator cap are generally used for pressurized reservoir conditions (see Fig. 6-14).

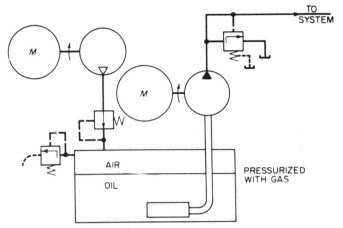

FIGURE 6-14 Reservoir pressurization with gas.

PUMP DESIGN TYPES

Pumping units are available for practically any conceivable use. In fact, it has been stated that more patents have been granted for pump designs than for any other mechanical device invented.

Hydraulic pumps are available for volumetric capacities of a few cubic inches per minute or for hundreds of gallons per minute. The pressure ratings of these pumps range from zero to 10,000 psi, and in some cases higher. Pump operating speeds are increasing as the pressure ratings go up and the volumetric capacities decrease.

Some aircraft and missile pumps operate at 12,000 rpm with a capacity of about 10 gpm and a pressure rating over 3000 psi. This would be nearly 20 hp for a pump weighing less than 10 lb. Imagine a 20-hp electric motor and its size and weight compared to the aircraft and missile unit (see Fig. 6-15).

Hydraulic pump operating pressures are steadily increasing in almost all fields of usage. More and more emphasis is being placed on better efficiencies, faster response, and greater accuracy and dependability with less maintenance. Because of these demands, pump manufacturers are constantly upgrading their designs and developing new pumping techniques to meet these new performance requirements.

Hydraulic pumps are positive displacement pumps which can be further divided into two categories: fixed delivery or variable displacement. A *fixed-delivery pump* provides a specific volume displacement per revolution which cannot be varied.

EXAMPLE

A fixed-delivery pump with a displacement of 2 in.³ per revolution would deliver 3600 in.³ of fluid at 1800 rpm.

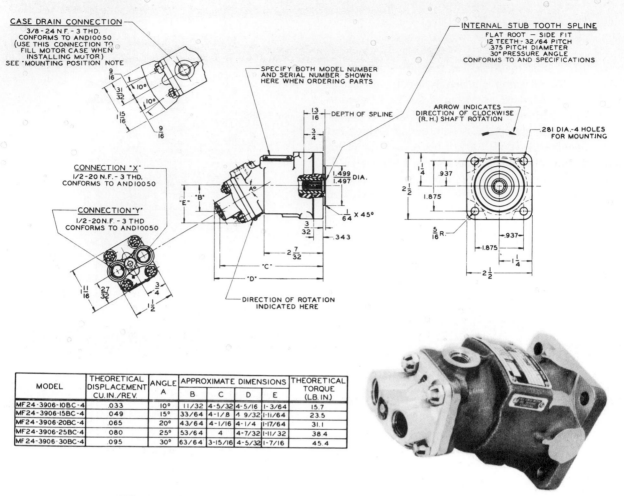

MODEL	THEORETICAL DISPLACEMENT CU. IN./REV.	ANGLE A	APPROXIMATE DIMENSIONS				THEORETICAL TORQUE (LB. IN.)
			B	C	D	E	
MF 24-3906-10BC-4	.033	10°	11/32	4-5/32	4-5/16	1-3/64	15.7
MF 24-3906-15BC-4	.049	15°	33/64	4-1/8	4 9/32	1-11/64	23.5
MF 24-3906-20BC-4	.065	20°	43/64	4-1/16	4-1/4	1-17/64	31.1
MF 24-3906-25BC-4	.080	25°	53/64	4	4-7/32	1-11/32	38.4
MF 24-3906-30BC-4	.095	30°	63/64	3-15/16	4-5/32	1-7/16	45.4

FIGURE 6-15 A 20-hp electric motor can be much larger than a 20-hp hydraulic pump. (Courtesy of Vickers, Incorporated, Troy, Mich.)

It is apparent from the example that the output volume of a fixed-delivery pump depends on the displacement per revolution and the speed (rpm) at which the pump is operated (see Fig. 6-16). As an example, the check ball constant-displacement pump of Fig. 6-16 works well in dirty environments because each piston has its own inlet and outlet check valve. When other pumps fail, due to contamination, cavitation, or vibration, check ball pumps show no loss of performance.

During its suction stroke, each piston is filled through an inlet check ball. During compression, the check ball seats and pressure in the piston chamber rises until it exceeds

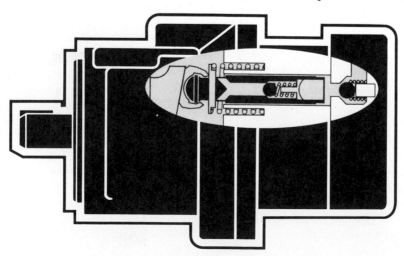

FIGURE 6-16 Fixed-displacement pump. (Courtesy of Dynex/Rivett Inc., Pewaukee, Wis.)

load pressure. The outlet check ball then unseats, and fluid is then pumped across the check, out of the chamber.

The piston barrel is stationary. A rotating wobble plate, keyed to the pump shaft, causes the pistons to reciprocate. The piston check valves eliminate valve plates, which can wear or score. With no valve plate-to-barrel wear, check ball pumps deliver consistent, high-volume efficiency.

The outlet checks do not unseat until pressure in the pumping chambers exceeds load pressure. This prevents cavitation damage sometimes caused by partially filled piston chambers in valve plate pumps.

A *variable-volume pump* is a pump in which the displacement per cycle can be varied (see Fig. 6-17). Pumps of this design provide infinitely variable volume from zero to the maximum volumetric capacity.

EXAMPLE

A variable-volume pump with a maximum displacement per revolution of 2 cu in. could deliver any increment of fluid between 0 and 3600 in.^{3} of fluid at a constant speed of 1800 rpm.

Some variable-volume pumps have the ability to provide infinitely variable volume in either direction through the pump. Pumps of this design are called reversible variable units or over-the-center pumps (see Fig. 6-18).

Hydraulic power pumps are also categorized by their construction type, such as lobe element, external gear, internal gear, screw, vane, axial piston, radial piston, and

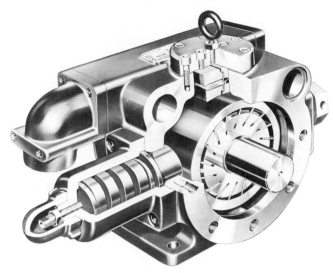

FIGURE 6-17 Varible pump. (Courtesy of Racine Hydraulics Division, Dana Corporation, Racine, Wis.)

reciprocating plunger. There are also a number of construction types that have variations from these basic designs.

FIXED-DELIVERY HYDRAULIC PUMPS

The simplest and most easily understood pump is a hand pump of the design commonly used in a hydraulic jack. This is the type that is used to raise barber and beauty chairs.

FIGURE 6-18 Variable piston pump with attached supercharge pump. (Courtesy of Dynapower, a Unit of General Signal, Watertown, N.Y.)

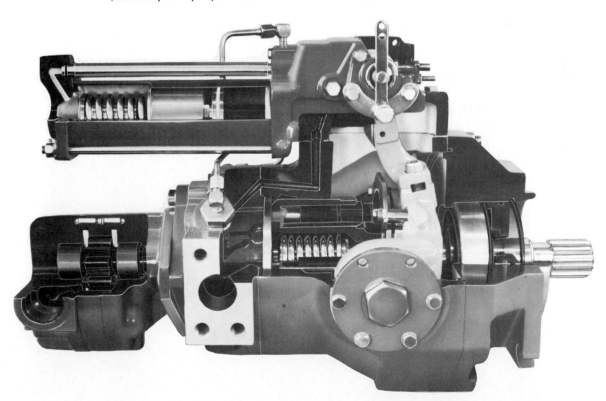

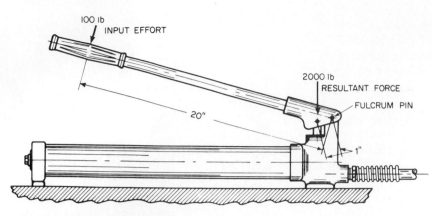

FIGURE 6-19 Principle of operation: hydraulic hand pump. [Courtesy of Duff-Norton (Ram-Pac).]

Hand pumps were among the first hydraulic pumps designed and, surprisingly, operate at rather high pressures.

Figure 6-19 shows the principle of operation. A lever arm is used to provide mechanical advantage with a person as the input power source. If 100 lb of force were applied on the 20-in. handle, a 2000-lb force would result on the pumping piston. The resultant force of 2000 lb is applied to a small-diameter pumping piston, as shown in Fig. 6-20. Since pressure is force per unit of area, a very high hydraulic pressure of 10,000 psi builds up and is transmitted undiminished through the conductor or hose to the large-diameter ram or cylinder to raise the load.

$$\frac{\text{Force of small piston}}{\text{Area of small piston}} = \frac{\text{force of large piston}}{\text{area of large piston}}$$

$$\frac{f = 2000}{a = 0.2} = \frac{F = 20,000}{A = 2.0}$$

$$10,000 \text{ psi} = 10,000 \text{ psi}$$

Thus the pressure at any given moment in a hydraulic system is proportional to the load. *It should also be obvious from the previous relationship that a pump does not pump pressure; it pumps fluid.* The pressure in a hydraulic system is due to the resistance of the load (see Fig. 6-21).

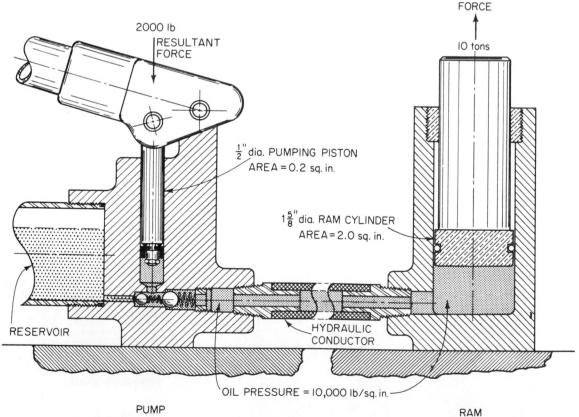

FIGURE 6-20 Hand pump system. [Courtesy of Duff-Norton (Ram-Pac).]

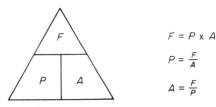

FIGURE 6-21 Force triangle.

$$F = P \times A$$
$$P = \frac{F}{A}$$
$$A = \frac{F}{P}$$

EXAMPLE

(a) What would the pressure (psi) be in the previous system if the load were reduced to 8 tons? (8 tons = 16,000 lb.)

(b) What handle effort would be required?

Solution:

(a) P (psi) $= \dfrac{F}{A}$

$$= \frac{16{,}000 \text{ lb}}{2 \text{ in.}^2}$$

$$= 8000 \text{ psi}$$

(b) Handle force (lb) × mech. ad. = total input force on small pumping piston

$$\frac{F \times 20 (\text{M.A.})}{\text{area piston}} = \frac{\text{force (lb) output}}{\text{area of large piston (in.}^2)}$$

Substituting, we have

$$\frac{F(\text{lb}) \times 20}{0.2 \text{ in.}^2} = \frac{16{,}000 \text{ lb}}{2 \text{ in.}^2}$$

Transposing gives

$$2 \text{ in.}^2 \times F(\text{lb}) = \frac{0.2 \text{ in.}^2 \times 16{,}000 \text{ lb}}{20(\text{M.A.})}$$

$$= \frac{3200}{20} = 160$$

$$F(\text{lb}) = \frac{160}{2}$$

(Handle) Input force = 80 lb

Because the small pumping piston cylinder is $\frac{1}{10}$ the area size of the large cylinder, it has to make ten 1-in. strokes for every inch of travel for the large output piston. This means that the pump must be stroked in order to raise the load any great distance.

In order to accomplish repetitive stroking without losing pressure, hydraulic pumps isolate the discharge stroke from the inlet stroke by means of internal valving. Figure 6-22 shows check valves for this purpose.

On the *inlet stroke*, as the handle is raised a vacuum

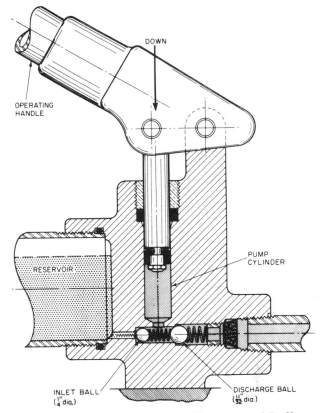

FIGURE 6-22 Inlet stroke. [Courtesy of Duff-Norton (Ram-Pac).]

is created in the cylinder beneath the pumping piston. Atmospheric pressure pushing on the fluid in the reservoir causes the fluid to overcome the slight spring force on the inlet check ball, and fluid enters the piston cavity. As soon as the upward travel of the handle stops, the inlet ball seats by the spring force.

On the *discharge stroke* the handle is forced downward with enough force to overcome the work resistance and transmit fluid under pressure through the discharge check ball and conductor into the large cylinder to raise the load. As the handle stops on the downward stroke, the spring force plus the pressure due to the suspended load causes the discharge check valve to close, and the pump is ready to repeat the cycle (see Fig. 6-23).

When it is necessary to lower the load, a needle or bleeder-type valve is used. When the valve is opened, fluid is forced back into the reservoir by gravity or force due to the load on the large cylinder (see Fig. 6-24).

Hand pumps are also available as double-acting and two-stage or two-speed units. Double-acting hand pumps deliver fluid to the system for each stroke of the handle. The two-speed hand pump in Fig. 6-25 uses two pumping pistons and a relief valve for each piston outlet. Two-speed hand pumps are popular for applications where actuator resistances and speeds vary.

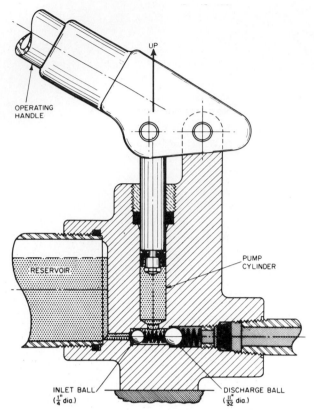

FIGURE 6-23 Discharge stroke. [Courtesy of Duff-Norton (Ram-Pac).]

HYDRAULIC POWER PUMPS

Hydraulic power-driven pumps are commonly used for almost all machines and devices using fluid power. There are many different construction types available.

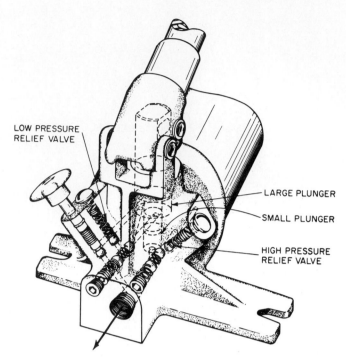

FIGURE 6-25 Two-speed hand pump. [Courtesy of Duff-Norton (Ram-Pac.)]

EXTERNAL GEAR PUMPS

A typical external gear pump consists of a drive gear and a driven gear enclosed within a precision machined housing. The close fit between the meshing gears and the housing maintains a seal between inlet and outlet sides of the pump. When the gears rotate, teeth unmesh on the inlet side, allowing hydraulic fluid from the reservoir to fill the resulting space. The fluid is then transferred around the periphery of

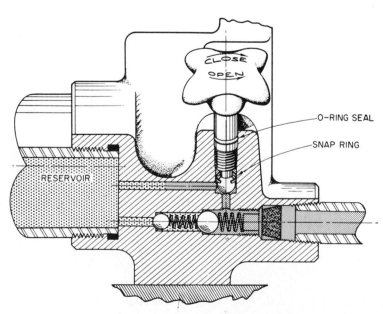

FIGURE 6-24 Bleeder valve. [Courtesy of Duff-Norton (Ram-Pac.)]

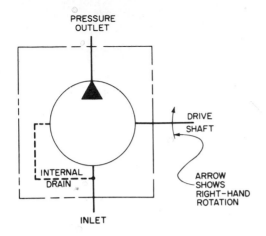

PRESSURE OUTLET

DRIVE SHAFT

INTERNAL DRAIN

ARROW SHOWS RIGHT-HAND ROTATION

INLET

FIGURE 6-26 Gear pump and ANSI symbol. (Courtesy of Webster Electric Company, Inc., Racine, Wis.)

both gears to the discharge side. As the gear teeth remesh on the discharge side, hydraulic fluid is forced through the discharge port and into the system (see Fig. 6-26).

The direction of rotation for all gear-type hydraulic pumps is determined by facing the shaft end of the pump. Manufacturers generally stamp an arrow into the pump housing indicating either left-hand or right-hand rotation. This is important in most pumps because integral passages for draining fluid from bearings, seals, and glands are connected back to the inlet side to insure flow for lubrication and seal. If the pump is operated in the wrong direction of rotation, the inlet and output ports reverse their primary function, and the pump may be destroyed. (Some gear pumps can be connected for either direction of rotation, and this would be indicated by the manufacturer's literature.)

right

true

ular in the low operating pressure ranges. They provide good life expectancy, are low in cost, have relatively good efficiency, and are less likely to be damaged by contamination.

External gear pumps may be either single or double units. Double pump or motor units can be used in series or in parallel hookups. More than two pumps can be driven from a common shaft.

INTERNAL GEAR DESIGNS

A popular construction type of pump or motor designed with internal gear teeth or motor is the gerotor unit (see Fig. 6-28). This pump consists of two elements, an inner gerotor and an outer gerotor. The inner element always has one less tooth than the outer.

The volume of the "missing tooth" multiplied by the number of driver teeth determines the volume of fluid pumped for each revolution. As the toothed elements, mounted on fixed centers but eccentric to each other, turn, the chamber between the teeth of the inner and outer elements gradually increases in size through about 180° of each revolution until it reaches its maximum size—equivalent to the full volume of the "missing tooth."

During the initial half of the cycle, the gradually enlarging chamber is exposed to the inlet port, creating a partial vacuum into which the hydraulic fluid flows. During the next 180° of the revolution, the chamber gradually decreases in size as the teeth mesh, and the fluid is forced out the discharge or outlet port and into the system. The gerotor unit illustrated in Fig. 6-29 is a popular design used in a number of fields of application as both pumps and motors. Gerotor-type pumps are available as single or double units.

The unit illustrated in Fig. 6-30 includes a thrust washer and dual check valves. The dual check valves provide a path for lubrication flow to the case to pass to the port with the lowest pressure. This is useful for fluid motor applications and in circuits where a motor may also function as a pump during certain machine cycles.

Another internal gear tooth design uses a gear-within-a-gear principle with a crescent that divides the flow and acts as a seal between the inlet and outlet sides of the pump. This unit operates equally well in either direction.

The pumping action is similar to other gear designs, except that both gears operate in the same direction. Gear-within-a-gear-type pumps are used advantageously where a

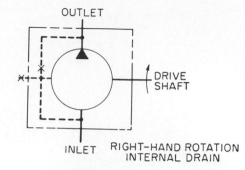

OUTLET

DRIVE SHAFT

INLET   RIGHT-HAND ROTATION INTERNAL DRAIN

FIGURE 6-28   Gerotor unit. (Courtesy of Double A Products Company, Manchester, Mich.)

drive shaft must go through the pump, as in automatic transmissions used in automobiles (see Fig. 6-61).

Internal gear pumps are produced with capacities up to 200 gpm. In the larger sizes they are generally applied as fluid transfer pumps.

SCREW PUMPS

There are three types of screw pumps which are commercially available. These three types are the single-screw, the two-screw, and the three-screw pumps. A single-screw pump

GEROTOR   STANDARD

FIGURE 6-29   Operation of a gerotor set. (Courtesy of Double A Products Company, Manchester, Mich.)

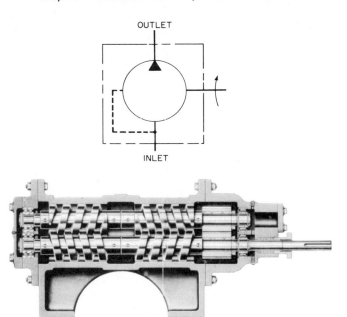

FIGURE 6-31   Screw pump and ANSI symbol. (Courtesy of Sier Bath.)

FIGURE 6-30   Gerotor design with integral check valves for draining lubrication oil to port with lowest pressure level. (Courtesy of Nichols Fluid Power Division, Parker Hannifin Corp., Sturtevant, Wis.)

consists of a spiraled rotor which rotates eccentrically in an internal stator. A two-screw pump consists of two parallel rotors with intermeshing threads rotating in a closely machined housing. These pumps use external or internal timing gears. A three-screw pump consists of a central drive rotor with two meshing idler rotors; the three rotors are surrounded by a closely machined housing.

The flow through a screw pump is axial and in the direction of the power rotor. When the inlet side of the pump is flooded with hydraulic fluid, a certain volume of the liquid that surrounds the rotors is caught as the rotors rotate. This fluid is pushed uniformly with the rotation of the rotors along the axis and is forced out the other end. In operation, when the power rotor is turning clockwise, the idler rotors are turning counterclockwise. This is true of all types of screw pumps.

The volume delivered by screw pumps is easy to maintain, because the fluid does not rotate but moves linearly. The rotors work like endless pistons which continuously move forward; thus there are no pulsations even at higher speed. Figure 6-31 shows the construction of the screw-type pump. The absence of pulsations and the fact that there is no metal-to-metal contact results in a very quiet operating pump.

Screw pumps are produced for volumetric capacities from 2 to 3400 gpm and for pressures up to 4000 psi.

Operating speeds of this design have been as high as 24,000 rpm.

The larger pumps are used as low-pressure, large-volume prefill pumps on large presses. A typical example of this would be the fast-closing phase of a cycle on a large hydraulic press as used in the plywood industry. These presses have wide openings to allow multiple stacking of the plywood sheets without difficulty. The press is then closed at a high rate of speed by using a 500-gpm screw-type pump. When the press is closed on the work, pressure unloads the 500-gpm screw pump and a low-volume, high-pressure pumping unit comes into operation to finish the pressing action.

Other applications include hydraulic systems on submarines and other uses where noise must be controlled.

VANE-TYPE PUMPS

Vane-type hydraulic pumps may be either fixed-delivery or variable-volume units. The fixed-delivery design usually has a pressure-balanced rotor which is slotted and contains the flat rectangular vanes. The rotor is splined to the pump shaft and rotates within a cam ring. Wear plates are fitted against both sides of the cam ring, thus making up the pumping element. The wear plates generally contain the fluid passages.

The *balanced vane* unit shown in Fig. 6-32 is popular for industrial applications requiring continuous operating cycles.

As the rotor is turned, centrifugal force causes the vanes to move out against the hardened and ground contour of the

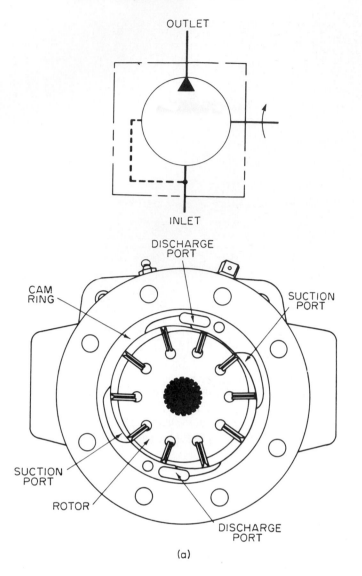

(a)

cam-shaped ring. Fluid is trapped between the rotating vanes when they pass over the inlet port. Because the vanes reciprocate in and out of the rotor as they rotate, while passing over the inlet port the vanes are extended out of the rotor and carry a maximum amount of fluid. As the vanes reach the outlet port, the cam ring contour forces them back into the rotor and the hydraulic fluid must go out the discharge side of the pump. Balanced vane units have two inlet ports and two discharge ports which are diametrically opposite.

Forces acting against the rotor are equal on diametrically opposite sides. These is no unbalanced force against the drive shaft assembly. Calculated areas within the side wear plates are supplied with discharge pressure to help keep the plates against the rotor and vane assembly with a suitable holding force regardless of the output pressure value. This means that both sides of the pump will be pressure balanced, and equal forces will result on both sides of the rotor.

When resistance builds up in the system and pressure increases, small passages in the wear plates allow fluid under pressure to operate beneath the vanes during their delivery period to keep them held tightly against the cam ring. This helps to increase the efficiency of the pump during the work phase of an operating cycle.

Fixed-delivery vane-type pumps are available as double pumps or two-stage pumps. Volumetric capacities of vane units are most popular up to about 60 gpm. Operating pressures are generally held to a maximum of 3000 psi and under. Figure 6-33 illustrates a balanced vane pump with two pumping cartridges driven by a common shaft.

FIGURE 6-32 (a) Vane pump and ANSI symbol. (Courtesy of Mobil Oil Corporation.) (b) Cross section of vane pump. (Courtesy of Abex Corporation, Denison Division, Columbus, Ohio.)

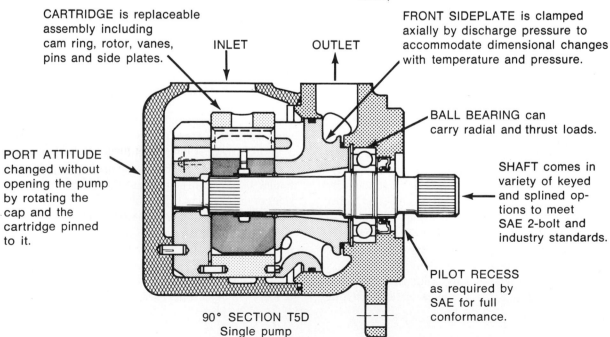

CARTRIDGE is replaceable assembly including cam ring, rotor, vanes, pins and side plates.

FRONT SIDEPLATE is clamped axially by discharge pressure to accommodate dimensional changes with temperature and pressure.

BALL BEARING can carry radial and thrust loads.

PORT ATTITUDE changed without opening the pump by rotating the cap and the cartridge pinned to it.

SHAFT comes in variety of keyed and splined options to meet SAE 2-bolt and industry standards.

PILOT RECESS as required by SAE for full conformance.

90° SECTION T5D
Single pump

FIGURE 6-33 Two vane cartridges in one housing driven by a common shaft. (Courtesy of Abex Corporation, Denison Division, Columbus, Ohio.)

The *unbalanced vane* type of hydraulic pump has a rotor which is eccentric to the housing cam ring. As the rotor containing the flat rectangular vanes is turned, hydraulic fluid entering the inlet port is trapped between the housing and vanes. When the vanes rotate to the outlet port, the space between the rotor and the housing cam ring lessens and the vanes are pushed back into the rotor. The fluid must then go out the discharge side of the pump and into the system.

Figure 6-34 shows a slipper-vane type of pump, which indicates the large number of varied designs that are available.

Most industrial unbalanced vane pumps manufactured are variable-volume units. This is easy to accomplish by making the cam ring housing adjustable from maximum eccentricity to a concentric position where the vanes are held equal distant between rotor and the ring through 360°.

Most generally, a mechanical spring force is used to hold the housing cam ring eccentric with the rotor, which delivers maximum volume until the system's work load causes a certain pressure at the outlet side of the pump. When this pressure level is reached, forces due to this pressure within the pumping element push the housing cam ring and overcome the force of the spring. This moves the housing cam ring to a center position relative to the rotor, and the pump delivery becomes nearly zero. Sufficient liquid is pumped to lubricate the moving members and to replace the leakage oil. The type of pump illustrated in Fig. 6-35 is biased to the extreme eccentric position with a mechanical spring. An adjustment is provided to permit the setting of a maximum flow value at minimum pressure. Maximum

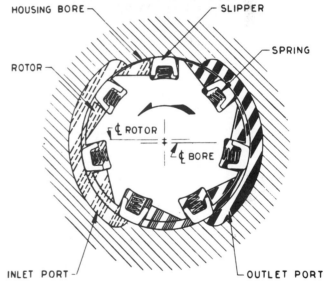

FIGURE 6-34 Unbalanced slipper vane type of pump.

pressure at minimum flow is adjusted with the jack screw that bears against the spring mechanism. Through the application of different spring forces the variable vane-type pump can be adjusted to reach zero delivery or "deadhead" at various pressure levels.

Hooke's law expresses the meaning that within the elastic limits of the spring material the ratio of the stress to the strain produced is a constant:

$$\frac{\text{Stress}}{\text{Strain}} = K$$

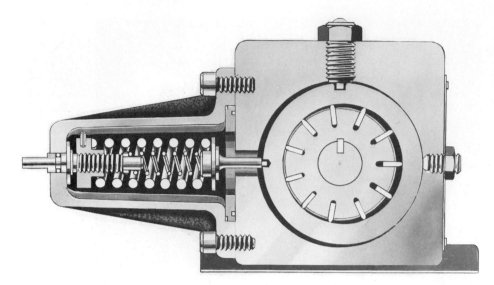

FIGURE 6-35 Variable vane pump with direct spring actuation. (Courtesy of Racine Hydraulics Division, Dana Corporation, Racine, Wis.)

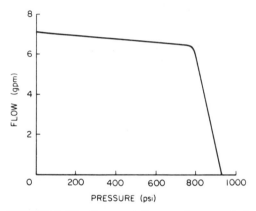

FIGURE 6-36 Pump delivery characteristics relative to type of spring. (Courtesy of Racine Hydraulics Division, Dana Corporation, Racine, Wis.)

The curves in Fig. 6-36 indicate the pump delivery characteristic relative to the type of spring being used.

One of the advantages of the variable-volume pump is its low power consumption. The amount of power that is used varies with the work load on the system. This concept is termed *pressure compensated,* which means that when the work resistance is the least the volume output is at a maximum. When the work resistance is the greatest, the pump delivery output is the least. It is important to repeat again: "Pressure in a hydraulic system is due to the resistance on the system." Also, "A hydraulic pump pumps the hydraulic fluid, not pressure." Thus a pressure-compensated pump is one that changes its volume output with changes in system pressure.

The variable-displacement pump shown in Fig. 6-37a consists of a housing (1), a rotor (2) with dual vanes (3), a cam ring (4), a pressure controller (5), a flow adjustment screw (6), and an automatic bleed valve (7).

With a standard pressure regulator as shown, the pump works on the following principle: the rotor (2) driven by the motor through the drive shaft rotates within the cam ring (4). The double vanes (3) are carried in the rotor, and are thrown outward by centrifugal force and pressed against the cam ring (4). The chambers (8) necessary to transport the fluid are each covered by two dual vanes (3), the rotor (2), the cam ring (4), and the port plates (9).

When the rotor (2) is turned away from the suction line, the chambers (8) increase in size and fill up with oil. When the highest volume is reached, the chambers (8) are separated from the suction side. If the rotor (2) is turned further, the chambers are connected to the pressure side, decrease in size, and transmit the fluid into the system via pressure line P.

Figure 6-37b shows a typical pressure control function. The circular cam ring (4) is held between the two control pistons (10) and (11). The third point of contact is the tuning screw (15). To ensure correct functioning of the pump during startup, the cam ring must be in an eccentric position.

The spring (12) behind the larger control piston (11) holds the ring in this position. As soon as pressure builds up in the system, the two pistons are pressurized. The fluid is channeled to the piston (11) via the control spool (14).

The larger piston (11) holds the cam ring (4) in its eccentric position. The pump then delivers oil until the maximum pressure is reached. The maximum pressure is set at the pressure regulator, which is flange mounted to the larger piston. The control spool (14) is held in a specific position by a spring (13).

When the hydraulic force on the control spool (14) exceeds the opposing force of the spring, the spool is moved against the spring. This allows the chamber behind the larger piston (11) to unload to tank. The small piston (10) is continually under pressure and is thus moved forward together

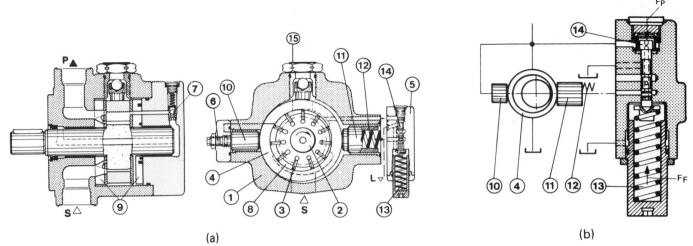

(a)   (b)

FIGURE 6-37 Pressure-compensated variable-delivery pump with pressure assist to bias spring. (Courtesy of Rexroth Corporation, Bethlehem, Pa.)

with the cam ring (4). The pump maintains pressure, and output is practically zero; only leakage oil is replaced. Power losses and heating of the fluid are therefore reduced to a minimum.

Should the system pressure fall below the set pressure, the spring (13) moves the control spool (14) into its original position. The control piston (11) is once more open to pressure and moves the cam ring into an eccentric position. Oil is again pumped into the system.

Many options are available for electrohydraulic proportional control. Manufacturers' descriptive literature is available to help in providing the most efficient circuit design for maximum performance and economy of operation.

PISTON HYDRAULIC PUMPS

Piston-type pumps are the oldest form of hydraulic pump. They have been used successfully since about 1900 for naval gun turrets, steering control systems, and other shipboard uses. Piston pumps are now being used in practically all fields.

The piston design itself lends to excellent control of clearances and use of materials in manufacture. At the present time, these units provide the highest degree of sophistication found in the many hydraulic pump designs that are available.

Piston pumps offer the highest volumetric efficiencies (up to 97%), higher pressure ratings (10,000 psi or higher), and operating speeds as high as 12,000 rpm. With their high operating efficiency less power is converted to heat, and better total system economy is provided with less size and weight.

Piston units are available for fixed delivery, variable

volume, and in variable reversible flow pumps (over-the-center). In general, there are three basic designs of piston pumps: radial, axial, and reciprocating plunger.

The radial hydraulic piston pumps operate on the principle of converting the rotary shaft motion to radial reciprocating piston motion (see Fig. 6-38).

Axial hydraulic piston pumps operate on the principle of converting rotary shaft motion to axial reciprocating piston motion (see Fig. 6-39).

Reciprocating plunger pumps may be horizontal or vertical piston pumps and operated by either a rotating or reciprocating power input shaft (see Fig. 6-40). They are gen-

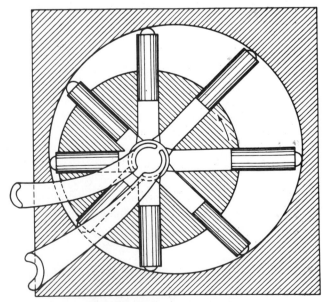

FIGURE 6-38 Radial action.

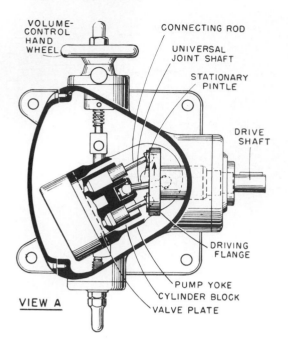

VOLUME-CONTROL HAND WHEEL
CONNECTING ROD
UNIVERSAL JOINT SHAFT
STATIONARY PINTLE
DRIVE SHAFT
DRIVING FLANGE
PUMP YOKE
CYLINDER BLOCK
VALVE PLATE
VIEW A

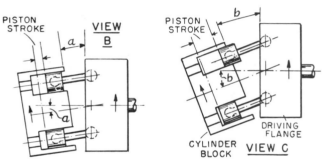

PISTON STROKE
VIEW B
a
PISTON STROKE
b
b
a
DRIVING FLANGE
CYLINDER BLOCK
VIEW C

FIGURE 6-39 Bent-axis structure.

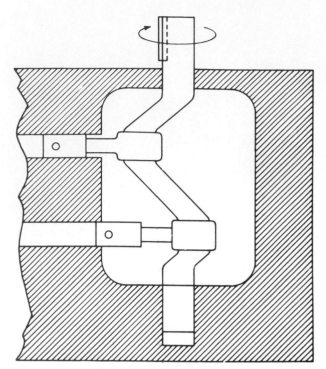

FIGURE 6-40 Reciprocating plunger.

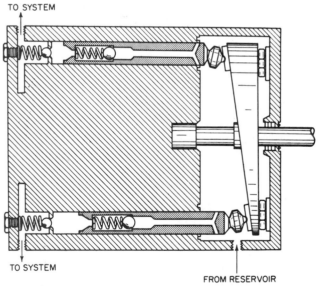

TO SYSTEM

TO SYSTEM

FROM RESERVOIR

FIGURE 6-41 Principle of check valve porting.

erally large piston-type pumps as used in central hydraulic power systems or for extremely large volume installations.

Piston hydraulic pumps may also be classified for their method of porting fluid between inlet and outlet. There are three common methods of porting fluid: valve plates, check valves, and pintle valve arrangements. As a general rule, units designed with pintle valves or valve plates are adapted to both pumps and fluid motors, but check valve units are available only as pumps (Fig. 6-41).

RADIAL PISTON PUMPS

The radial piston pump classification is based on the line of piston motion, which is perpendicular to the centerline of the input shaft (see Fig. 6-43).

Check Valve Design

An example of the check valve design is shown in Fig. 6-42. When the pump shaft is rotated, fluid from the reservoir enters the inlet side of the pump. Fluid is then ported from the flooded crankcase through the inlet checks and into the piston chamber as the piston travels inward during one-half of one shaft revolution. During the other one-half of one revolution, the eccentric bearing forces the piston outward, causing fluid delivery to the system through the discharge check valve. This design is a fixed-delivery pump and is available in a number of different volumetric capacities.

It is also important to observe that each piston is essentially an independent pumping unit, which is generally true with all check valve piston pump designs. These units

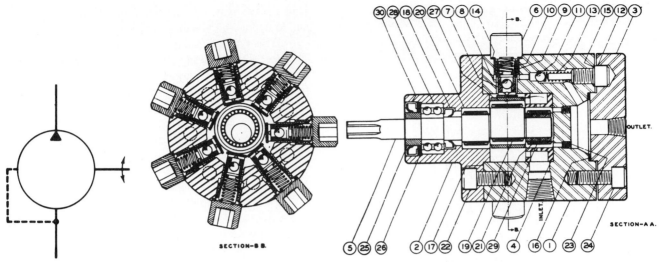

FIGURE 6-42   Check valve design and ANSI symbol. (Courtesy of Racine hydraulics Division, Dana Corporation, Racine, Wis.)

can be assembled to supply fluid independently to several circuits simultaneously, if desired. The check valve radial piston pump previously shown can be operated at pressures up to 10,000 psi.

Pintle Valve Arrangement

The radial piston pump shown in Fig. 6-43 uses the pintle valve method of porting fluid. It is available as a fixed-delivery, variable-volume, or reversible variable-volume unit. The volumetric capacity of the fixed-stroke type varies with the size of the unit and the rotational speed.

Input power applied to the pump shaft rotates the cylinder, piston, and rotor assembly. Centrifugal force keeps the pistons against the thrust rings to rotate the rotor with the cylinder. During the lower one-half of one revolution, the pistons move outward to allow fluid to enter the piston cavity through the inlet port and pintle arrangement. During the upper one-half of one revolution, the pistons are thrust inward to discharge the fluid from the piston cavity through the pintle and out into the system.

Pumping pistons (see Fig. 6-44) have a rounded head and are fitted in the cylinder at a slight angle. This causes the pistons to roll and reciprocate simultaneously. Both motions are uniformly accelerated and decelerated for smooth action.

The reversible variable-volume unit shown in Fig. 6-45 provides an infinitely variable volume of fluid flow in

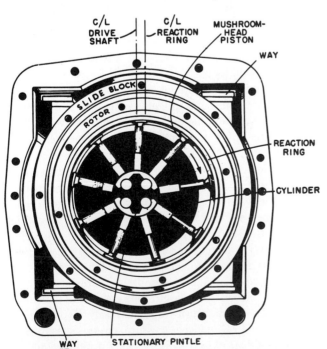

FIGURE 6-43   Radial piston pump. (Courtesy of The Oilgear Company, Milwaukee, Wis.)

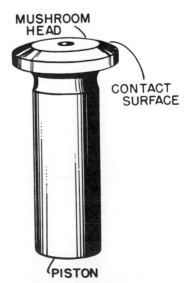

FIGURE 6-44   Piston. (Courtesy of The Oilgear Company, Milwaukee Wis.)

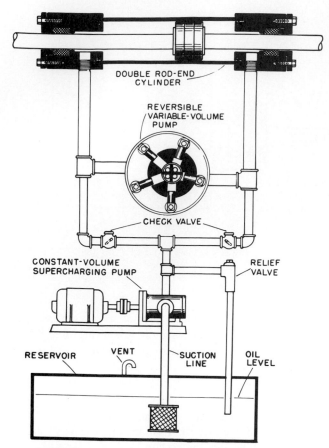

FIGURE 6-45 Variable reversible radial pump. With a double-rod-end cylinder, which has the same displacement on both ends, the reversing-pump circuit can theoretically consist of nothing but the pump, cylinder, and suitable pipe connections. The oil pumped from one end of the cylinder is delivered to the other. Speed is controlled by adjusting the pump stroke, and no reversing valve is needed. Actually, a small fixed-volume pump—usually a gear pump—and reservoir is also required to make-up main pump leakage losses and keep the entire system under positive pressure so that no air can get in. This "super-charging" gear pump, the make-up check valves, and the relief valve are sometimes built into the main pump. A rotary fluid motor may be used instead of a cylinder as shown. This type of circuit is used for variable-speed transmissions, ship steering equipment, and windlass drives. (Courtesy of Mobil Oil Corporation.)

either direction through the pump. The internal porting arrangement of the pump facilitates flow in either direction as the slide block race is positioned from side to side. The pump shaft rotates in one direction constantly; only the fluid changes direction. This type of hydraulic pump can be controlled manually, mechanically, by fluid, electrically, or

with a servo system to discharge the exact amount of fluid and to operate a cylinder or fluid motor in either direction.

This type of system is called a *closed-circuit system*. The closed-circuit design, as shown in Fig. 6-45, eliminates the need for a large storage volume of oil. Although this feature is more important to mobile equipment than it is to general industrial machinery, a large number of closed-circuit pumps are used in industrial hydraulics. Even though this circuit requires special pumps (normally piston pumps) and pump controls, it does offer several advantages.

Figure 6-46a shows a typical circuit using ANSI (American National Standards Institute) symbols. This is how it works. A single hydraulic pump is used to drive a single hydraulic motor. The closed circuit has little or no significance for cylinder actuators, which displace different volumes during extension and retraction. The reason for this is that the oil which passes through the motor actuator is returned directly to the low-pressure side of the pump. For proper operation, the pump must receive the same quantity of oil at its inlet as it is pumping from its outlet.

The closed circuit is always used in conjunction with a small supercharge circuit. The supercharge circuit consists of a small fixed-displacement pump (usually 15% of the displacement of the main pump), a small oil reservoir (sometimes the pump housing), and the necessary filters and heat exchangers.

During operation the main pump control can cause the pump's displacement to go over center, which means that the main pump can pump high-pressure oil from either of its two main ports. It can cause clockwise or counterclockwise flow of fluid through the closed-circuit plumbing. This, in turn, will allow the motor actuator to operate in either direction of rotation. The port that serves as the high-pressure inlet to the actuator will determine the high-pressure leg, while the low-pressure leg will be determined by the actuator's outlet port.

The supercharge circuit always works on the low-pressure leg of the main circuit, pumping freshly filtered oil into the circuit through the makeup check valve network while bleeding off a percentage of oil through the hot-oil bleed valve. This hot oil is then cooled by a heat exchanger and stored in the small reservoir before returning to the main system. The pressure in the low-pressure leg is maintained at a value between 100 and 300 psi by the supercharge relief valve. (Other pressure values may be dictated by special requirements.) The pressure setting of the supercharge relief is determined by the requirements of the pump and/or motor actuator and the operating conditions of the system.

In closed circuits, pressure, flow, and directional control are all achieved by the controlling element of the pump. The crossport reliefs are incorporated only to protect the actuator from load-induced pressure peaks. They cannot function as a main system relief valve, since this would, in short order, cause severe overheating of the circuit.

The advantages of a closed-circuit pump system are that high-horsepower systems are compact, and they operate with a minimum amount of excess storage oil. The systems are highly efficient since the pump control must be designed to supply only the oil flow required by the actuator at the load-induced pressure. The pump controls direction, acceleration, deceleration, and maximum speed and maximum torque of the motor actuator, thus eliminating the need for pressure and flow control components. (Figure 46b shows a complete assembly.)

The major disadvantage of closed-circuit systems is that a single pump can operate only a single output function. In addition, this type of hydraulic drive is generally usable only with motor actuators, with the circuit of Fig. 6-45 as an exception, where a double rod cylinder provides equal flow in each direction of travel.

AXIAL PISTON PUMPS

Axial piston hydraulic pumps constitute many different versions of converting rotary shaft motion to an axial reciprocating motion of a piston. Both check valve and valve plate methods of porting fluid are used with axial piston pumps.

Check Valve Design

The check valve design illustrated in Fig. 6-47 is rotatable in either direction. The rotating input motion is converted to a reciprocating axial piston motion by the cam plate, which is keyed to and revolves with the shaft. One face of the cam plate is at an angle to the centerline of the shaft.

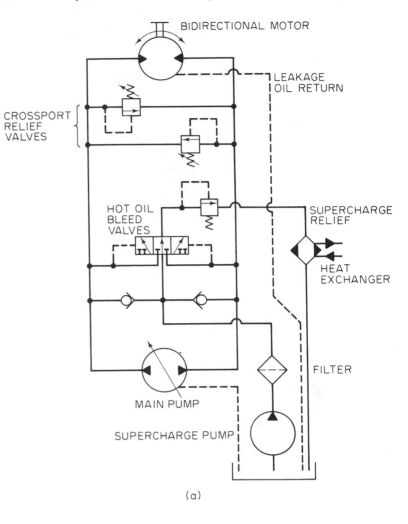

(a)

(b)

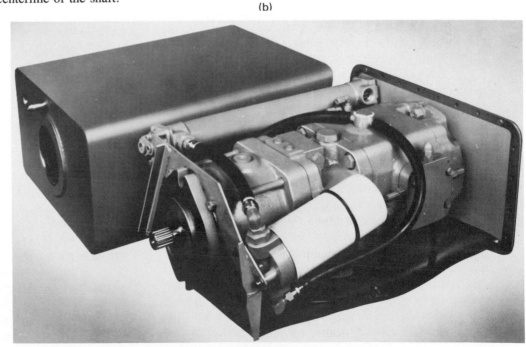

FIGURE 6-46 (a) Typical closed-circuit hydrostatic transmission using ANSI symbols to show functional operation. (Courtesy of Rexroth Corporation, Bethlehem, Pa.) (b) Integrated hydrostatic transmission. (Courtesy of Abex Corporation, Denison Division, Columbus, Ohio.)

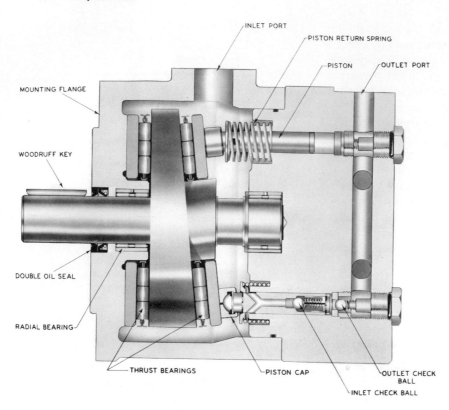

FIGURE 6-47 Check valve axial piston pump. (Courtesy of Dynex/Rivett Inc., Pewaukee, Wis.)

As the input shaft and cam plate assembly are rotated, the *thin section* of the cam slides under the piston cap. The piston return spring forces the piston toward the inlet side, causing a low-pressure condition within the tightly sealed piston cavity. This causes the inlet check valve to open and allow fluid to enter the low-pressure area of the piston cavity.

When the cam plate rotates 180°, the *heavy section* slides under the piston cap, which pushes the piston toward the discharge side. The pressurized fluid forces the discharge valve open, and the fluid goes out into the system. As the thin section again slides under the piston cap, the process is repeated.

Each piston contained in the cylinder barrel of this pump design is an independent pumping unit. This arrangement provides for multiple circuitry from the use of one pump.

The variable pump concept for the check valve axial piston design is shown in Fig. 6-48. Volume is varied by bypassing a portion of the displaced fluid back to the inlet side on each piston stroke. A cam-operated bypass valve connects each piston discharge cavity with a passage back to the inlet side. The pumping action is the same as in the fixed delivery pump, except for the bypass feature.

The cycling phase relationship between the cam plate and the bypass cam can be varied during pump operation. If both the cam plate and the bypass cam are phased so that the bypass valves are open during the full 180° of rotation when the pistons are moving toward the discharge side, the pump output is zero. The fluid merely circulates through the open passages back to the inlet side. However, if the bypass cam is phased with the full 180° of rotation when pistons are being returned by the springs on the inlet stroke, the bypass valves will be closed on the discharge stroke and all the fluid will be delivered to the system. The bypass cam can be adjusted for any increment of overlapping to provide infinitely variable volume to the system.

This type of variable-volume pump delivers flow in only one direction and is used with open-circuit systems (see Fig. 6-49). Notice that in this circuit a directional control is used to change the direction of the fluid motor rotation.

The check valve design can also be used for multiple circuitry. It is also possible to install both fixed-delivery and variable-volume features in one pump (see Fig. 6-50).

Valve Plate Design

Figure 6-51 shows examples of the valve plate porting design of axial piston pumps. These pumps are available as fixed-delivery, variable-volume, and as reversible variable-volume units.

In this design a valve plate with two kidney-shaped ports is used. One port allows fluid to enter the piston chamber and the other directs the fluid out into the system.

As the pump shaft and cylinder barrel rotate, the pistons maintain a fixed pumping action due to the cam angle. During one-half of one revolution the pistons move in a direction away from the kidney port, and fluid enters to fill the cavity of the cylinder bore. On the next one-half rev-

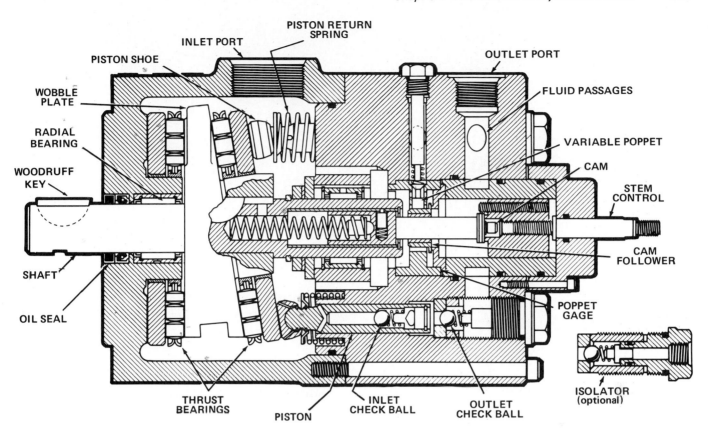

PV4000 SERIES

(a)

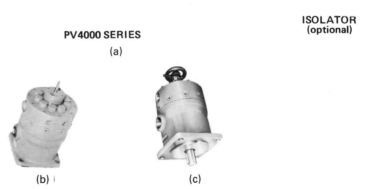

(b)   (c)

FIGURE 6-48 (a) Isolation unit to permit independent discharge from selected pistons; (b) rear view of pump; (c) shaft end view of pump. (Courtesy of Dynex/Rivett Inc., Pewaukee, Wis.)

olution the pistons are forced toward the kidney port and fluid is forced out into the system.

The design of valve-plate-type axial piston pump shown in Fig. 6-52 consists mainly of a two-section housing which establishes a fixed angularity. The rotating elements maintain the fixed-angle relationship during operation. This angle determines the distance through which a piston travels in its bore during shaft rotation. Because of the angular relationship between cylinder block and drive shaft, pistons reciprocate in their respective bores as they rotate with the cylinder block. Each piston accepts fluid during one-half of one revolution and forces it out during the other one-half

revolution. The inlet and outlet ports located in the valve block lead into separate kidney-shaped slots which are mated with corresponding valve plate slots. The valve plate slots are also in phase with port openings in the cylinder barrel bores. Narrow walls separate inlet from outlet.

Units of the valve plate design type are readily available as variable-volume units or as reversible variable-volume units. When the cam angle or angular relationship between cylinder barrel and drive shaft can be adjusted on only one side of center, the pump is a variable-volume unit. This means that the pump delivers flow out one specific port. However, if the pumping angle can be adjusted on

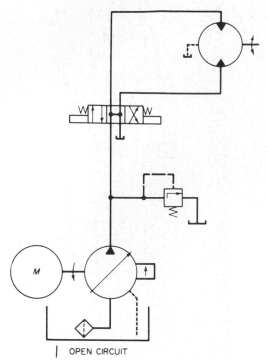

FIGURE 6-49 Open-circuit hydraulic system.

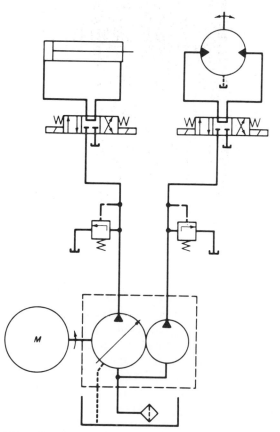

FIGURE 6-50 Multiple circuit from one pump.

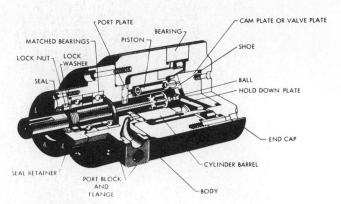

FIGURE 6-51 Valve plate: fixed-delivery axial piston pump. (Courtesy of Abex Corporation, Denison Division, Columbus, Ohio.)

both sides of center (over-the-center), the pump can deliver variable flow from either port for reversing fluid motors or cylinders (see Figs. 6-53 and 6-54).

CONTROLS FOR VARIABLE-DISPLACEMENT PUMPS

Pressure Compensation

Pressure compensation is the most widely used control for a variable-displacement pump. The inherent reduction in flow as pressure increases is a useful way to control energy input in many circuits and provides the most efficient use of the energy. The automatic action is dependable and repetitively accurate when using direct spring control as shown in Fig. 6-35 or pilot-operated control as shown in Fig. 6-37.

Controls for Multiple Pressure Levels

Pilot with Direct Solenoid Control. As we study pressure control valves we will find that selective pilot pressure sources can be used to add force values to a mechanical spring. In the upper view of Fig. 6-55 a pilot flow passage is provided from the outlet port to the bias spring chamber. A restriction is included in this passage. A solenoid-actuated two-way valve can divert this pilot flow to the tank faster than it can enter through the restricted passage. In this mode, with the solenoid deenergized, the bias spring determines the maximum pressure level to the outlet port.

If the solenoid is energized as shown in the lower view of Fig. 6-55, the passage to tank is blocked. Pressurized fluid can assist the bias spring. This will increase the pressure until the pilot fluid can pass through the small integral pilot relief valve at the pressure level established by the bias spring in the pilot relief valve.

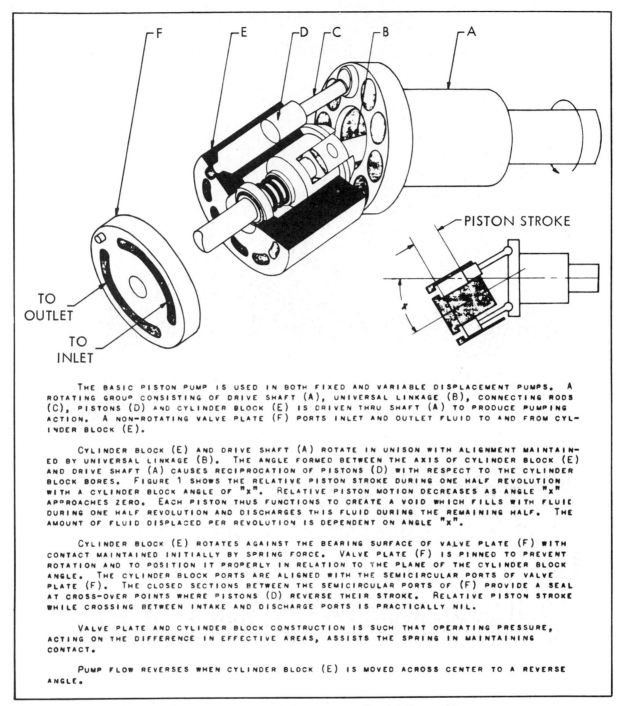

F E D C B A

PISTON STROKE

TO OUTLET

TO INLET

THE BASIC PISTON PUMP IS USED IN BOTH FIXED AND VARIABLE DISPLACEMENT PUMPS. A ROTATING GROUP CONSISTING OF DRIVE SHAFT (A), UNIVERSAL LINKAGE (B), CONNECTING RODS (C), PISTONS (D) AND CYLINDER BLOCK (E) IS DRIVEN THRU SHAFT (A) TO PRODUCE PUMPING ACTION. A NON-ROTATING VALVE PLATE (F) PORTS INLET AND OUTLET FLUID TO AND FROM CYLINDER BLOCK (E).

CYLINDER BLOCK (E) AND DRIVE SHAFT (A) ROTATE IN UNISON WITH ALIGNMENT MAINTAINED BY UNIVERSAL LINKAGE (B). THE ANGLE FORMED BETWEEN THE AXIS OF CYLINDER BLOCK (E) AND DRIVE SHAFT (A) CAUSES RECIPROCATION OF PISTONS (D) WITH RESPECT TO THE CYLINDER BLOCK BORES. FIGURE 1 SHOWS THE RELATIVE PISTON STROKE DURING ONE HALF REVOLUTION WITH A CYLINDER BLOCK ANGLE OF "X". RELATIVE PISTON MOTION DECREASES AS ANGLE "X" APPROACHES ZERO. EACH PISTON THUS FUNCTIONS TO CREATE A VOID WHICH FILLS WITH FLUID DURING ONE HALF REVOLUTION AND DISCHARGES THIS FLUID DURING THE REMAINING HALF. THE AMOUNT OF FLUID DISPLACED PER REVOLUTION IS DEPENDENT ON ANGLE "X".

CYLINDER BLOCK (E) ROTATES AGAINST THE BEARING SURFACE OF VALVE PLATE (F) WITH CONTACT MAINTAINED INITIALLY BY SPRING FORCE. VALVE PLATE (F) IS PINNED TO PREVENT ROTATION AND TO POSITION IT PROPERLY IN RELATION TO THE PLANE OF THE CYLINDER BLOCK ANGLE. THE CYLINDER BLOCK PORTS ARE ALIGNED WITH THE SEMICIRCULAR PORTS OF VALVE PLATE (F). THE CLOSED SECTIONS BETWEEN THE SEMICIRCULAR PORTS OF (F) PROVIDE A SEAL AT CROSS-OVER POINTS WHERE PISTONS (D) REVERSE THEIR STROKE. RELATIVE PISTON STROKE WHILE CROSSING BETWEEN INTAKE AND DISCHARGE PORTS IS PRACTICALLY NIL.

VALVE PLATE AND CYLINDER BLOCK CONSTRUCTION IS SUCH THAT OPERATING PRESSURE, ACTING ON THE DIFFERENCE IN EFFECTIVE AREAS, ASSISTS THE SPRING IN MAINTAINING CONTACT.

PUMP FLOW REVERSES WHEN CYLINDER BLOCK (E) IS MOVED ACROSS CENTER TO A REVERSE ANGLE.

FIGURE 6-52   Bent axis: axial piston pump: fixed delivery. (Courtesy of Vickers, Incorporated, Troy, Mich.)

Proportional Pump Stroke Control

Electronic proportional pump stroke control offers a useful method of establishing the maximum pressure level of the output flow from a pump. This control can be directly interfaced with a microprocessor, programmable controller, computer mechanism, and/or remote manual control.

A proportional direct-current solenoid or torque motor converts a dc signal into a force output. A typical proportional solenoid operates on a 24-V dc voltage and typically varies its force output in the control range 150 to 700 mA.

The cross-section illustration (Fig. 6-56) represents a pump control that uses a proportional solenoid to vary pump displacement. The pump responds from minimum to max-

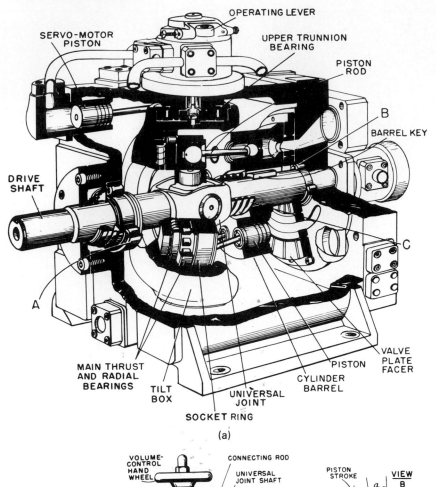

(a)

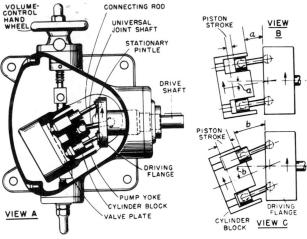

(b)

FIGURE 6-53 (a) Component parts of an axial piston variable-volume pump that is reversible to provide infinitely variable volume in either direction, controlled; (b) manual version of controlling pump delivery using the same bent-axis concept. (Courtesy of Mobil Oil Corporation.)

imum displacement proportional to the current of a 24-V dc command signal. The displacement of the pump is shown in the minimum position with no electrical signal supplied to the solenoid. The pump is held in this position by the force of the feedback spring and system pressure working over the smaller area of the positioning piston.

As dc current is supplied to the proportional solenoid, the solenoid pushes on the pilot spool with a specific force. When the current, and therefore the force, are high enough to move the pilot spool against the small spring (which adjusts the beginning of regulation), pilot pressure is exposed to the large diameter of the positioning piston. Because of the area difference of the positioning cylinder, the pump begins stroking toward maximum displacement. However, stroking of the pump also causes the feedback spring to be compressed, which increases its force. When feedback spring force exceeds the limited force of the proportional solenoid the spool will be moved back to its orig-

(a)

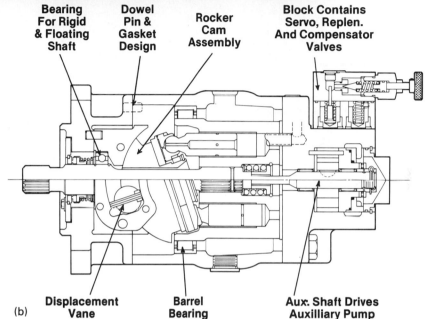

Bearing For Rigid & Floating Shaft

Dowel Pin & Gasket Design

Rocker Cam Assembly

Block Contains Servo, Replen. And Compensator Valves

Displacement Vane

Barrel Bearing

Aux. Shaft Drives Auxilliary Pump

(b)

FIGURE 6-54 (a) Overcenter axial piston pump with auxiliary pump; (b) cross-sectional view showing valve assembly and auxiliary pump drive. (Courtesy of Abex Corporation, Denison Division, Columbus, Ohio.)

inal position. This will cause the large diameter of the positioning piston to be disconnected from the pressure source and vented to tank.

The net result is that the pilot spool modulates the pressure on the large area of the positioning piston, so that the precise balance of feedback spring force and proportional solenoid force is maintained. The pump stays at this displacement position until the current supply is changed, which changes the force output of the proportional solenoid.

The control, as described, can be used to change the velocity of the output actuator, while the maximum force (or torque) output would be controlled by a separate pressure control piped into the hydraulic circuit. There are, however, many other variations of pump controls which use a proportional solenoid to adjust both displacement and the maximum pressure capabilities of the pump. For instance, load-sensing control is available with the adjustment of both the main flow orifice and the pilot relief established by two proportional solenoids. This allows both velocity and force output of the hydraulic system to be interfaced with electronic controls.

FIGURE 6-55 Solenoid-controlled, two-pressure governor. (Courtesy of Racine Hydraulics Division, Dana Corporation, Racine, Wis.)

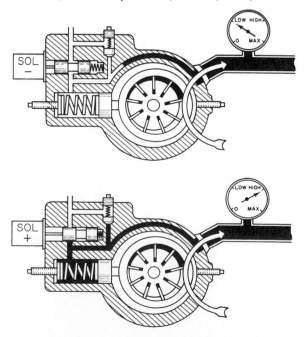

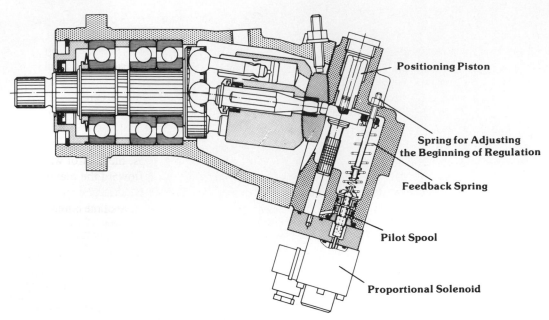

Positioning Piston

Spring for Adjusting
the Beginning of Regulation

Feedback Spring

Pilot Spool

Proportional Solenoid

FIGURE 6-56 Proportional pump stroke control (Courtesy of Rexroth Corporation, Bethlehem, Pa.)

Servo Pump Controls

A servo pump control provides an amplification of a small or low-value signal to an amplitude adequate for the major work at hand, which is usually the movement of the pump displacement mechanism. This may require significant force.

The proportional control can have many characteristics of a servo control. For our purposes we deal with hydromechanical or hydroelectric servo controls where a valving structure can accept a relatively small energy input, such as that of the human hand, and amplify it with a power assist to control massive forces. Automotive power steering is a good example. A power piston is used to move the displacement element as shown in Fig. 6-56. A relatively small directional control valve can direct fluid to the piston to cause movement. The moving member of the pump mechanism is fastened to a valve mechanism to cancel the initial movement but in a new physical position. A relationship is established between the input signal movement and the resulting mechanical power movement. The input signal can be human, mechanical, electrical, or electronic in nature, with appropriate interface devices.

The term *master/slave action* has been used to describe a servo system. The input or master signal controls the slave or power delivery member. A sensing device is often used to feed back information showing that the slave unit has provided the action dictated by the master signal. This may be a signal showing physical position, torque, pressure level, temperature level, or any other value sought by the servomechanism. Note in Fig. 6-53 that the operating lever is actuating a small servo valve. The servo valve directs fluid

to the servomotor piston. This in turn moves the tilt box to vary the pump output flow. As it moves the tilt box it also moves a feedback connection to the servo valve which cancels the signal in this new position. It will stay in that position until another movement is desired and the operating lever is moved again.

Constant-Horsepower Control

A constant-horsepower control (sometimes referred to as a horsepower-limiter control) is intended for use with a prime mover which drives the pump at a constant speed. The prime mover can be either an electric motor or an internal combustion engine with a speed governor. The purpose of the pump control is to keep the prime mover working at its maximum torque capabilities, or better still, at a constant-horsepower level. Remember:

$$\text{hp} = \frac{T \times \text{rpm}}{5252}$$

where T is the torque in lb-ft.

To draw constant horsepower from the prime mover, the pump must maintain the mathematical product of flow and pressure at a constant value. This means that if the flow output is high, the operating pressure must be low. Similarly, when pressure increases, flow must decrease. Since the operating pressure level of a system is dictated by the load conditions, the flow must vary with changes in load-induced pressure if we want to maintain the product of flow and pressure at a constant value.

The constant-horsepower control senses the load-induced pressure in the system and regulates pump displacement accordingly. The pump control holds the pump at its maximum displacement until the pressure level reaches the point at which regulation begins. During regulation, the pump supplies as much flow as possible for the input power available.

The graph of Fig. 6-57 shows the pressure and flow relationships for a constant-horsepower control. The solid black (curved) line represents the theoretical constant-horsepower relationship between flow and pressure, while the dashed line depicts the actual characteristics of the pump control. In reference to the graph, the pump is at full displacement until pressure A is reached. The slight slope of the curve between zero pressure and point A represents only a loss in flow because of pump leakage. Once regulation

begins, the pump flow decreases quickly as pressure increases from point A to point B. Further increase in pressure (between points B and C) decreases pump flow more gradually until the control reaches the minimum flow value.

The pump control must be used with a main system relief valve capable of relieving this minimum pump flow. Once the end of regulation is achieved, the slightest increase in system pressure will open the relief valve and bypass the minimum pump flow to tank. Needless to say, if load-induced pressure drops, the control will follow the curve in the reverse direction (toward the maximum flow). You can see that the two straight lines, which represent the actual pressure versus flow curve of the pump, closely approximate the actual horsepower curve.

The cross-sectional view of a bent-axis design open-circuit pump (Fig. 6-58) can be used to explain the operation of any constant-horsepower control, such as that shown in Fig. 6-59. During startup, control spring 1 holds the pump on full displacement and the pump begins delivering a flow of fluid to the system. As the flow encounters resistance, system pressure builds on the small area of the positioning piston which holds the pump on full displacement, and on the small piston sensing the pressure in the system. As long as system pressure working over the area of the sensing piston is not high enough to move the pilot spool against the small spring (which establishes the beginning regulation), the large area of the positioning piston will be vented to the pump housing through the pilot spool.

When system pressure exceeds the pressure setting for the beginning of regulation, the sensing piston will push on the control rod and shift the pilot spool. The pilot spool will then direct system pressure to the large area of the positioning piston. Under this condition, the area differential of the positioning piston will cause the pump to begin destroking.

Destroking of the pump will cause control spring 1 to be compressed, which, in turn, will increase the mechanical

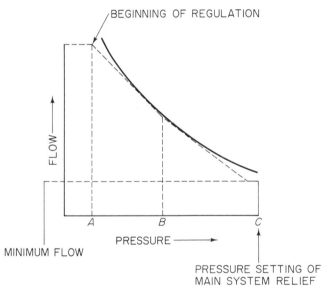

FIGURE 6-57 Constant-horsepower performance curves.

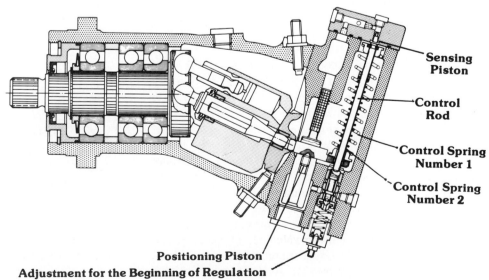

FIGURE 6-58 Pump structure for constant-horsepower capabilities. (Courtesy of Rexroth Corporation, Bethlehem, Pa.)

FIGURE 6-59 Variable-volume vane pump with constant-horsepower control. (Courtesy of Rexroth Corporation, Bethlehem, Pa.)

force on the control rod and the sensing piston. When the spring force balances the force due to pressure working over the area of the sensing piston, the pilot spool will modulate the pressure on the large diameter of the positioning pistons. In this way, the pump will destroke only to the point at which the spring force of control spring 1 balances the hydraulic force on the sensing piston. Consequently, control spring 1 establishes the slope of the initial pressure versus flow curve of the pump.

When the pump destrokes to approximately 50% of its maximum displacement control, spring 2 begins being compressed. This further increases the pressure needed on the sensing piston to cause displacement of the pilot spool. The second, more gradual, slope of the pump's pressure versus flow characteristics is shaped by the combined force of the two control springs.

Load-Sensing Controls

The typical load-sensing control circuit shown in Fig. 6-60 can use a pump similar to that shown in Fig. 6-55 or any similarly equipped variable-displacement pump. The basic requirement of a load-sensing pump is the ability to provide an established minimum pressure, usually 200 psi or less, to ensure a source of pilot pressure and keep the circuit "live." A pilot pressure is added to this minimum pressure value. This pilot pressure is supplied through a shuttle valve network. The shuttle valve senses the highest pressure and blocks the lower pressures.

As an example, if valves H and J are in a neutral position in Fig. 6-60, the cylinder lines to motor K and cylinder L are open to the tank. Shuttle valve E will have

no source of pilot pressure. Compensator valves C and D will block flow to valves H and J at the value of springs P and Q, respectively. This is usually at a value of 150 psi or less.

By energizing valve H (which has 150 psi at the inlet because of valve C) flow is directed to motor K. The pressurized supply line will provide pilot fluid through shuttle valve G to assist spring P. Flow is also available to shuttle valve E. The flow is directed to the maximum pressure control N on pump A. The resistance of the load in psi is added to spring P of valve C and control N of the pump. The pressures are increased only by the amount created by the load. The pump output pressure is then the established minimum pressure plus the load. The flow will be enough to satisfy the flow controlled by valve H. Valve H is a solenoid, pilot-operated proportional valve. It functions as both a directional and flow control valve. Valve J is similar. Valve J is shifted and cylinder L extends at the desired flow rate while motor K is turning at a controlled rate. Pilot flow passes through shuttle F and E to control N. The pressure level at N (maximum pressure) is set at pilot valve R. This pressure level passes through valve D to valve J and cylinder L. Pressure load at valve H is the work load on K plus spring P. Valve C limits the pressure feeding into valve H to the minimum established by P plus the load.

Many variations of this circuit provide economical power transmission functions with minimum loss through heat generation. Response time is excellent. Shock loading is held to a minimum.

Valves H and J are connected in parallel. Additional valves appropriate to the supply capabilities of pump A can be added. The shuttle network senses the greatest pressure need up to the maximum established by pilot valve R. Control valves C and D limit flow and pressure to the directional valves at a value reflecting the bias spring plus the sensed branch circuit load. Load at motor K does not affect operation of cylinder L. Cylinder L is isolated from the motor as long as both function within the flow and pressure capabilities of pump A. Pump A is usually of the variable-delivery type using vanes or pistons.

FIXED-DELIVERY PUMP SELECTION FACTORS

1. Volumetric capacity (gpm)
2. Operating speed (rpm)
3. Pressure rating (psi)
4. Type of duty (continuous or intermittent)
5. Type of mounting
6. Shaft size and type (splined or keyed)

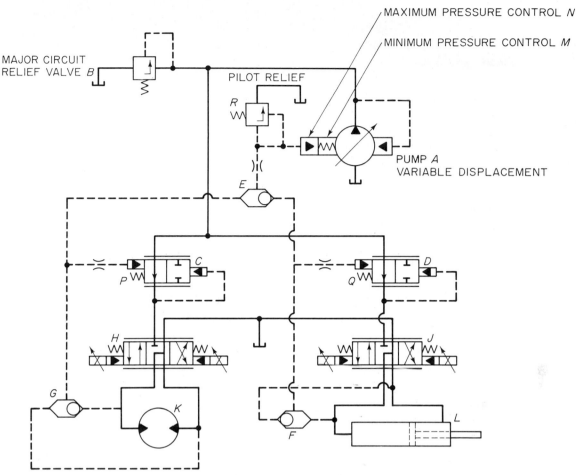

FIGURE 6-60 Load-sensing circuit.

7. Cost
8. Type of fluid
9. Operating temperatue (°F)
10. Design type (gear, vane, or piston)
11. Size and weight
12. Operating conditions (shock, loads, etc.)
13. Environmental factors (temperature, dust, etc.)
14. Type of drive (direct coupled, belt and pulley, etc.)
15. Inlet characteristics (atmospheric pressure, etc.)

VARIABLE-VOLUME PUMP SELECTION FACTORS

1. Volumetric capacity (0 to maximum gpm)
2. Operating speed (rpm)

3. Pressure range (0 to maximum psi)
4. Type of volume control (manual, pressure, servo)
5. Type of duty (continuous or intermittent)
6. Type of mounting
7. Shaft size and type (splined, keyed)
8. Cost.
9. Type of fluid
10. Operating temperature (°F)
11. Design type (vane, piston)
12. Size and weight
13. Operating conditions (shock, loads, etc.)
14. Environmental factors (temperature, dust, etc.)
15. Type of drive (direct coupled, pulley and belt, etc.)
16. Inlet characteristics (atmospheric pressure, etc.)

Figure 6-61 illustrates typical pump structures.

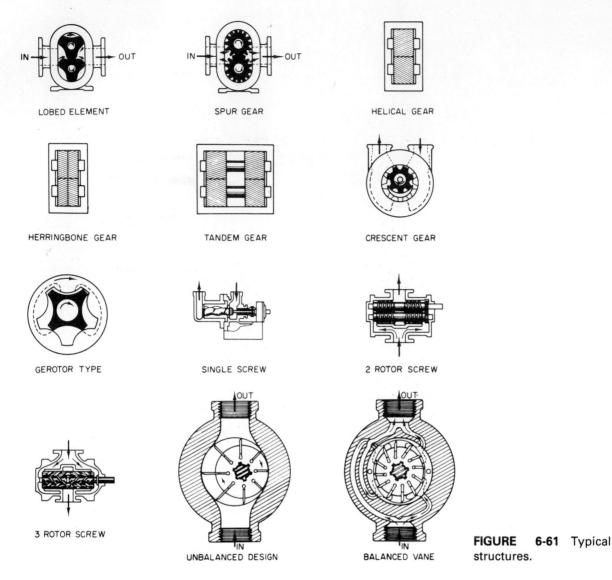

IN → ← OUT LOBED ELEMENT

IN → ← OUT SPUR GEAR

HELICAL GEAR

HERRINGBONE GEAR

TANDEM GEAR

CRESCENT GEAR

GEROTOR TYPE

SINGLE SCREW

2 ROTOR SCREW

3 ROTOR SCREW

UNBALANCED DESIGN

BALANCED VANE

FIGURE 6-61 Typical pump structures.

QUESTIONS

1. Name several prime movers that are used to impart energy to hydraulic pumps.
2. Why are electric motors especially popular for supplying energy to a hydraulic system?
3. What is a power takeoff?
4. When are gear reductions used to drive hydraulic pumps?
5. What precaution must be taken regarding the side loading of pumps by belts, chains, or other devices?
6. How much hydraulic horsepower would a pump produce when working at 4000 psi and delivering 12 gpm? What size (in horsepower) electric motor would be selected to drive this pump if it were 90% efficient?
7. What are the two main classes of pumps?

8. Name the three popular construction types of nonpositive displacement pumps.
9. What is meant by *staging* centrifugal pumps?
10. What is a positive displacement pump, and how does it differ from a centrifugal pump?
11. How is the volumetric efficiency of a positive displacement pump determined?
12. How is the theoretical output of a positive displacement pump determined?
13. What is mechanical efficiency, and how is it determined?
14. Explain how atmospheric pressure forces hydraulic oil up into the pump inlet.

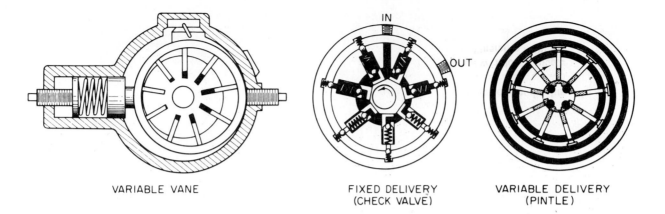

VARIABLE VANE

FIXED DELIVERY
(CHECK VALVE)

VARIABLE DELIVERY
(PINTLE)

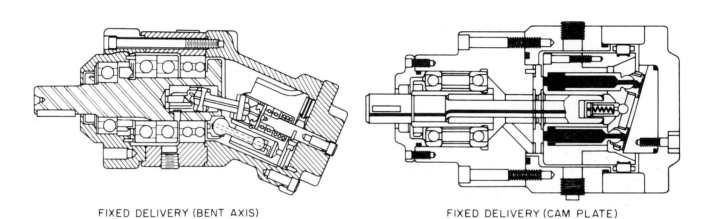

FIXED DELIVERY (BENT AXIS)

FIXED DELIVERY (CAM PLATE)

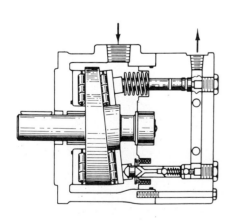

FIXED DELIVERY (CHECK VALVE)

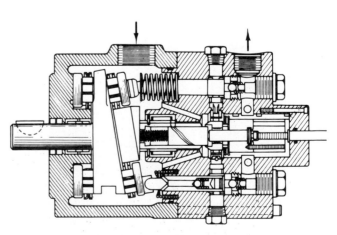

VARIABLE VOLUME (CHECK VALVE)

FIGURE 6-61 (continued).

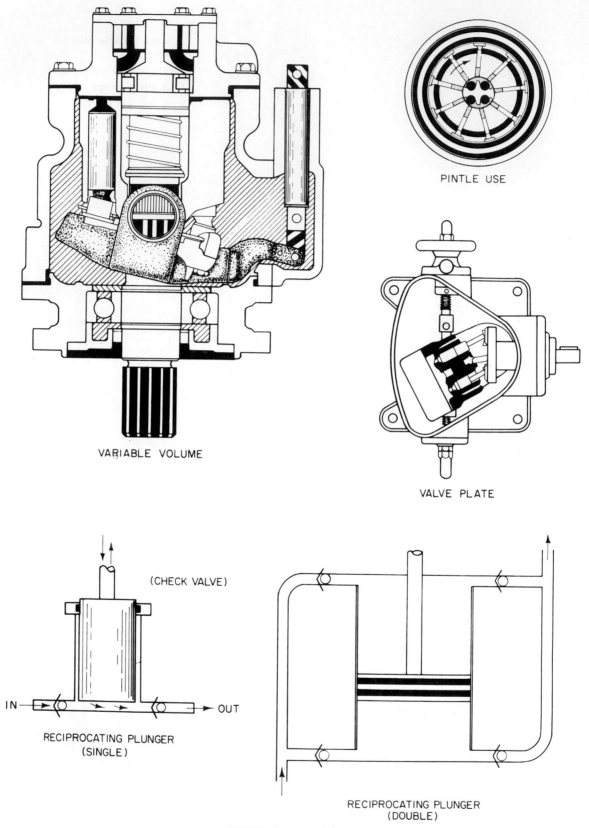

VARIABLE VOLUME

PINTLE USE

VALVE PLATE

(CHECK VALVE)

IN ——→ OUT

RECIPROCATING PLUNGER
(SINGLE)

RECIPROCATING PLUNGER
(DOUBLE)

FIGURE 6-61 (continued).

15. What is the recommended velocity for good inlet conditions when a pump is operating in normal atmospheric conditions?

16. List several reasons for pressurizing the hydraulic pump inlet.

17. What might happen to the inlet supply to a hydraulic pump if the air breather on a reservoir becomes dirty and clogged?

18. Describe the operation of a fixed-delivery hydraulic pump. Describe the operation of a variable-volume hydraulic pump.

19. List several construction types of fixed-delivery hydraulic pumps.

20. When is it necessary externally to drain a hydraulic pump or motor unit?

21. Name three designs of external gear pumps.

22. Name two designs of internal gear pumps.

23. Describe the operation of a two-stage pump.

24. Why is the operation of a screw-type pump quiet?

25. Describe the operation of a balanced vane pump.

26. Why is an unbalanced vane pump used when variable-volume characteristics are important?

27. What is meant when a variable-volume vane pump "dead-heads"?

28. Explain the main differences between radial, axial, and reciprocating plunger piston pumps.

29. What are the three common methods of porting fluid from inlet to outlet in piston-type pumps?

30. Which methods of displacing fluid from inlet to outlet in a piston unit allow operation as either a pump or a motor?

31. What is meant when a piston pump is said to be variable and reversible?

32. What kind of a pump is used in a closed-circuit system?

33. What is the purpose of a replenishing pump?

34. What is multiple circuitry using a single pump?

35. List the important considerations in the selection of a pump for any given application.

7

The Control of Hydraulic Power

Fluid power is power transmitted and *controlled* through use of a pressurized fluid. In the definition above the "controlling" factor in the transmission of fluid power not only is important but provides a real challenge to the ingenuity of system designers. Through slight modification in the designs of standard components, many new techniques in the art of controlling hydraulic power are being discovered.

Primarily, controlling hydraulic power involves the pressure, direction, and volume of fluid flow. Controlling fluid pressure concerns the maximum system operating pressure, or a pressure-sensing device that controls the direction or volume of the fluid. Because of the flexibility of applying hydraulic controls, they are often named because of what they do rather than what they actually are. Therefore, it is important to know the primary function of the various controls available and how they operate. Knowing the operational characteristics of a control leads to the discovery of new techniques in applying the control to any given hydraulic circuit.

PRESSURE CONTROLS

Relief Valves

One of the most important pressure controls is the relief valve or safety valve. Its primary function is to limit system pressure. Relief valves usually establish or limit a working pressure. Safety valves operate only when there is a circuit malfunction. A relief valve may function in dual capacity as both relief and safety device. There are two basic designs, which are the directly operated or inertia type, and the pilot-operated design (see Fig. 7-1).

The direct type of relief valve has two basic working port connections. Additional ports may be provided for convenience in making system connections. System pressure opposes the poppet, which is held on its seat by an adjustable spring. The adjustable spring is set to limit the maximum pressure that can be attained within the system. When pressure exceeds the spring setting, the poppet is forced off its seat and excess fluid in the system is bypassed back to the reservoir. When system pressure drops to or below established set value, the valve automatically reseats.

The pilot type of relief valve is a two-stage design sometimes referred to as a compound-type valve. The in-line type may have three port connections. The main pressure line can pass straight through the body in the three-port design, eliminating the need for one tee connection, and the secondary port is connected to the reservoir. In certain designs the main spool may be fitted with a piston-like skirt. An orificed passage is provided to direct fluid at a controlled rate from the primary high-pressure area to the upper side of the piston-like skirt or into the bias cavity, which contains the spring that provides minimum pressure

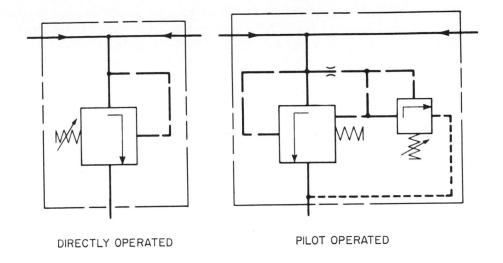

DIRECTLY OPERATED   PILOT OPERATED

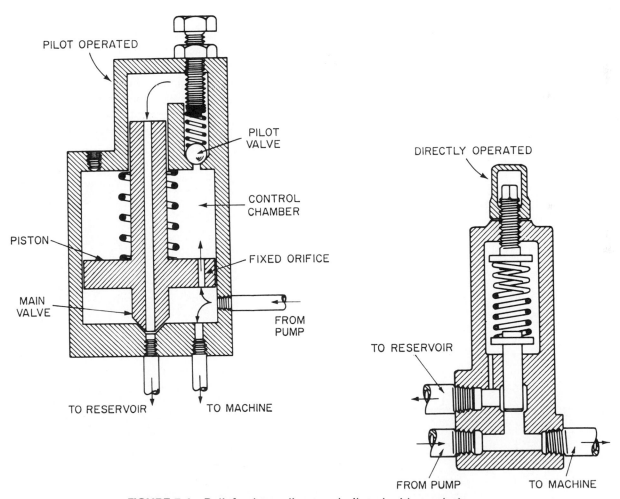

FIGURE 7-1   Relief valves: direct and piloted with symbols.

value (plus the additive from the flow through the orifice). This orifice may be in the piston section of the spool itself or in a separate passage in the valve body. System pressure operates on both sides of the piston, causing it to be pressure-balanced. Because the main spool is pressure-balanced, any increase in system pressure will operate on both sides of the piston, and the spool remains seated with a relatively light bias spring.

From the chamber on top of the piston (see Fig. 7-1) there is a fluid passage that leads up to the small pilot valve. (A pilot valve is applied to operate another valve or control.) The pilot valve is also held on its seat by a spring that is adjustable in this specific design.

When system pressure exceeds the setting of the pilot valve bias spring, the small pilot valve opens, permitting escape of pressurized pilot fluid from the top side of the main spool piston. This action causes the main valve to open because of the greater force below the piston. The orifice through the spool piston is smaller than the passage to drain, so the relief valve remains open until the system pressure drops and allows the pilot valve to close. The main spool again becomes pressure-balanced and assumes its normally closed position.

A choice between the directly operated and the pilot-type relief valve depends primarily on the application. When the pressure level within a circuit must remain relatively constant, the pilot type is generally used. This design gives less pressure differential during its normal cycle of opera-

tion. This cycle involves its cracking point to fully opened and then reseating when the pressure drops to normal.

Pilot-operated relief valves are popular in the machine tool field, where more exact pressure levels are maintained. This design also provides for greater flexibility in control.

Pilot-operated relief valves are especially important in circuits where rapid operations can create shock waves. The pilot relief valve can usually open much faster than an equivalent-size direct spring-loaded relief valve with a heavy spring and a mass to get into motion. Some variable-displacement pumps cannot respond as fast as a relief valve. For this reason a relief valve is a desirable and often essential component in systems using variable displacement pumps. A vent line or drain connected to the chamber on top of the main spool piston and fitted with an on-off valve could be used to operate the relief valve remotely.

This technique is used for bypassing pump delivery back to the reservoir through the relief valve at the relatively low bias pressure created by the spring during the idle time of a machine. As long as the vent line is open to the reservoir, the relief valve is held open at a low pressure, and flow from the pump is bypassed or circulated back to the reservoir, for the relief valve now provides the path of least resistance to the reservoir.

Another technique shown in Fig. 7-2 is dual pressure control. Note that the remote pilot control can be separate from the piloted relief valve. The schematic view shows the integral pilot and the piloted spool in one envelope. The

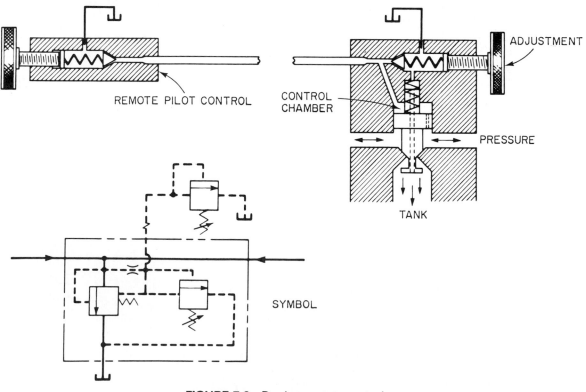

FIGURE 7-2 Dual pressure control.

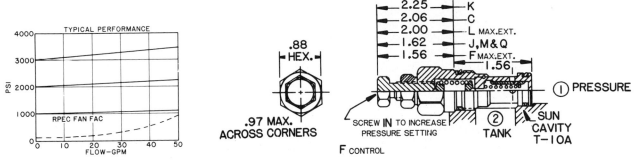

FIGURE 7-3 Cartridge-type structure and typical performance curves. (Courtesy of Sun Hydraulics Corporation, Sarasota, Fla.)

remote pilot control is shown as an independent unit in its own envelope. It is interconnected with the piloted segment with suitable tube, pipe, or hose. The two pilots are in parallel. Pilot pressure will be as per the unit set at the lowest pressure.

The cartridge valve of Fig. 7-3 is small in physical size but capable of handling flow to 50 gpm. Note the slight change in pressure as flow increases from minimum to 50 gpm at the set pressure values. This type of valve can be installed in a machined bore close to the point of control

with appropriate drilled passages for input pressure and return to tank.

A similar cartridge valve is shown in Fig. 7-4. A second passage is introduced for the vent or remote control function. Note the vent pressures as they relate to flow through the cartridge.

By connecting the remote pilot control to the external tank line of the integral pilot in the piloted relief valve assembly the two pilots would be in series. Thus the resistance to flow of pilot fluid would be the summation of the

FIGURE 7-4 (a) Ventable cartridge type of relief valve; (b) pressure-drop curves with valve relieving; (c) pressure-dop curves with valve vented (Courtesy of Sun Hydraulics Corporation, Sarasota, Fla.)

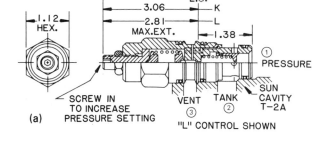

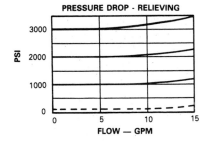
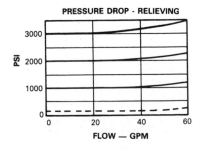
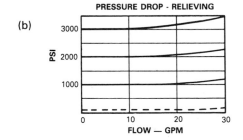
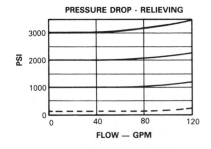
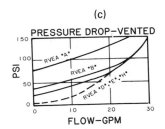
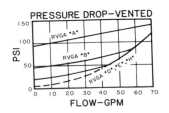

two pilot valve values. As an example, the main adjustment in the piloted assembly can be set for a desired minimum pressure. Any appropriate pressure above that value can be set by the pilot valve, which is in series via the connection to the outlet port of the integral pilot valve.

Figure 7-5 shows a cartridge-type piloted relief valve. An envelope (1) manifold mounts to a subplate into which the plumbing is connected. A poppet (3) is contained within a cross-drilled cylindrical housing which is held in place by a threaded sleeve (2).

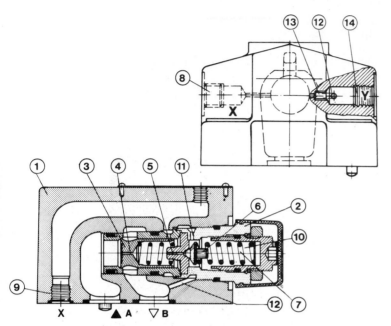

FIGURE 7-5 Piloted-type cartridge relief valve in subplate-mounted envelope. (Courtesy of Rexroth Corporation, Bethlehem, Pa.)

The orificed pilot supply port through hole 4 provides a fluid supply to the control chamber. Passage 5 is closed by poppet 6. Port 8 is connected to the port identified as X. This port is the parallel connection such as that shown in Fig. 7-2.

Spring chamber 10 can be internally connected to port B through passage 12. Passage 12 can be blocked. Flow from chamber 10 can be connected through passage 13 to connection 14, identified as the Y port. Any remote pilot to the Y port is in series with the integral pilot assembly consisting of cone 6 and spring 7. Ports A, B, X, and Y can usually be connected to the subplate. Ports X and Y can often be connected to auxiliary ports on the valve body (1).

The pilot assembly of the upper view of Fig. 7-6 is located in the cap of the valve. The piloted poppet and its sleeve assembly (1) are concentrically located in a bore in the body of the valve. A removable orifice (2) is threaded into the valve body. Fluid passing through orifice 2 is directed through passage 4 to a poppet seat (15). It also passes through passage 5 and orifice 3 to the control chamber in

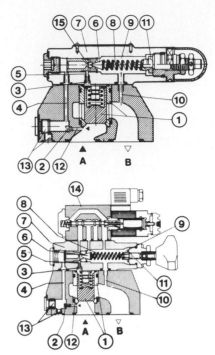

FIGURE 7-6 Subplate-mounted piloted relief valve wth cartridge-type major poppet structure. (Courtesy of Rexroth Corporation, Bethlehem, Pa.)

which the main spool poppet spring is located. Cone 6 biased by spring 8 can control maximum pressure. Fluid in chamber 9 can be internally drained to port B through hole 10. External connection 11 is used if line 10 is plugged. Connection 11 is then the Y port. Connection 13 can be external through the threaded port X or through the subplate.

The lower view of Fig. 7-6 shows a solenoid-actuated, spring-return, directional valve. This valve can be normally open or normally closed (as shown). Valve 14, as installed, is normally closed because of the spool positioning by the spring. The passage to cone 6 is blocked by the pilot venting valve spool. Energy to the solenoid of valve 14 directs flow from line 4 back to tank internally through passage 10 or externally through port 11.

The remote pilot of Fig. 7-2 could be a proportional-type unit such as that shown in Fig. 7-7. Solenoid 2 of Fig. 7-7 is a direct-current device. Current to the windings causes the armature assembly (7) to increase push on spring 6. This increases the mechanical force holding poppet 5 against seat 4 in body 1. A linear variable differential transformer (3) is used to electronically monitor the position of the solenoid armature. This provides the feedback signal to ensure the desired pressure level as controlled by this pilot assembly. Figure 7-8 shows a typical amplifier card used to control a valve of this type.

A command voltage at the input to the electronic card can produce a resulting current to the solenoid winding at

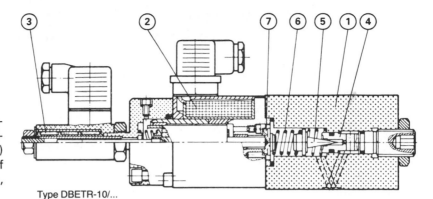

FIGURE 7-7 Pilot valve with proportional solenoid and linear variable differential transformer (LVDT) feedback assembly. (Courtesy of Rexroth Corporation, Bethlehem, Pa.)

Type DBETR-10/...

FIGURE 7-8 External view of interface card showing adjustments and test terminals. (Courtesy of Rexroth Corporation, Bethlehem, Pa.)

a desired value and at a controlled rate. Figure 7-9 shows the card circuit when the LVDT (linear variable differential transformer) is employed. The card circuit shown in Fig. 7-10 does not include the electronics associated with the LVDT.

The relief valve assemblies of Figs. 7-11 and 7-12 are quite similar to those of Fig. 7-6. The solenoid in the lower view of Fig. 7-6 is digital in nature, that is, it is either on or off and there is no control of the rate at which the solenoid armature moves.

The proportional solenoid used to actuate the pilot of the valve shown in Fig. 7-11 can be controlled as to the rate at which pressure rises and the rate at which pressure decays as well as pressure level.

The assembly shown in Fig. 7-12 establishes two fixed pressures and is infinitely adjustable between the two fixed values. The pressure relief cartridge establishes maximum pressure. The spring on the main poppet assembly establishes minimum pressure.

In summary, the direct-operated design type can be used as a pilot valve, adjustable or nonadjustable. It can be used to control overpressures in simple circuits at relatively low pressures.

The piloted type is usually faster, particularly in the larger sizes. They are also used to dampen hydraulic shock and to prevent overpressures in long lines or in branch circuits which may be subject to local intensification.

Relief valves must be properly sized and placed in the hydraulic system to provide maximum protection at all times. Generally, the main relief valve is installed close to the pump. Other relief or safety valves may be placed throughout the system if needed.

Once the relief valve is adjusted for the work cycle of a machine's operation, it should not be tampered with. Generally, the relief valve is set about 10% higher than the normal pressure necessary to perform the actual work done by the hydraulic circuit, unless it is used to establish the working pressure level, a case which might be encountered in a press or clamp circuit.

Sequence Valves

A sequence valve's primary function is to direct flow in a predetermined sequence. It is a pressure-actuated valve similar in construction to a relief valve. The sequence valve operates on the principle that when main system pressure overcomes the spring setting, the valve spool moves up, allowing flow from the secondary port (see Fig. 7-13).

A sequence valve may be direct or remote pilot-operated. These valves are used to control the operational cycle of a machine automatically or in response to a manual input signal. Sequence valves may be directly operated as shown in Fig. 7-13 by directing pressurized fluid from the primary port to the area beneath the directional spool opposite the bias spring. The pressure in the primary line is effective on the end of the spool. This pressure will urge the spool against

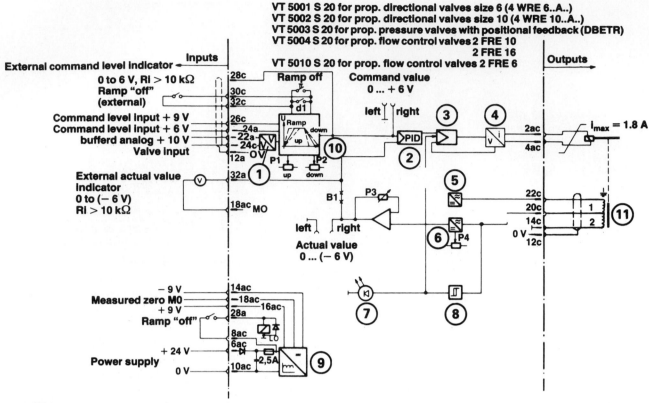

VT 5001 S 20 for prop. directional valves size 6 (4 WRE 6..A..)
VT 5002 S 20 for prop. directional valves size 10 (4 WRE 10..A..)
VT 5003 S 20 for prop. pressure valves with positional feedback (DBETR)
VT 5004 S 20 for prop. flow control valves 2 FRE 10
2 FRE 16
VT 5010 S 20 for prop. flow control valves 2 FRE 6

Cautions:
- Turn off Power before unplugging amplifier card.
- Measurements to be made with a high impedance multimeter.
- Regulated zero volt output measured zero (M0) is raised + 9 V with respect to 0 V input supply voltage, therefore, M0 may not be connected to 0 V of supply voltage.
- The ground connection of the inductive positional transducer must not be connected to 0 V of the supply voltage.
- Radio transmitters may not be used within 3 ft (1 m) of this card.
- Input signals may only be switched using contacts suitable for currents < 1 mA.
- Input signal lines and lines to the inductive positional transducer must be shielded. Leave one end of the shield open and connect the card end to 0 V of the supply voltage.
- Do not run the solenoids wires in the vicinity of power lines.

P1 ramp up 0.03 to 5 sec.
P2 Ramp down 0.03 to 5 sec.
P3 Sensitivity (solenoid)
P4 Zero Point Adjustment

1 Differential Input Amplifier
2 PID Proportional Integral Differential
3 Current Regulator
4 Power Amplifier
5 2.5 kHz Oscillator
6 Demodulator
7 LED Cable Break
8 Cable Break Detcedor
9 Voltage Regulator
10 Ramp Generator
11 Inductive positional Transducer (L.V.D.T.)

FIGURE 7-9   Circuit for interface card with linear variable differential transformer (LVDT) feedback used on pilot valve. (Courtesy of Rexroth Corporation, Bethlehem, Pa.)

the main bias spring and at the preset value of the spring allow a passage from the primary to the secondary port.

For remote operation it is necessary to close the passage used for direct operation by plugging or some other mechanical blockage and provide a separate pressure source as required for operation of the spool in the remote operation mode.

Figure 7-14 shows an automatic sequencing operation using the system's pressure as an actuating signal. When the clamp cylinder moves out and clamps the workpiece, resistance is met and pressure in the system increases. When the pressure reaches the spring setting in the sequence valve, it opens, and flow is directed out of the secondary port to operate the work cylinder.

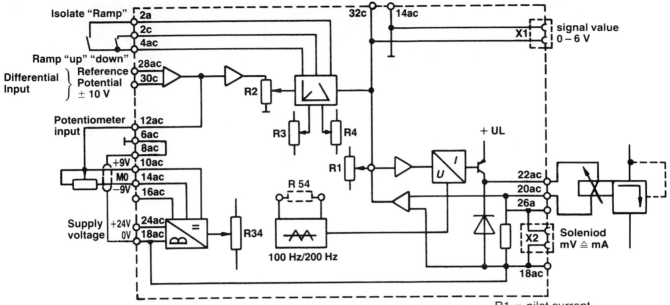

FIGURE 7-10 Electronic interface card for use with relief valves equipped with proportional solenoid operators. (Courtesy of Rexroth Corporation, Bethlehem, Pa.)

R1 = pilot current
R2 = maximum current
R3 = ramp time up
R4 = ramp time down

Note:
On model "K" (with terminal strip), terminal 2a is not used, ramp times may not be separately isolated and 6 V signal output terminals are not used.

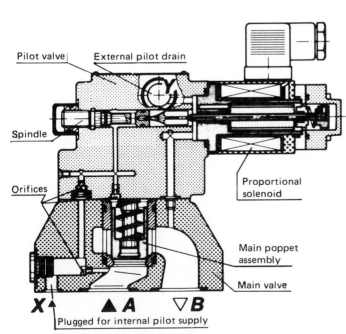

FIGURE 7-11 Proportional solenoid control for piloted relief valve. (Courtesy of Rexroth Corporation, Bethlehem, Pa.)

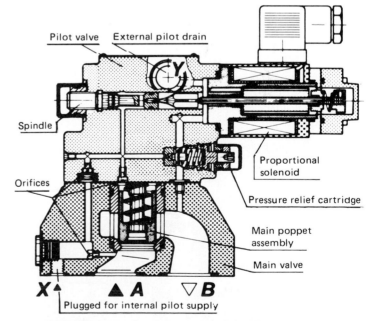

FIGURE 7-12 Proportional solenoid control for piloted relief valve with maximum pressure control. (Courtesy of Rexroth Corporation, Bethlehem, Pa.)

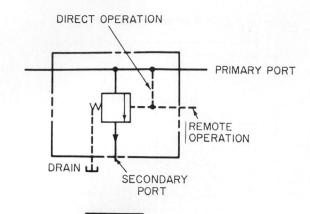

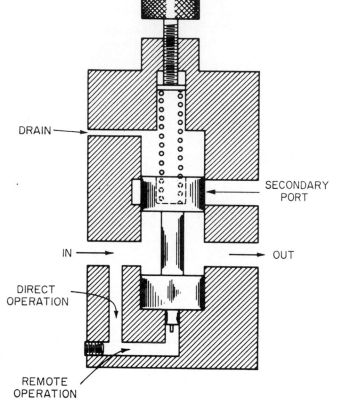

FIGURE 7-13 Sequence valve.

A good rule to remember with almost all pressure control valves is: "When a pressure control valve operates— if the flow from the secondary port performs work or is pressurized—the valve must be externally drained."

It is obvious that trapped fluid in the spring chamber would not allow the valve spool to move upward, because hydraulic fluids are practically incompressible. It is, therefore, necessary to allow clearance flow or internal valve leakage to drain back to the reservoir unrestricted.

Sequence valves may also be pilot operated. The drain from the pocket housing spring D in Fig. 7-14 can be used to control the poppet by selectively adding pressurized fluid to assist spring D in holding the poppet closed.

The sequence valve of Fig. 7-15 includes an orificed passage (5) to supply pressurized fluid to the main poppet bias spring chamber. A small pilot assembly is provided in the cap (4). Internal signal is sensed through line 2. Sequence valves are usually internally piloted and externally drained. As pressure rises in the primary port A it is sensed through line 2 and orifice 1 and pushes spool 3 against spring 7.

As spool 3 moves, it creates a passage from the bias spring chamber of poppet 6, allowing oil to pass to tank through port Y or Y_1 faster than it can pass through orifice 5 so that poppet 6 can open, creating a passage from A to B.

These valves are multipurpose structures. They can provide different functions by inserting or omitting certain plugs or closures. Thus the plug at (9) prevents flow of pilot fluid to port B when used as a sequence valve. Line 10 offers a path to port Y at the subplate interface if desirable. Port X is rarely used when a sequence function is desired as the signal is usually from port A rather than an external source.

Unloading Valves

An unloading valve is used to permit a pump to operate at minimum load (see Fig. 7-16). The unloading valve operates on the principle that pump delivery is diverted to the secondary port and back to the reservoir when sufficient pilot

The Control of Hydraulic Power

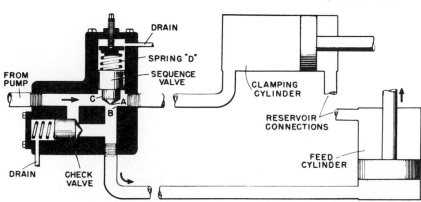

FIGURE 7-14 Clamp and sequence. (Courtesy of Mobil Oil Corporation.)

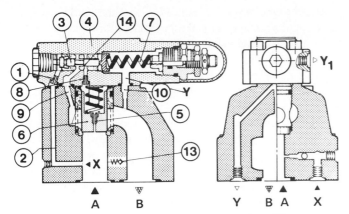

FIGURE 7-15 Piloted sequence valve. (Courtesy of Rexroth Corporation, Bethlehem, Pa.)

pressure is applied to move the spool against the spring force. The valve is held open by pilot pressure until the pump delivery is again needed by the circuit. The pilot fluid applied to move the spool upward becomes a static system. In other words, it merely pushes the spool upward and maintains a static pressure to hold it open. When the pilot pressure is relaxed, the spool is moved down by the spring, and flow is directed through the valve and into the circuit.

The unloading valve is useful in systems having one or more fixed delivery pumps to control the amount of flow at any given time. It is especially popular for use in feed and traverse circuits where rapid approach, feed, and return strokes are needed.

Unloading valves also help to prevent heat buildup in a system, which is caused by fluid being discharged over the relief valve at its pressure setting. A well-designed hydraulic system uses the correct amount of fluid for each phase of a given cycle of the machine operations. When pressure builds up during the feed phase of the cycle, pilot pressure opens the unloading valve, causing the large pump to bypass its 15 gpm back to the reservoir. The check valve isolates the high pressure side from the low.

A typical application of an unloading valve is shown in Fig. 7-17. The total delivery of the two pumps in the figure is 20 gpm. This would provide 20 gpm for rapid advance, 5 gpm for feed, and 20 gpm for rapid return (see Fig. 7-18).

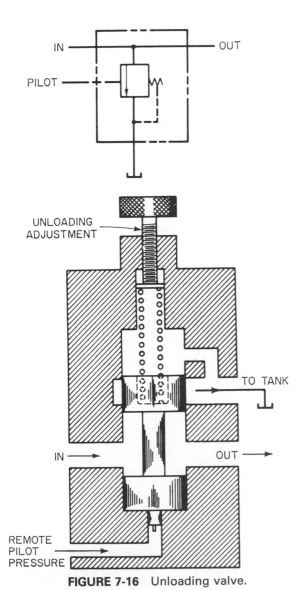

FIGURE 7-16 Unloading valve.

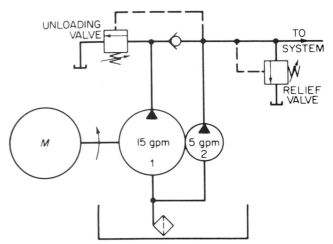

FIGURE 7-17 Application of unloading valve.

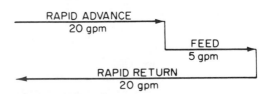

FIGURE 7-18 Machining cycle for an unloading circuit.

EXAMPLE

What size electric motor drive would be needed in the "unloading circuit" if the rapid advance operating pressure is 250 psi and the feed operating pressure is 2000 psi? (Rapid return is the same psi as rapid advance; assume pumps to be 100% efficient.)

Solution:

$$hp = gpm \times psi \times 0.000583$$

$$Rapid\ advance\ hp = 20 \times 250 \times 0.000583$$

$$= 2.91\ hp$$

$$Feed\ hp = 5 \times 2000 \times 0.000583$$

$$= 5.83\ hp$$

A standard 5-hp electric motor could be used if the duty cycle were intermittent. Generally, it is good practice to select the next-highest-hp motor, which would be $7\frac{1}{2}$ hp.

This sample problem shows that using an unloading valve to eliminate the large pump during the high-pressure feed portion of the cycle reduces the input horsepower. Without the unloading valve, the horsepower input would be

$$hp = 20 \times 2000 \times 0.000583$$

$$= 23.32\ hp$$

By using the unloading valve, only 5 hp input is needed, about 18 hp being saved. This saves in the cost of the electric motor and prevents hp from turning into heat during the feed portion of the cycle.

A great number of hydraulic systems in all fields of usage are faced with the common problem of fast approach or high volume at low pressure, then low feed strokes during the high pressure work phase of the cycle. It is always the prime target of the circuit designer to conserve horsepower and prevent system heat buildup. If an unloading circuit is not feasible, a more sophisticated approach is necessary, using variable-volume pumps.

Unloading valves are used with accumulator circuits. The circuit of Fig. 7-19 shows how a relief valve can be vented by a pilot assembly that accepts a signal from the system beyond a check valve. The pump is protected by the relief valve to prevent overpressure in the conventional manner. In addition, a plunger with a slightly larger diameter than the seated pilot poppet is connected to the circuit beyond the check valve. As the accumulator is being charged there is equal area on each face of the plunger so that the only force difference is the pressure difference created by the bias spring in the check valve. As system pressure reaches the value of the pilot spring adjustment and flow passes by the poppet to tank, the pressure on the face of the plunger

adjacent to the poppet is reduced so that the plunger snaps the poppet open and vents the pump flow through the relief valve piston at the value created by the bias spring on the piston (usually less than 100 psi). As system pressure decays because of usage or leakage, the force against the poppet decreases so that the poppet snaps closed and directs pump flow through the check valve to the circuit.

This automatic unloading and reloading continues in the normal power usage pattern of the machine and its "demand" circuit. It is called a demand circuit because of the relaxation of the pump until there is a demand for fluid and associated power after the accumulator is depleted to a predetermined pressure value. The physical position of the plunger, piston, and check valve are shown in Fig. 7-20.

Counterbalance Valves

A counterbalance valve is used to maintain back pressure to prevent a load from falling. Common applications include vertical presses, loaders, lift trucks, and other machines that must position or hold suspended loads.

The counterbalance valve operates on the principle that fluid is trapped under pressure until pilot pressure, either direct or remote, overcomes the spring force setting in the valve. Fluid is then allowed to escape, letting the load descend under control (see Fig. 7-21).

This same type of pressure control valve is used as a "braking valve" for decelerating heavy inertia loads. It can be applied to either linear or rotary motion.

When the counterbalance valve is used on large vertical presses, it may be important to analyze the source of pilot operating pressure. Figure 7-22 shows a comparison between direct pilot and remote pilot operation.

Through the application of Pascal's law P (psi) = force/area = 800 psi, a pressure of 800 psi is needed to support the heavy load. If pilot pressure is taken directly, then the counterbalance valve will operate at about 800 psi or slightly higher because of inertia and friction. In the other case, where remote pilot pressure is taken from the pressure line at the top of the cylinder, a choice of the operating pressure can be made for the valve. A counterbalance valve is a normally closed valve and will remain closed until acted upon by the remote pilot pressure source. Therefore, a much lower spring force can be selected to allow the valve to operate at a lesser pilot pressure. It should also be noted that the press member cannot move downward unless flow from the pump is directed into the top of the cylinder; this is a normal function of the machine.

A number of the pressure control valves manufactured can be used as relief, sequence, unloading, counterbalance, or braking valves by making some modification in the drains, pilot connections, or other slight changes. Most designs offer unlimited applications because of their designed flexibility.

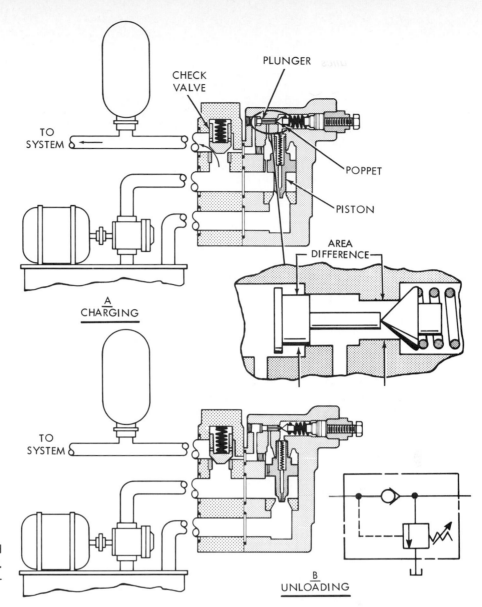

CHECK
VALVE

PLUNGER

POPPET

PISTON

TO
SYSTEM

AREA
DIFFERENCE

A
CHARGING

TO
SYSTEM

B
UNLOADING

FIGURE 7-19 Valving to control the unloading of an accumulator. (Courtesy of Vickers, Incorporated, Troy, Mich.)

FIGURE 7-20 Cross-sectional view of valving for controlling the unloading of an accumulator. (Courtesy of Vickers, Incorporated, Troy, Mich.)

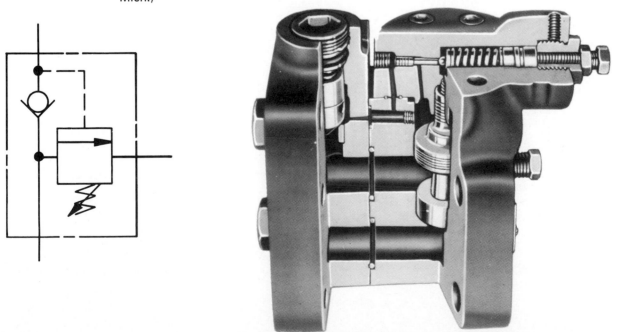

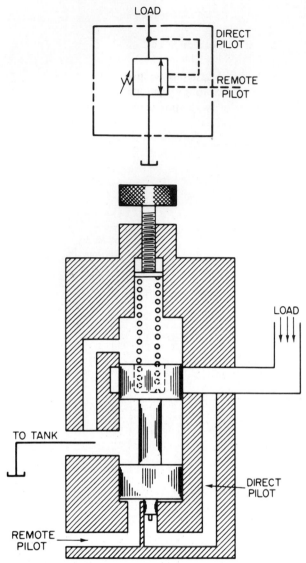

FIGURE 7-21 Counterbalance valve.

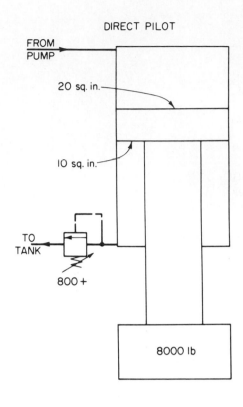

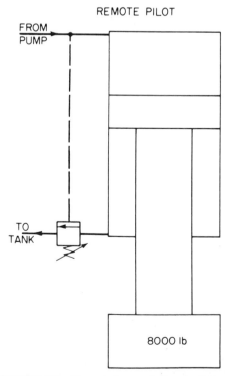

FIGURE 7-22 Comparison of direct pilot and remote pilot operation.

As an example, the cross-sectional view of the valve in Fig. 7-21 illustrates two pilot sources. The direct pilot source is connected to a reduced-diameter piston. The diameter ratio of the big piston to the small piston could be 10:1. The bias spring could be set to open the passage with a pressure of 2500 psi holding the load. Thus the potential intensification is minimized. Because of the 10:1 ratio it is only necessary to have a pressure of 250 psi on the remote pilot line to open the valve. In this mode a safety function is incorporated within the load control device.

Because of the potential of broken lines creating a hazard for overcenter and suspended loads, it is often mandatory that counterbalance or holding valves be manifolded to the cylinder or motor or be in cartridge format such as that shown in Fig. 7-23. Area ratios are also available in the cartridge structure.

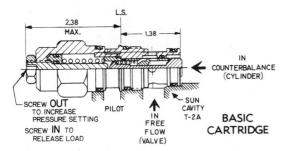

FIGURE 7-23 Cartridge-type counterbalance valve. (Courtesy of Sun Hydraulics Corporation, Sarasota, Fla.)

Pressure-Reducing Valves

A pressure-reducing valve is used to limit its outlet pressure. Usually these valves are available in various operating ranges and are adjustable. Their application is important where limited pressure must be controlled for such things as clamping light metal objects, or other light duty uses.

Reducing valves are also used for the operation of branch circuits, where pressure may vary from the main system pressures. The pressure-reducing valve is normally an open-type valve. A free flow passage is provided through the valve from inlet to secondary outlet until a signal from the outlet side tends to throttle the passage through the valve. The valve operates on the principle that pilot pressure from the controlled pressure side opposes an adjustable bias spring normally holding the valve open. When the two forces are equal, the pressure downstream is controlled at the valve setting. Thus, it can be visualized that if the spring has greater force, the valve opens wider; if the controlled pressure has greater force, the valve moves toward the spring and throttles the flow (see Fig. 7-24).

Reducing valves are widely used as compensators for variables in work load or supply pressure when it is desirable to maintain a uniform pressure drop across an orifice or throttle mechanism. The reducing valve of Fig. 7-25 is an integral part of a flow control valve. The reducing valve is called a hydrostat in this installation. It is a spring-biased normally open two-way valve. Flow from the pump can pass directly to the throttle control. The bias spring load sets the pressure difference across the throttle. If the throttle is closed, the pressure rises to the value of the bias spring and the path from the pump is closed.

When the throttle is opened, flow passes to the load. Any resistance to movement of the flow created by the load is added to the value of the bias spring so that the net pressure drop across the throttle is always uniform regardless of flow or work-load-induced pressure.

The reducing valve of Fig. 7-26 uses pilot flow to the spring chamber to add to the spring value up to the pressure value that opens the pilot poppet, at which time a maximum reduced pressure value is established.

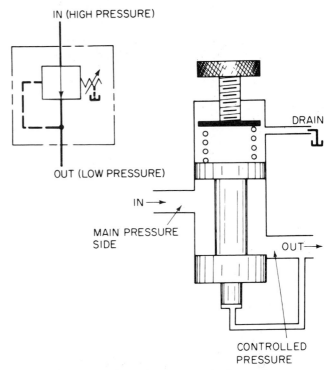

FIGURE 7-24 Pressure-reducing valve

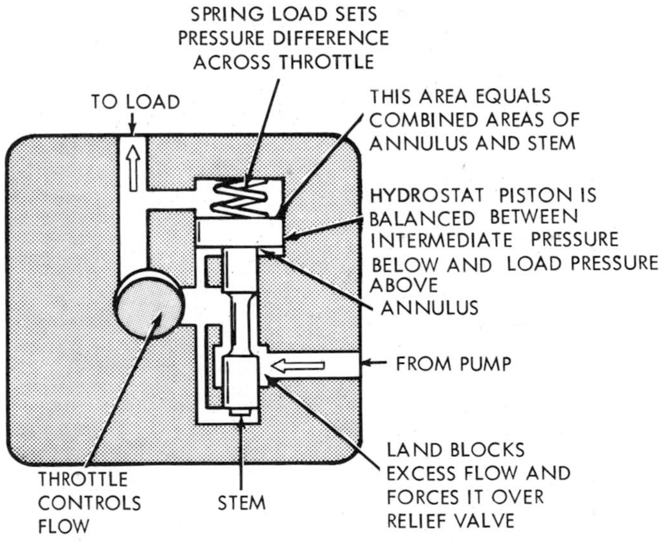

FIGURE 7-25 Pressure-reducing valve used as "hydrostat" in pressure-compensated restrictor-type flow control. (Courtesy of Vickers, Incorporated, Troy, Mich.)

A sandwich structure is provided in the direct spring-loaded reducing valve shown in Fig. 7-27. It can be used to reduce pressure to a directional pilot control.

A cartridge structure such as that shown in Fig. 7-28 permits the integration of the valve into manifold plates or multiple valve assemblies and ensures quick and easy replacement or servicing.

Figure 7-29 illustrates a pilot-operated reducing valve

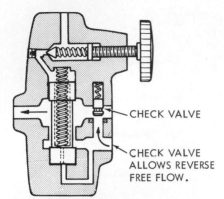

FIGURE 7-26 Piloted pressure-reducing valve with integral check. (Courtesy of Vickers, Incorporated, Troy, Mich.)

which uses a small reducing valve compensator to restrict flow to the pilot assembly to a predetermined value. Note the enlarged view of the compensated flow valve used to ensure an appropriate flow of pilot fluid to the cone assembly that provides the desired secondary pressure (see Fig. 7-31).

Pump pressure is available to port 2. Flow control 3 allows a flow (usually less than 50 in.^{3}/min) to the face of poppet 6 and through to chamber 4. Normally open two-way assembly 1 is biased open by the spring and pressure from passage 4. Ball assembly 7 allows flow from the secondary area to the tank by cone 6 if excessive pressure builds up in output port A as a result of external forces or leakage. Spring 5 and adjustment 9 in cap 10 determine secondary pressure values.

When branch circuits are used for clamping heavy

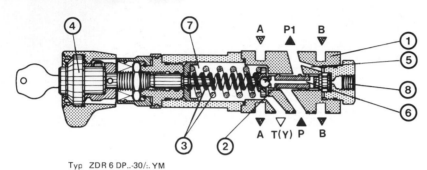

Typ ZDR 6 DP..-30/:. YM

FIGURE 7-27 Sandwich-type pressure-reducing valve. (Courtesy of Rexroth Corporation, Bethlehem, Pa.)

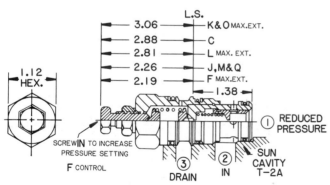

FIGURE 7-28 Cartridge-type pressure-reducing valve. (Courtesy of Sun Hydraulics Corporation, Sarasota, Fla.)

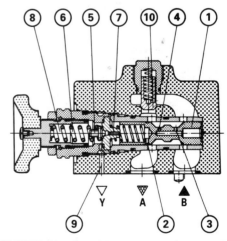

FIGURE 7-29 Cartridge-type pressure-reducing valve in subplate-mounted envelope. (Courtesy of Rexroth Corporation, Bethlehem, Pa.)

of cartridge design inserted into a housing which can be subplate mounted to an interface plate or manifold assembly. Pressure at port B flows through orifice 3 and 4 to act as a controlled damping agent to provide proper rate of spool movement. Orifice 2 directs pilot fluid into the main spool spring chamber. Channel 7 directs fluid to the poppet (5) biased by spring 6 adjusted by knob assembly 8. Conduit 9 directs fluid to port Y, which can be connected to a tank or to another remote control device in series with cone 5. Check valve 10 permits return flow through the valve.

Figure 7-30 illustrates a pilot-operated reducing valve

work pieces or for other important uses, the circuit designer must analyze the safety aspects of his design. There have been cases where pump failure due to the work cycle caused accidental unclamping of heavy pieces, which caused serious damage. Sometimes a separate clamp system, independent of the main hydraulic system, should be considered.

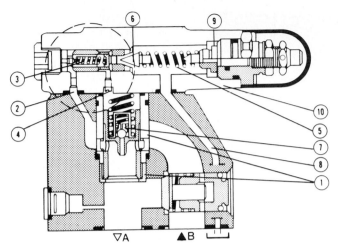

(a)

(b)

FIGURE 7-30 Piloted pressure-reducing valve with primary pilot pressure source. (Courtesy of the Rexroth Corporation, Bethelehem, Pa.)

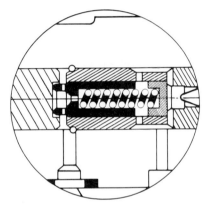

FIGURE 7-31 Pressure-compensated supply orifice for pilot section of pressure-reducing valve. (Courtesy of the Rexroth Corporation, Bethlehem, Pa.)

(c)

DIRECTIONAL CONTROL VALVES

Directional control valves have a primary function to direct or prevent flow through selected passages. They are used to extend, retract, position, or reciprocate hydraulic cylinders and other components for linear motion.

When rotary motion is required, directional controls are used to operate fluid motors in either direction with smooth reversals. When directional controls are used to operate other controls, they are called pilot directional valves.

Figure 7-32a illustrates how a spool-type directional control can be integrated into a body with cored passages so that load-holding check valves and cylinder line relief valves can be included in one rugged assembly for use on construction machinery where heavy-duty units are essential. A pilot flow can be used to actuate the valve spool.

FIGURE 7-32 (a) Cross section of piloted directional control valve; (b) Cross section of lever-actuated directional valve; (c) Exterior view of mobile-type bank valve. (Courtesy of Commercial Shearing, Inc., Youngstown, Ohio.) (d) In-line shutoff valve. (Courtesy of Republic Manufacturing Co.)

(d)

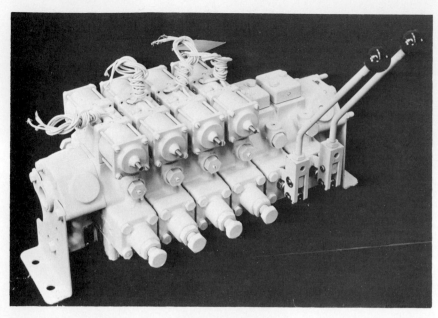

FIGURE 7-33 Combination solenoid and manually operated directional control assembly. (Courtesy of Racine Hydraulics Division, Dana Corporation, Racine, Wis.)

Figure 7-32b illustrates a similar unit with a spool extension which can be actuated by mechanical means. Figure 7-32c shows how several spool assemblies can be manifolded together to reduce need for interconnecting lines which can be potential leak sources.

Thus directional control valves range from a simple shutoff valve (Fig. 7-32d) to a complex multiple-segment,

banked valve package (Fig. 7-33) which is used for operating a number of circuits from one hydraulic power source. Generally speaking, each linear or rotary circuit in an open-circuit-type system is operated independently by a directional control valve or segment of a multiple control assembly.

Directional control valves (Fig. 7-34) are used to direct the fluid through selected passages to or from actuators

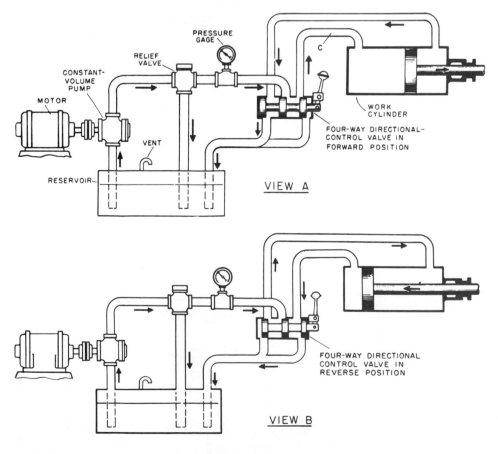

FIGURE 7-34 Constant-volume hydraulic system. Elements common to many hydraulic systems are shown: Reservoir, pump, piping, control valves, and hydraulic cylinder, or "motor." The hydraulic fluid is the "life blood" of such systems. In view *A*, the four-way directional-control valve is in its forward position so that hydraulic fluid is delivered by the pump to the left side of the piston, causing it to move to the right on its work stroke. Hydraulic fluid from the right side of the piston flows through the four-way valve to the reservoir. A relief valve controls maximum pressure. In view *B*, the four-way valve is in its reverse position, and hydraulic fluid is delivered by the pump to the right side of the piston, causing it to move to the left on its retraction stroke. Hydraulic fluid from the left side of the work cylinder flows through the four-way valve to the reservoir. (Courtesy of Mobil Oil Corporation.)

FREE FLOW

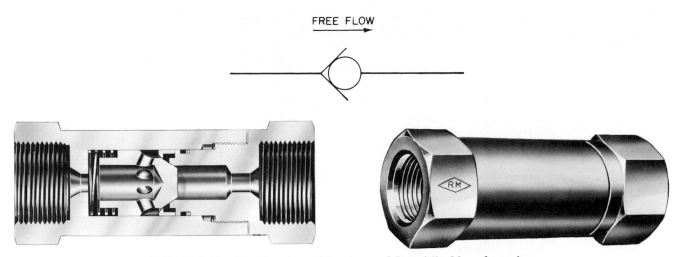

FIGURE 7-35 Check valves. (Courtesy of Republic Manufacturing Co.)

during the work cycle. Valves may be classified as one-way, two-way, three-way, and four-way valves with manual, mechanical, pilot, solenoid, or servo actuators. Many of these valves are standard units with a combination of solenoid, pilot, mechanical, and manual actuators. There is a valve or combination of valves for every purpose, with special ones available for handling exotic fluids, high temperatures, high pressures, or to solve problems concerning weight or space.

One-Way Valves

The simplest directional valves are the one-way valves or check valves (see Fig. 7-35). These units allow flow in one direction through the valve and are usually designed for in-line or right-angle connections. They are used primarily to allow free flow around other controls when actuating cylinders and fluid motors are being reversed (see Fig. 7-36).

Check valves are also built with various spring forces to act as low-pressure relief valves or to create a *pressure drop* between the system and the return line to the reservoir, for use as pilot pressure (see Fig. 7-37).

Two-Way Valves (Straightway Designs)

Two-way valves, or on-off valves, are used primarily as simple shutoff valves. These valves are of the seating action design which utilizes a gate, disk, ball, or other device to obstruct the flow path. Generally, fluid flow can be in either direction through the valve (see Fig. 7-38).

Gate valves are generally restricted to low-pressure applications, whereas globe- and ball-type valves are used for high pressures. Gate valves are used as shutoff valves and should be operated either fully opened or fully closed. Otherwise, high flow rates or hydraulic shock may crack or break off the wedge-like gate. Limit switches are often installed on gate valves, especially when they are used as

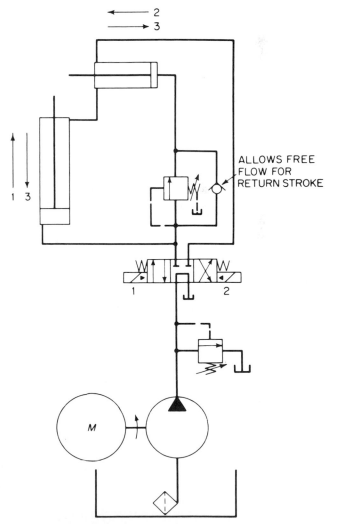

ALLOWS FREE FLOW FOR RETURN STROKE

FIGURE 7-36 Check valve application.

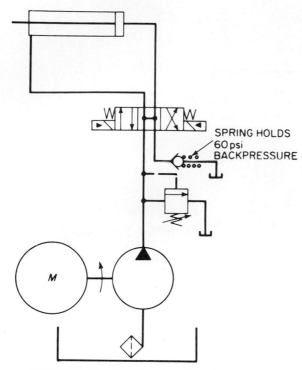

FIGURE 7-37 Check valve for pilot pressure.

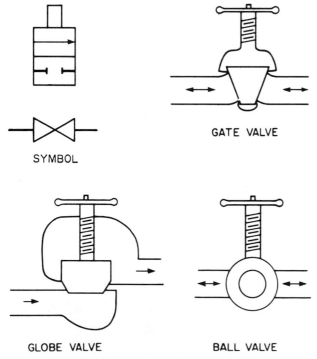

GLOBE VALVE   BALL VALVE

FIGURE 7-38 Two-way valves.

cutoff valves for suction lines. The limit switches may be so arranged in the electrical circuit so that the machine cannot be started up until the valve is wide open. When the gate valve is shut off it may engage a limit switch, which

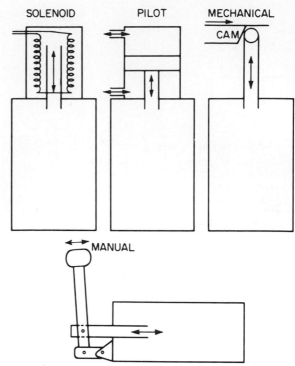

FIGURE 7-39 Solenoid, pilot, mechanical, and manual spool two-way and poppets.

will light a warning light on the control panel and prevent starting the pump motor until it is opened fully to allow fluid to the pump suction. Globe and ball designs may be adjusted to provide flow regulation where accuracy is not of prime importance. They are manufactured in almost all standard port configurations and are used to connect or isolate component parts of a hydraulic system.

Two-way valves may also be of the spool or poppet design and are available as normally opened or as normally closed valves. They are also available as solenoid, pilot, mechanical, or manually actuated (see Fig. 7-39).

Pilot-Operated Check Valves

A pilot-operated check valve is a two-way valve which allows free flow in one direction but prevents reverse flow until actuated by pilot pressure or other means. They are available with either single or double checks (see Fig. 7-40).

These valves are applied as a safety device to prevent an accidental occurrence of a cycle, or phase of a cycle, out of sequence. In other words, they may be used as a safety interlocking device. They are also used for controlling the sequence of a machining cycle and to prevent a load from dropping (see Fig. 7-41).

Three-Way Valves

A directional control valve whose primary function is alternately to pressurize and exhaust *one* working port is called

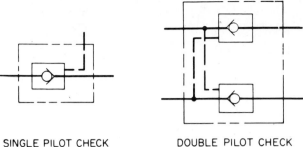

SINGLE PILOT CHECK   DOUBLE PILOT CHECK

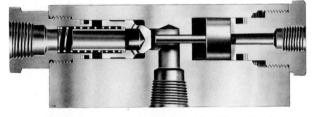

FIGURE 7-40 Double and single pilot check. (Courtesy of Republic Manufacturing Co.)

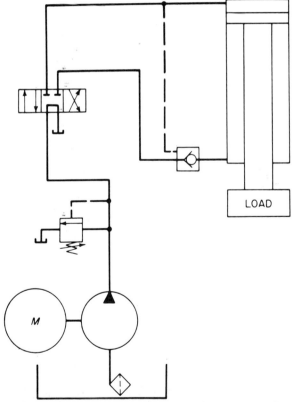

FIGURE 7-41 Use of pilot-operated check valve.

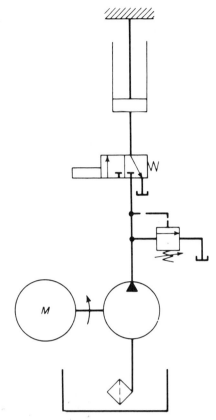

FIGURE 7-42 Three-way valve circuit.

a three-way valve. Generally, these valves are used to operate cylinders or fluid motors which need power in only one direction (see Fig. 7-42). Single-acting cylinders are used when loads are extended or positioned by fluid and retracted by some other means, such as a spring, or by the weight of the load itself.

Three-way directional valves are available for manual, mechanical, pilot, solenoid, or servo actuation. There are three basic designs, which include the spool, poppet, and shear seal types (see Fig. 7-43).

Three-way valves may be two-position or three-position. Most commonly they have only two positions, but in some cases a neutral position may be needed. Figure 7-44 shows the operations of a spool type, two-position, three-way valve.

The three-way-valve working ports are inlet from the pump, outlet to the cylinder, and exhaust to reservoir. These ports are generally identified as follows: P = pressure; C = cylinder; and T or R = tank or reservoir. Sometimes the letters A or B are used to identify the ports connected to an actuating cylinder or fluid motor.

When port identification cannot be readily found, blow low-pressure air through the various ports and shift the valve from one position to the other to determine port connections. (Make sure that clean, dry air is used.)

A three-way valve may be used to direct pressurized flow from one system to another; for example, a tractor hydraulic system may be used alternately to actuate a front-end loader or a plow control mechanism. The diversion-

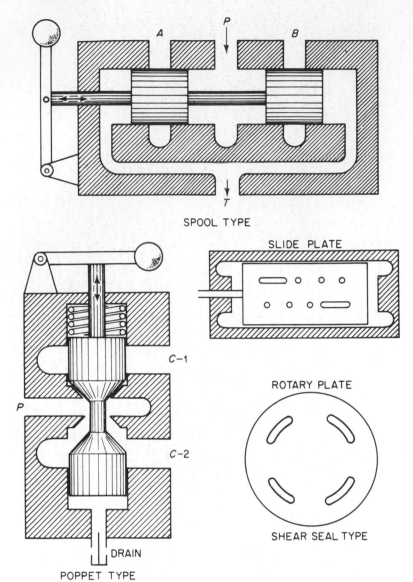

SPOOL TYPE

SLIDE PLATE

C—1

P

C—2

ROTARY PLATE

SHEAR SEAL TYPE

DRAIN

POPPET TYPE

FIGURE 7-43 Spool, poppet, and shear seal three-way valves.

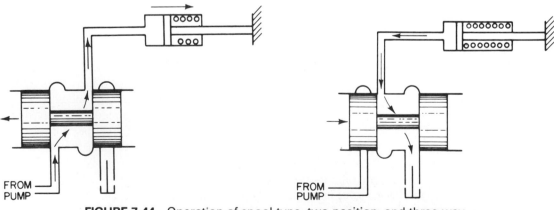

FROM PUMP

FROM PUMP

FIGURE 7-44 Operation of spool type, two-position, and three-way valves.

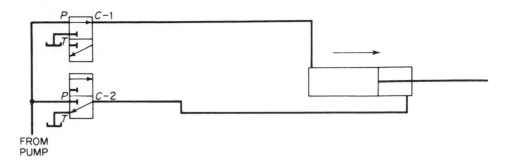

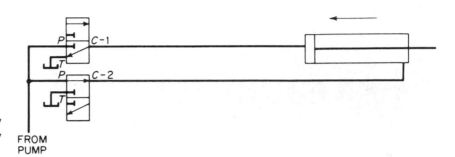

FIGURE 7-45 Two three-way valves can operate as a four-way valve.

type three-way valve will direct the pump delivery to the system that is to function at a specific time.

In cases of emergency or other reasons, two three-way valves can be used to reciprocate a double-acting cylinder or for reversing fluid motors (see Fig. 7-45).

Four-Way Valves

The type of the valves generally used to operate cylinders and fluid motors in both directions are four-way valves. These valves are manufactured as two-, three-, four-, five-, and even six-position valves in the spool, poppet, or shear seal designs. The primary function of a four-way directional control valve is alternately to pressurize and exhaust two working ports (see Fig. 7-46).

These valves are also available with a choice of actuators, including manual, mechanical, solenoid, servo, pilot, or in any combinations. Four-way valves can easily be converted to three-way valves by plugging one of the working ports.

Two-position four-way valves are used to reciprocate and hold an actuating cylinder in one position. They are also used on machines where fast reciprocating cycles are needed. Since the valve actuator moves such a short distance to operate the valve from one position to the other, this design is used for punching, stamping, and for other machines needing fast action.

Figure 7-47 shows a two-position four-way directional control valve holding a cylinder in the "up" position. Ac-

cording to fluid power standards, the connecting lines are shown in the position with the spring. It means that the valve assumes this position unless actuated manually by an operator. This applies for any means of actuation.

When fast actuation is needed, electrical solenoids, pilot fluid, and/or mechanical springs are used to shift the four-way valve from one position to the other (see Fig. 7-48).

When electricity is supplied alternately to the two solenoids, magnetic force causes an iron core to be pulled into the magnetic field. Since the valve actuator is attached to the iron core, it shifts back and forth as the solenoids are energized and deenergized.

It is possible to actuate a four-way valve with both a solenoid and mechanical spring plus a manual override. In Fig. 7-49 the solenoid is energized to position the valve in one direction; then when deenergized the spring force moves it back, thus reciprocating the valve spool. The manual lever can also be used to override the solenoid or spring in an emergency.

Pilot fluid is also an excellent means of positioning control valves. The fluid may be either a liquid or a gas. Air-piloted hydraulic valves provide rapid operation, freedom from fire hazard, and independent power source for the pilot function.

Figure 7-50 shows a four-way valve where the rate of shifting the spool from one side to the other can be controlled by a needle valve. Fluid entering the pilot pressure port on the X end flows through the check valve and operates against the piston. This forces the spool to move

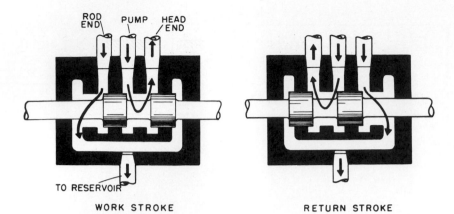

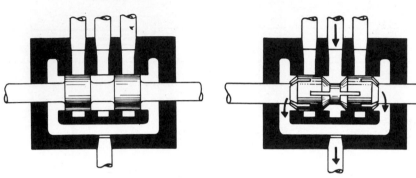

WORK STROKE

RETURN STROKE

NEUTRAL

OPEN-CENTER TYPE
VALVE IN NEUTRAL

FIGURE 7-46 Operation of four-way valve. (Courtesy of Mobil Oil Corporation.)

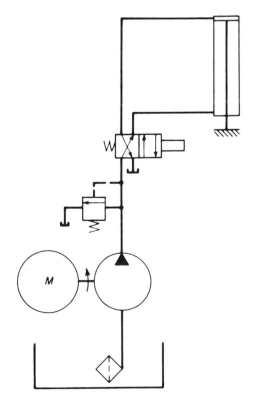

FIGURE 7-47 Two-position four-way directional control valve holding cylinder in up position.

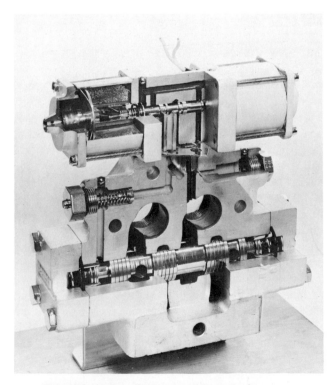

FIGURE 7-48 Solenoid-actuated four-way valve. (Courtesy of Racine Hydraulics Division, Dana Corporation, Racine, Wis.)

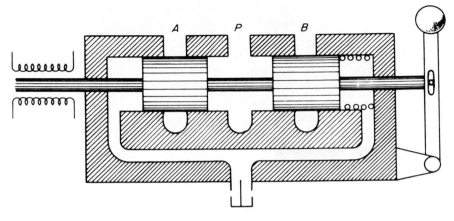

FIGURE 7-49 Four-way valve with solenoid and mechanical spring plus manual override.

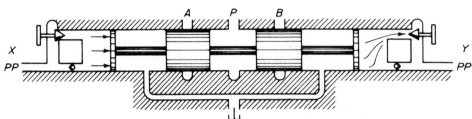

FIGURE 7-50 Four-way valve: needle valve to control shifting.

toward the opposite position. Fluid in the Y end passed through the adjustable needle valve and is exhausted. The amount of fluid bled through the needle valve controls how fast the valve will shift. Some machines, where heavy work loads are encountered, cannot be reversed too rapidly. Pilot-controlled valves, such as the one described, provide flexibility to adjust machine reversals.

Figure 7-51 shows a typical example of using air as

the pilot fluid. In this case, limit switches are used to energize and deenergize the air pilot valve solenoids. Their operation is similar to any off-on electric switch, such as the three-way type used between a house and garage. Either switch can be operated to supply electricity.

Limit switches are generally located on a machine to provide the proper stroke length and are actuated automatically by the movement of the machine. As solenoids A and

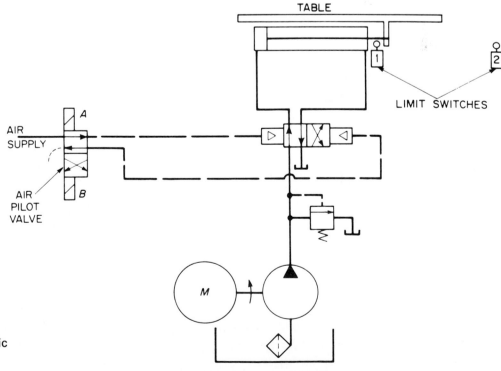

FIGURE 7-51 Air pilot hydraulic circuit.

B of the pilot valve are energized and deenergized by the limit switches, air is alternately supplied to either end of the main hydraulic four-way valve to reciprocate the machine table automatically.

In recent years, air controls have become available to replace electrical solenoids, switches, and other devices to increase service life and eliminate potential fire, explosion, and shock hazards. Figure 7-52 shows a four-way hydraulic control valve being actuated automatically by air components.

Air controls are clean and safe to use, having a minimum of moving parts. Therefore, air controls eliminate much of the maintenance generally found with electrical devices. Burned-out solenoids, fire hazards from switches arcing, and wear of moving parts are common problems with electrical controls.

Directional controls may also be actuated with hydraulic fluid obtained directly from the system or from an auxiliary circuit. In the circuit shown in Fig. 7-53, as the cylinder reciprocates, the pilot directional valve is operated by mechanical means to supply fluid to either end of the main four-way valve. This causes the machine to reciprocate automatically, and it will not stop until the pump is shut off. Automatic cycling of a machine is accomplished with controls that allow the electric motor to operate continuously because of the high-starting-torque requirements (Fig. 7-54).

Directional control valves with a center position allow the machine to stop without shutting down the entire system.

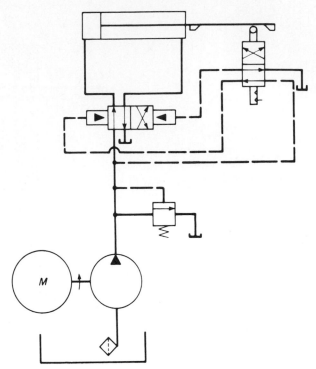

FIGURE 7-53 Circuit diagram.

Valves with three or more positions provide flexibility for starting, stopping, reversing, holding, positioning, speed control, stabilizing, and many other operational characteristics.

Typical center positions are shown in Fig. 7-55. The most common varieties are the open; closed; tandem; A, B to R; and P to A, B. A variety of center configurations provides great flexibility for circuit design.

In an open-center circuit, all ports are open to each other in the center position (Fig. 7-56). When a three-position, open-center-type valve is used in a hydraulic circuit, pump flow is directed to the reservoir when the valve is in the center. The other two positions are used to reciprocate the cylinder (see Fig. 7-57).

Open-center valves help to prevent heat buildup in the system by allowing the pump flow to go back to the reservoir at a minimum pressure during the idle time of a machine. However, it should be understood that the hydraulic fluid from the pump always follows the path of least resistance, and no work can be done by any part of the system depending on the pump so long as the valve remains in the center position. This means that pilot valves or other auxiliary devices using pressure energy cannot function. Air pilot or separate hydraulic pilot sources are often used with open-center systems.

In a closed-center circuit, all ports are closed to each other in the center position (Fig. 7-58). When a three-position, closed-center-type, four-way valve is used in a hydraulic circuit, pump flow must go over the relief valve when it is not being used to move a cylinder or fluid

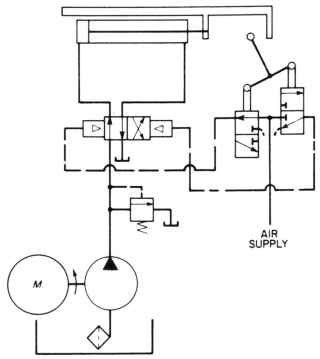

FIGURE 7-52 Hydraulic four-way operated valve with air controls.

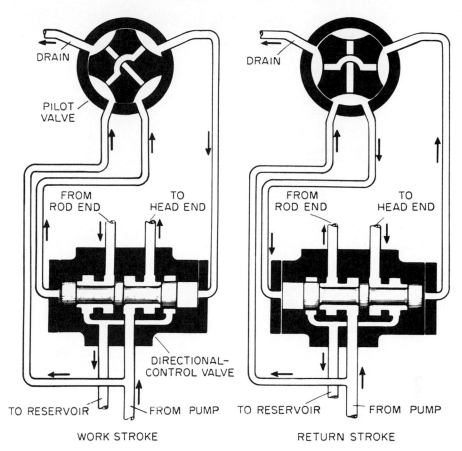

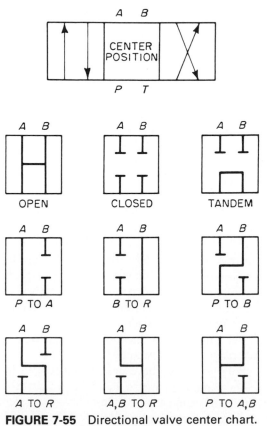

FIGURE 7-54 Method of shifting the directional control valve using a rotary pilot valve which is cam-operated by machine members. (Courtesy of Mobil Oil Corporation.)

FIGURE 7-55 Directional valve center chart.

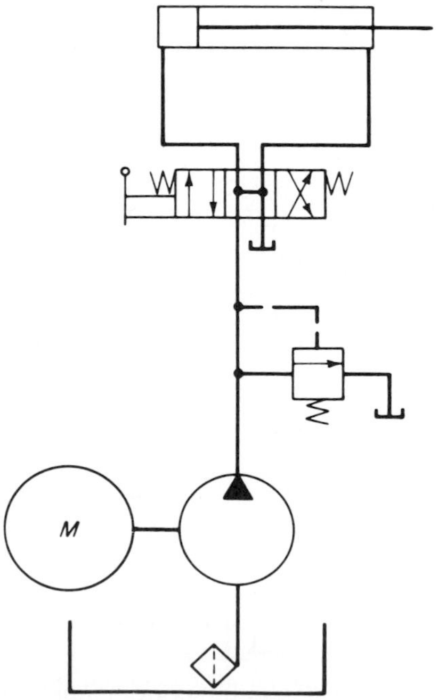

FIGURE 7-56 Open-center circuit.

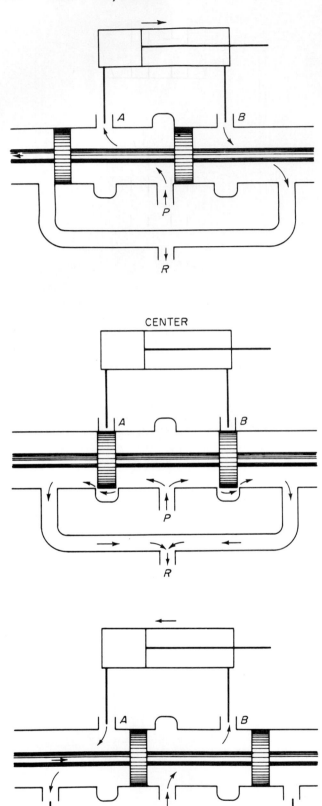

FIGURE 7-57 Three-position open-center valve: hydraulic circuit.

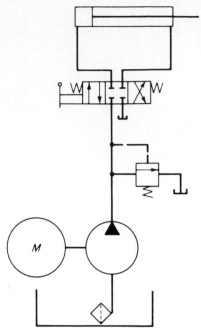

FIGURE 7-58 Closed-center circuit.

motor. A variable-displacement pressure-compensated pump must automatically reduce delivery rate under the same circumstances.

Closed-center versions are used when multiple circuits or functions must be accomplished from one hydraulic power source. There are many techniques used to prevent heat buildup when closed-center valves are used. Venting or unloading through a pressure control is one method which was discussed previously. Variable-displacement pumps with pressure-actuated controls are another economical method of avoiding heat and horsepower loss.

Tandem center valves (Fig. 7-59) direct the pump flow out the reservoir port with the other two working ports closed when in the center position. The application of this design may be to hold a cylinder or fluid motor under load or to permit the pump flow to be connected to a series of valves for multiple circuitry.

In addition to the flexibility in types of centers, directional valves are built with many other special features, such as integral restrictors, load-holding check valves, and relief valves (see Fig. 7-60). They are also available as banked units providing several valves in one package. Banked valves are commonly used on mobile equipment, where several circuits are operated from a central hydraulic power source (see Figs. 7-61 through 7-64).

Three- and four-way directional controls are available in a wide range of sizes and pressures.

Spool-type valves have been manufactured in sizes above 3-in. nominal pipe size and for pressures in excess of 3000 psi. The maintenance and installation considerations often lead to a decision to use a multiplicity of poppet valves

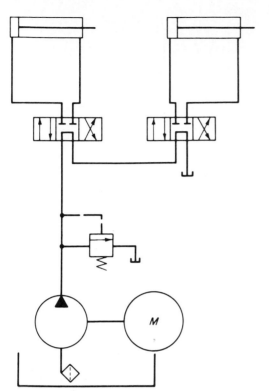

FIGURE 7-59 Tandem center four-way valves.

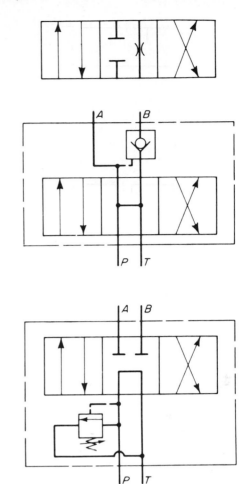

FIGURE 7-60 ANSI symbols of integral restrictor, load holding check valve and relief valve.

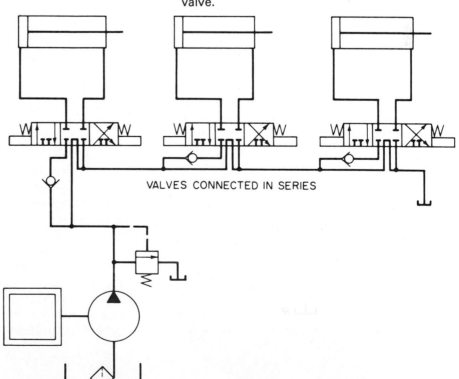

VALVES CONNECTED IN SERIES

FIGURE 7-61 Valves banked for multiple circuitry.

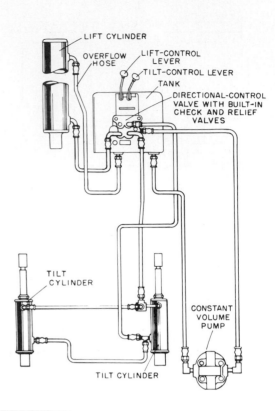

FIGURE 7-62 Lift truck and circuit showing directional control valve with built-in relief and load-handling check valve. (Courtesy of Mobil Oil Corporation.)

FIGURE 7-63 Backhoe and front-end loader. (Courtesy of Formrite Tube.)

FIGURE 7-64 Banked solenoid valves. (Courtesy of Racine Hydraulics Division, Dana Corporation, Racine, Wis.)

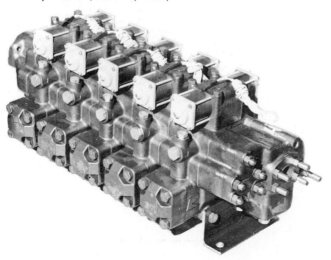

to provide the needed function as an option for flow needs above approximately 100 gpm or above 2-in. nominal pipe size. Poppet valves are precisely controllable and tolerate pressures above 3000 psi. A wide range of pilot controls are available. The term *logic circuitry* is often given to the use of a multiplicity of poppets with appropriate pilot circuitry to perform large flow, high-pressure power transmission tasks. More information on logic circuitry appears in Chapter 9.

The flow capacity of a directional control valve depends on the design type and the size of ports. Most generally, valves are selected on the basis of their port size. However, pressure drop across the working ports of a valve is important. If the valve is too small for the flow, excessive pressure drop causes power losses. If the valve is too large for a given flow, it adds excessive cost and more size and weight to the circuit.

Pressure drop is generally determined by the valve designer through actual bench testing. These data are available from the manufacturer.

A number of the control valve manufacturers use the flow coefficient Cv to express valve capacity and flow characteristics. The Cv coefficient of a directional control valve is defined as the flow of water at 60°F in gallons per minute at a pressure drop of one pound per square inch across the valve.

The following formula based on the Darcy equation is generally useful for determining valve size.

$$Cv = \frac{Q}{\sqrt{\Delta P/SG}}$$

where Cv = capacity factor (definition above)

Q = gallons per minute

ΔP = pressure drop

Sp. gr. = specific gravity (water = 1 = 62.4 lb/ft³)

$$\left[sp.\ gr. = \frac{\text{fluid specific weight (lb/ft}^3)}{62.4} \right]$$

The flow in gallons per minute of any liquid having a viscosity close to that of water at 60°F can be determined as follows:

$$Q = Cv \sqrt{\Delta P \left(\frac{62.4}{\rho} \right)}$$

where ρ is the specific weight of fluid (lb/ft³).

The pressure drop ΔP can be computed from the same formula changed to

$$\Delta P = \frac{\rho}{62.4} \frac{Q^2}{(Cv)^2}$$

There have been many improvements in valve designs which have increased the flow capacity of controls so that smaller pipe ports can be used. Table 7-1 can serve as a guide for typical size and capacity of valves

TABLE 7-1
VALVE SIZE AND CAPACITY

Valve (port) pipe size (in.)	Valve capacity nominal rating (gpm)
$\frac{1}{8}$	Up to 2
$\frac{1}{4}$ and $\frac{3}{8}$	Up to 8
$\frac{1}{2}$ and $\frac{3}{4}$	Up to 20
$\frac{3}{4}$ and 1	Up to 40
$1\frac{1}{4}$ and $1\frac{1}{2}$	Up to 85
2	Up to 180
3	Up to 320

(if in doubt, consult the manufacturer.) This chart will be considered by many as ultraconservative. Many valves are rated for much higher flows and pressures because of careful attention by the designer to flow forces and shock characteristics.

The advent of the proportional solenoid actuator for pressure, flow, and directional control valves as well as for various pump controls has had a distinct impact on the fluid power industry. A typical digital solenoid pilot valve (Fig. 7-65) can be actuated by ac or dc solenoids. Once energized, the solenoid shifts the valve spool quickly. Rapid change of pressurized fluid flow is subject to shock conditions of a magnitude related to the flow and pressure involved. Body 1 is fitted with spool 3. Springs (4) return the spool to center when the solenoids (2) are deenergized. Push pin 5 transmits linear force from armature 6 to spool 3.

A similar direct-current solenoid pilot valve is manifolded to a piloted body assembly as shown in Fig. 7-66a. It could also be an alternating-current solenoid.

Piloted spool 2 is fitted in a bore in valve body 1. Spring 3 returns the spool to neutral when the pilot pressure is relaxed.

The pilot valve (4) of Fig. 7-66a is fitted with a spool which blocks pilot pressure in neutral and directs both end caps to tank. This allows spring 3 to mechanically center the piloted spool.

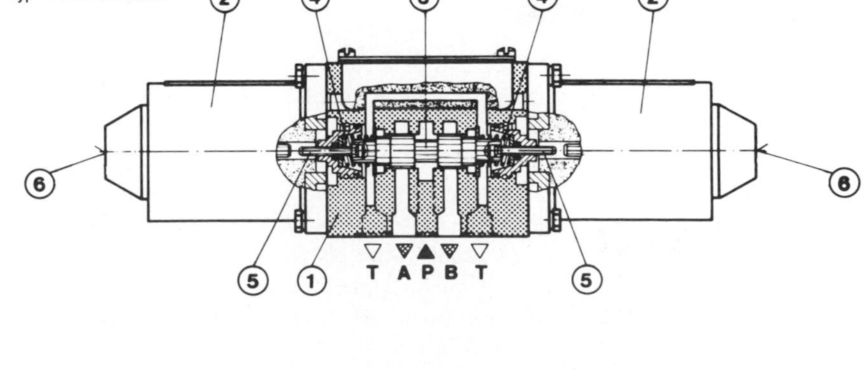

FIGURE 7-65 Closed-center, spring-centered double-solenoid-actuated four-way valve. (Courtesy of Rexroth Corporation, Bethlehem, Pa.)

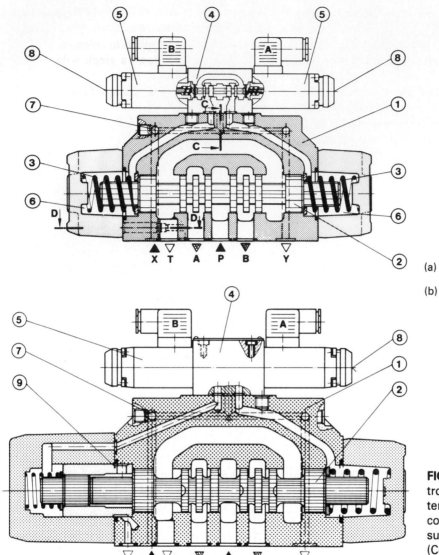

(a)

(b)

Type 4 WEH 25..H..50/..

FIGURE 7-66 (a) Solenoid-controlled, pilot-operated, spring-centered four-way valve; (b) solenoid-controlled, pilot-operated, pressure-centered four-way valve. (Courtesy of Rexroth Corporation, Bethlehem, Pa.)

The pilot valve (4) in Fig. 7-66b is fitted with a spool which blocks the return to tank port and connects pressure to both end caps. Piston assembly 9 is bigger in diameter than spool 2. The piston (9) is urged to neutral, where it is physically stopped by a shoulder in the valve spool bore. The small piston concentrically nested in assembly 9 can shift the spool to the right when pressure is relaxed in the right spring cavity and pilot pressure is applied to the left spring cavity. This directing of pilot pressure is a function of the pilot spool in valve 4. Similarly, the end of spool 2 can move piston assembly 9 to the left when pilot pressure is relaxed in the left cavity and applied to the right cavity. When the pilot spool is centered by deenergizing solenoids 5, the pilot spool again centers. With pressure to both end cavities of piloted spool 2 there is an action to urge piston assembly 9 to the right. It can only go as far as the neutral position because of the shoulder stop. The small concentric

piston is of lesser area than spool 2, so that it is blocked by the force generated against the end of spool 2. Note the drain connected to the cavity between the face of piston assembly 9 and the left end of spool 2. This takes care of the area differential between piston assembly 9 and spool 2. This drain, entitled L, must pass freely to the reservoir.

Internal or external pilot connections can be provided through line 7 to the pilot body. An orifice may be found in this line used to control the shifting speed of piloted spool 2. Line X is normally designated as pilot pressure source. Line Y is normally designated as pilot drain. Line Y is often returned to tank independently so that pressures in the tank line cannot accidentally shift the main spool (2). A manual push assembly on the pilot valve (8) is used for machine-setup purposes.

Spool 3 of Fig. 7-67a within body 1 is machined to provide a smooth gradual flow of liquid from input P to A

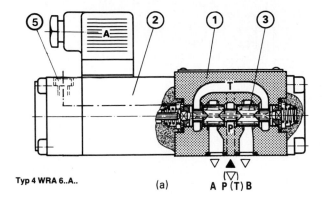

Typ 4 WRA 6..A..   (a)   A P (T) B

or B and from A or B to tank. The valve shown in Fig. 7-67b uses two solenoids. Springs 4 return the spool to neutral when no energy is supplied to solenoid 2.

Figure 7-68a shows a single-solenoid, spring-offset design which uses one solenoid to position the spool as it pushes against the bias spring (5) to provide the desired flow and direction of flow. The attached LVDT (linear variable differential transformer) in Fig. 7-68a can provide information as to spool position, which in turn will determine direction and rate of flow of the controlled liquid.

Proportional solenoid 2 is matched to springs 4 of Fig. 7-67b. Full current to the coil of the direct-current propor-

FIGURE 7-67 (a) Single-solenoid four-way valve with proportional solenoid actuation; (b) double-solenoid four-way valve with proportional solenoid actuation. (Courtesy of Rexroth Corporation, Bethlehem, Pa.)

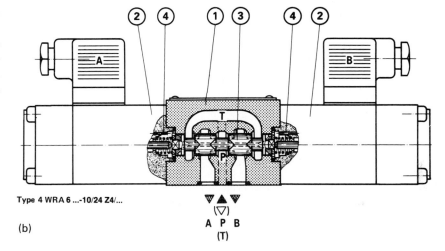

Type 4 WRA 6 ...-10/24 Z4/...

(b)   A P B (T)

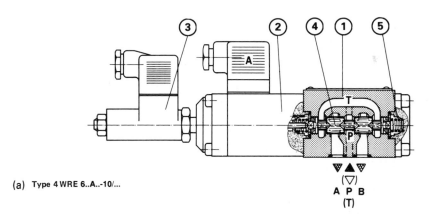

(a)   **Type 4 WRE 6..A..-10/...**   A P B (T)

(b)

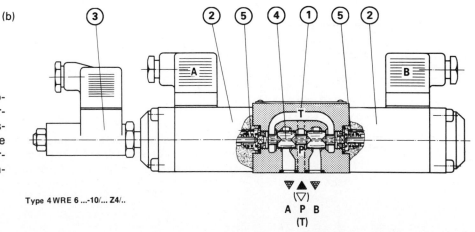

FIGURE 7-68 (a) Proportional solenoid with linear variable differential transformer feedback assembly; (b) double-solenoid valve with feedback assembly. (Courtesy of Rexroth Corporation, Bethlehem, Pa.)

Type 4 WRE 6 ...-10/... Z4/..   A P B (T)

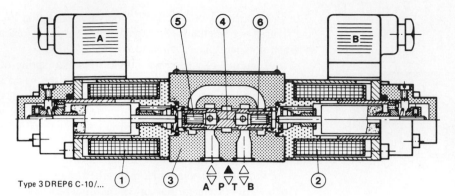

Type 3 DREP6 C-10/...

FIGURE 7-69 Double-proportional-solenoid pilot valve with hydraulic feedback. (Courtesy of Rexroth Corporation, Behtlehem, Pa.)

tional solenoid (2) shifts the spool (3) full stroke. The current flow to solenoid 2 can be controlled electronically as to the distance spool 3 is shifted and the time frame in which the shift is accomplished.

Valves of Fig. 7-68 include a linear variable differential transformer (LVDT) (3) attached to the proportional solenoid structure so that as the solenoid armature is moved it is sensed in the transformer, and thus the precise position of the spool can be electronically monitored and repositioned if necessary to provide a desired flow rate.

The valve of Fig. 7-69 employs a force feedback system to position the spool (4). Pistons 5 and 6 are fitted into spool 4. The cavity at the inner end of piston 5 is connected to port A. The cavity at the end of piston 6 is connected to port B. Current flowing to solenoid coil 1 causes the armature to push on piston 5, moving spool 4, directing a pressurized flow to port B. The load at port B is reflected

on piston 6 and will urge spool 4 back toward neutral at a predetermined pressure level. Energy to coil 2 moves spool 4 to direct fluid to port A and to piston 5, which functions in a similar manner.

A valve such as that illustrated in Fig. 7-69 can be assembled to a piloted body as shown in Fig. 7-70a. Spring 9 creates the load as solenoid 1 or 2 is subjected to a controlled current flow.

Piloted spool 8 will move in the bore of body 7 a distance proportional to the energy supplied to solenoid 1 or 2. Decay of signal source will allow spring 9 to center the spool (8) at the rate of signal decay to the solenoid.

Any pressure signal source can supply fluid to the end cavity of the spool (8) to move the spool and compress spring 9. Thus a proportional relationship is established wherein flow rate and directional control is infinitely variable.

Acceleration and deceleration rates are easily con-

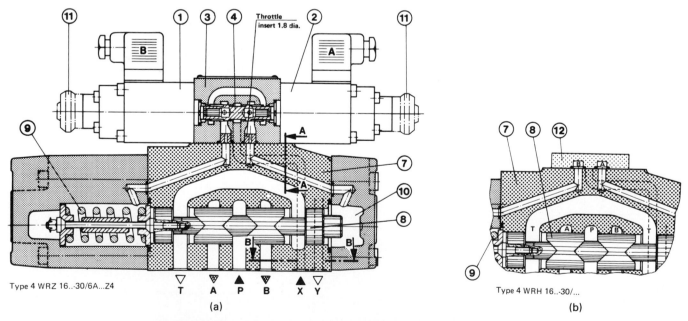

Type 4 WRZ 16..-30/6A...Z4

(a)

Type 4 WRH 16..-30/...

(b)

FIGURE 7-70   (a) Double-proportional-solenoid pilot valve and piloted spring-centered four-way valve; (b) with interface cap to permit pilot from remote source. (Courtesy of Rexroth Corporation, Bethlehem, Pa.)

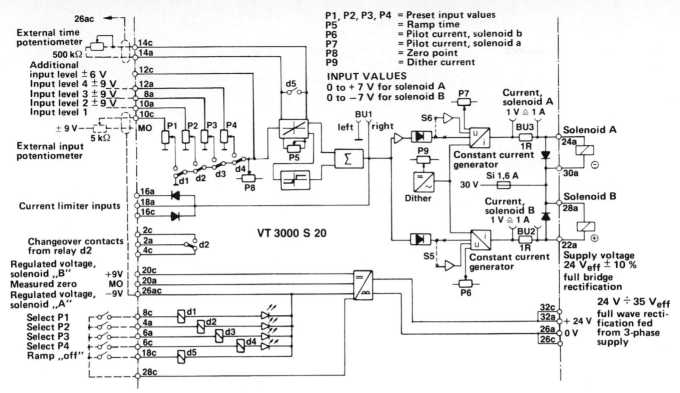

FIGURE 7-71 Interface card to match external signal source to the proportional solenoids. (Courtesy of Rexroth Corporation, Bethlehem, Pa.)

trolled and shock loading can be minimized or virtually eliminated as machine members can be accelerated or decelerated at the natural harmonic frequency values. Figure 7-70b shows provision for pilot.

A typical electronic interface card circuit is shown in Fig. 7-71. Ramp time for acceleration and deceleration is easily programmed. Typical voltage is 24 V dc.

FLOW CONTROL VALVES

A flow control valve has the primary function of controlling the rate of flow. Basically, the amount of fluid delivered to a given actuating cylinder or fluid motor determines the speed in which it will do the work.

When fixed-delivery pumps are used for a hydraulic system, the size of the pump is selected to provide a maximum of speed. Flow control devices are used to obtain any amount less than maximum speed. This means that fluid energy developed in the hydraulic system that cannot be used for useful work may be converted into heat. See Table 7-2 flow control formulas.

EXAMPLE

An actuating cylinder must extend 12 in. in the first second and 6 in. in the next second. What is the maximum gpm needed if the cylinder has a 2-in. bore?

TABLE 7-2
FORMULAS FOR APPLYING FLOW CONTROLS

p = pressure setting, maximum pressure valve (psi)

p' = pressure, cylinder inlet (psi)

p'' = pressure, cylinder outlet (psi)

V = velocity, piston (in./min)

Q = discharge, pump (in.^{3}/min)

q = flow through flow control valve (cu./in./min)

q' = discharge out of maximum pressure valve (cu./in./min)

A = area, piston, inlet side, (in^2.)

A' = area, piston, outlet side, (in.^{2})

F = work load, opposing motion (lb)

l = pressure drop at maximum pressure valve (psi)

l' = pressure drop at flow control valve (psi)

Meter-in	Meter-out	Bleed-off
$p' = \dfrac{F}{A}$	$p' = p$	$p' = \dfrac{F}{A}$
$p'' = 0$	$p'' = \dfrac{pA - F}{A'}$	$p'' = 0$
$V = \dfrac{q}{A}$	$V = \dfrac{q}{A'}$	$V = \dfrac{Q - q}{A}$
$q' = Q - q$	$q' = Q - \dfrac{Aq}{A'}$	$q' = 0$
$l = p$	$l = p$	no flow
$l' = p - p'$	$l' = p''$	$l' = p'$

Note: Formulas based on 0 pressure drop through piping and open valves.
Source: Courtesy of Vickers, Incorporated, Troy, MI.

Solution:

$$\text{Area} = 0.7854 \times \text{diameter}^2$$

$$= 3.14 \text{ in.}^2$$

$$Q \text{ (gpm)} = \frac{A \times \text{length (in.)} \times 60 \text{ sec}}{231 \times \text{time (sec)}}$$

$$\text{gpm to extend first second} = \frac{3.14 \text{ in.}^2 \times 12 \text{ in.} \times 60 \text{ sec}}{231 \text{ in.} \times 1 \text{ sec}}$$

$$= 9.7$$

$$\text{gpm for next 1 second} = \frac{3.14 \text{ in.} \times 6 \text{ in.} \times 60 \text{ sec}}{231 \text{ in.}^3 \times 1 \text{ sec}}$$

$$= 4.8 \text{ gpm/sec}$$

Thus, it can be seen that a 9.7-gpm pump would be needed for the maximum speed during the 12-in. stroke, whereas only 4.8 gpm would be needed for the 6-in. stroke. The pump would be selected on the basis of the 9.7 gpm. Therefore, at the end of the first second, 4.9 gpm would have to be diverted from the system, or a pressure-compensated pump, if used, would need to reduce output flow accordingly.

The speed of the cylinder for the 6-in. stroke can be controlled in three different ways:

1. Meter the fluid flow into the cylinder.
2. Meter the fluid flow out of the cylinder.
3. Bleed off the excess fluid flow and discharge it back to the reservoir.

If the meter out (2) is used, fluid could be discharged through the control with a maximum pressure drop. In other words, the fluid leaving the system via the control valve may have maximum pressure on one side and atmospheric pressure on the reservoir side, depending on load conditions. Metering the fluid into the cylinder (1) will reflect system pressure on the inlet and work load resistance on the outlet.

Flow and pressure drop across a control are related, and as flow in the system increases, pressure drop increases as the square of the flow changes. The reciprocal is also true in that as the pressure drop increases, flow increases in proportion to the square root of the change in pressure. When flow versus pressure drop is plotted for a fixed orifice, the curve in Fig. 7-72 results.

In spite of the inefficiencies in applying flow control devices, they are useful and important fluid power components. Variable-volume pumps can only be used as a flow control means if a pump is limited to one controlled cylinder or motor. If several flow control valves are used in parallel, a compensated variable-delivery pump will automatically adjust to the needed volume if it has sufficient capacity to actuate the multiple units at maximum value. If two or more

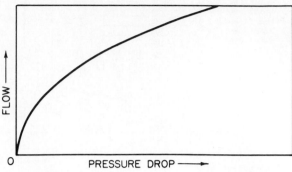

FIGURE 7-72 Flow versus pressure drop across a fixed orifice.

operations are being done at the same time, a separate power source is needed for each operation. A fixed-delivery pump sized for the maximum speeds could supply several circuits, with each actuator speed controlled by a flow device.

Restrictors

Restricting devices may be an orifice plate or a check valve with a small hole drilled through it. These devices are classified as noncompensated flow controls (see Fig. 7-73).

In some cases, simple on-off valves are drilled with a small hole to provide an orifice control (see Fig. 7-74). These are the simplest types of flow control devices. They are used to create a fixed rate of pressure drop to operate pilot valves, to stabilize a system with a slight back pressure, to prevent control overrides, and for decompressing systems under high pressure.

A needle valve is a variable restrictor device which allows the orifice size to vary by adjustment (see Fig. 7-75). Some designs incorporate a check valve and a needle valve which provides adjustable flow in one direction and free flow in the opposite direction. Flow control devices,

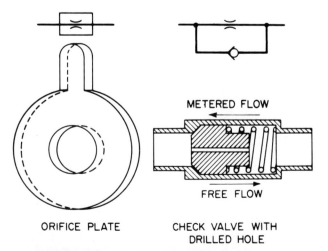

ORIFICE PLATE CHECK VALVE WITH DRILLED HOLE

FIGURE 7-73 Restricting devices: ANSI symbol, orifice plate, and check valve.

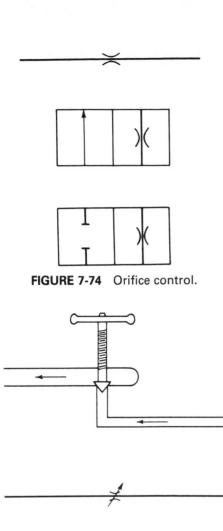

FIGURE 7-74 Orifice control.

CONTROLLED FLOW

FREE FLOW
FIGURE 7-75 Needle valve.

cuits are operated by a central hydraulic power source. Flow controls are used to help balance the entire system and to counter for uneven loads which normally cause fluid energy to seek the path of least resistance. They are also used to prevent gravity loads from dropping too rapidly.

Pressure-Compensated Flow Controls

Because flow through an orifice or valve varies as the square root of the pressure drop across it, any change in pressure at the control inlet or outlet changes flow through the valve. Pressure-compensated flow controls adjust automatically to pressure changes and maintain a constant pressure drop from inlet to outlet, thus providing constant flow.

Principle of Pressure-Compensated Flow Control

Assume that the pump capacity in Fig. 7-76 is 5 gpm and that the orifice is sized to pass 3 gpm. When the load is zero, 3 gpm is going into the system and 2 gpm is going back to the tank over the bypass valve. Since the bypass poppet has 1 in.^{2} of area and the spring has a force of 50, gauge A will read 50 psi. Gauge B will read zero because of no load. There will be a 50-psi pressure drop across the metering orifice.

If the system load increased, causing gauge B to read 100 psi, this same increase in pressure would be added to the bypass valve spring through the pilot line. This means that gauge A would now have to read 150 psi to bypass the extra 2 gpm back to the tank. Since the pressure difference between gauge A and gauge B remains the same (50 psi pressure drop), the flow across the metering orifice remains the same. Thus any increase or decrease in load will either

such as the needle valves and the combination needle and check valves, are used for controlling cylinder and fluid motor speeds where small adjustments are essential. However, when the work load in a system varies, system pressure also varies with each change in resistance, and precise accuracy in speed control cannot be maintained. This is because flow increases with pressure drop across the fixed orifice, and each time the system pressure changes the flow also changes. In some applications these designs are used to obtain equal speeds with two or more cylinders or fluid motors.

Noncompensated flow controls are also used on some mobile equipment where higher pressures and low volume are required. Generally, in mobile applications several cir-

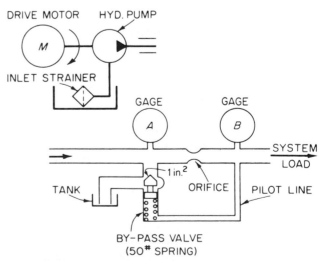

FIGURE 7-76 Principle of pressure compensation.

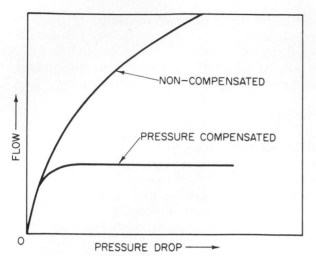

FIGURE 7-77 Curves comparing pressure drop.

add or subtract to the spring, maintaining constant pressure drop and flow across the metering orifice (Fig. 7-77).

One type of pressure-compensated flow control regulates or restricts fluid flow entering the valve. The other type bypasses the excess fluid back to the reservoir. In both types a constant pressure drop across the metering orifice provides constant flow (see Fig. 7-78).

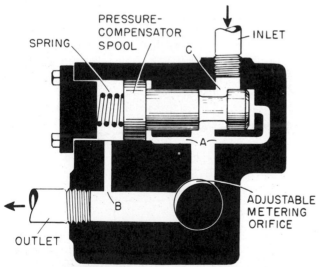

FIGURE 7-78 Restrictor-inlet-type pressure-compensated flow control.

Fluid entering the control at the inlet port is restricted at the orifice. Pressure builds up, causing the compensator spool to move toward the spring force, which is generally about 40 lb. When the spool moves in the direction of the spring, fluid is restricted across the inlet port.

When fluid passes across the orifice, back pressure builds up on the spring side of the compensator and the compensator spool is forced to the right, which opens the restriction at the valve inlet.

Thus, the compensating spool is pressure-balanced, regardless of any changes in system pressure. Changes in pressure affect both sides of the compensator because of the equal areas. In this case the spring creates a constant pressure differential from the inlet to the outlet. In other words, the inlet pressure is always slightly higher, only because of the spring force.

The bypass type of flow control valve maintains constant flow across the metering orifice and bypasses excess fluid back to the reservoir at a pressure only slightly above the system pressure.

$$A_1 = A_2 + A_3$$

In this design type, shown in Fig. 7-79, the compensator spool is held closed by the spring in A_1. Fluid can still get past the spool and across the metering orifice to cause back pressure to build up on the A_1 side. Since the areas are equal on both ends of the compensator spool, it becomes pressure-balanced. However, since the spring force is relatively light, and the spool is pressure-balanced, a slightly greater force on the inlet side opens the compensator to bypass the excess fluid. With the compensator spool balanced, any change in system pressure does not affect the flow rate across the metering orifice. If the pressure on the outlet side exceeds the pilot relief valve setting, it opens and causes fluid to bleed off the spring side of the compensator, and with the spool unbalanced it opens wide to bypass the system fluid directly to the reservoir.

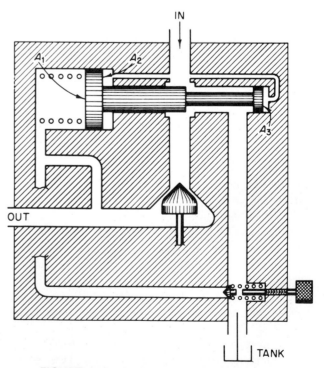

FIGURE 7-79 Bypass type of flow control.

This design has one main advantage over the restricted inlet type. The bypass design allows excess fluid to bypass back to the reservoir at a slightly higher pressure than the system's working pressure. However, the restricted inlet design causes the surplus fluid to bypass over the main relief valve at its pressure setting. This means that during periods when any portion of the pump delivery is restricted it blows across the main relief valve at its maximum pressure setting.

Metering-In Circuits

Metering-in circuits control the flow of fluid before it enters the working cylinder or fluid motor (see Fig. 7-80).

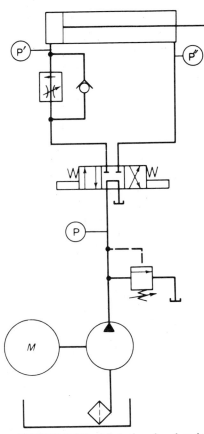

FIGURE 7-80 Metering-in circuit.

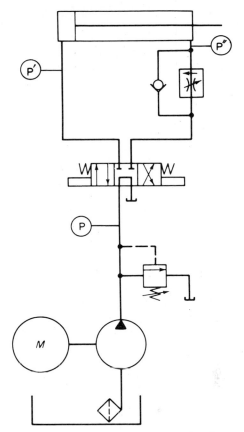

FIGURE 7-81 Metering-out circuit.

Metering-Out Circuits

Metering-out circuits control the flow exhausting from a working cylinder or fluid motor (see Fig. 7-81). Bleed-off circuits control the fluid by bleeding off the excess not needed by the working cylinder or fluid motor (see Fig. 7-82). The speed control circuit selected depends on the load characteristics of the cylinder or fluid motor.

Metering-in circuits are generally used when the load characteristics are constant and positive. If the load is erratic or negative, the cylinder or fluid motor will have jerky motion.

The bypass type of flow control is used for meter-in circuits and has an integral relief valve which provides protection between the work cylinder or motor and the control. This type of flow control is much more efficient than the inlet restricting type for meter-in, because the bypass feature allows fluid to be exhausted to the reservoir at just slightly higher pressure than that necessary to do the work. With the inlet restricting type, pump delivery not used would discharge over the main relief valve at maximum pressure.

Metering-in circuits are used on surface grinders, welders, milling machines, and other machine tools where fine speed control is essential. Bypass meter-in controls are usable only with one pump and one basic actuator because of the integrated relief valve type of compensation.

Metering-out circuits are best where negative loads may occur, because back pressure is maintained on the exhaust side of the actuator, preventing erratic motion. This type of speed control circuit operates satisfactorily for drilling, boring, reaming, or tapping operations.

When cylinders with large piston rods are used at high pressure, intensified back pressures may occur. Relief valve protection should be used for such conditions.

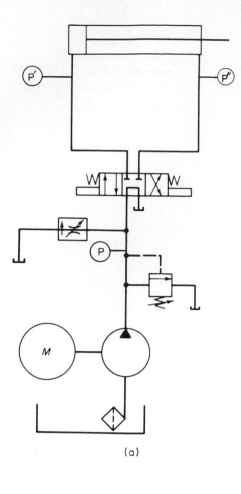

(a)

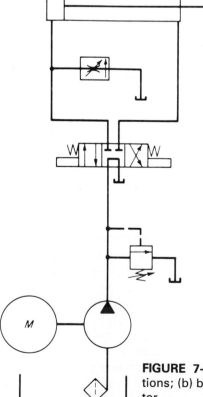

FIGURE 7-82 (a) Bleed-off for both directions; (b) bleed-off for inlet to cylinder or motor.

(b)

$$P = \frac{F}{A}$$

$$= \frac{30,000}{5}$$

$$= 6000 \text{ psi}$$

Figure 7-83 shows that an intensification may result in very high pressures on the rod side of cylinders; if it is not protected with a relief valve, a serious accident might occur. If a relief valve is installed ahead of a meter-out flow control valve, it may negate the flow control function. The use of a meter-in flow control is mandatory in many circuits because of this factor.

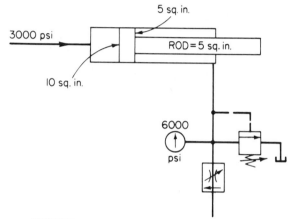

FIGURE 7-83 Intensified back pressure.

Bleed-off circuits provide less accuracy in speed control, because they do not compensate for any change in fluid losses due to pressure change. In other words, pump slip increases with pressure, and so do other internal losses which are not sensed by the compensator of a flow control used with bleed-off circuits.

This method of speed control does have an advantage with regard to efficiency, because system pressure is only high enough to accomplish the work and fluid does not discharge over the main relief valve.

Deceleration Flow Control

A deceleration flow control valve gradually reduces flow rate to provide deceleration of heavy loads (see Fig. 7-84). In Fig. 7-85, the deceleration valve gradually restricts the flow from the rod end of the cylinder to slow down the heavy moving table as the machine starts the feed phase of the cycle. The angle of the cam on the end of the cylinder piston rod and the designed features of the valve determine the time to decelerate.

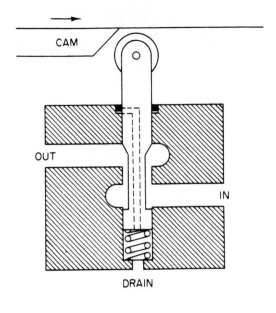

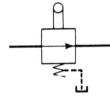

FIGURE 7-84 Deceleration valve.

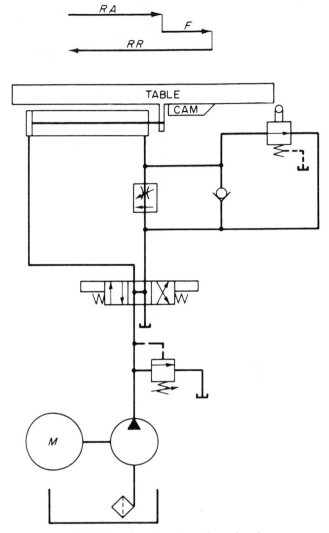

FIGURE 7-85 Deceleration circuit.

There is a tremendous amount of flexibility in hydraulic components to control speeds, feeds, positions, pressures, and various motions. The discussion in this chapter has barely scratched the surface.

Hydraulic pumps are designed with flow adjustment mechanisms as integral parts of the pump itself. Directional controls are available with metering orifices built into any given flow passage through the valve. One combination control device might have a directional control, a flow control, and other features all built into one package.

QUESTIONS

1. What are the three important elements that are controlled in a hydraulic system?
2. What is the function of a relief valve in a hydraulic system?
3. What are the two general types of relief valves?
4. Which type of relief valve is used as a means of unloading a pump during the idle time of machine?
5. What rule generally applies to the pressure setting of a relief valve?
6. What is the function of a sequence valve?
7. What is the general rule for draining pressure control valves?
8. What is the function of an unloading valve?
9. Is an unloading valve directly or remotely pilot-controlled?

10. What is the advantage of using an unloading circuit when feeds and speeds of a machine must be varied?
11. What is the function of a counterbalance valve?
12. When is a pressure-reducing valve used in a hydraulic system?
13. Where is pilot oil taken to operate or control a pressure-reducing valve?
14. What are directional control valves used to operate other control valves called?
15. What is the simplest type of directional control valve?
16. Name three types of on-off valves.
17. What is the purpose of the pilot-operated check valve?

18. When are three-way valves used in a hydraulic circuit?

19. How are directional control valves actuated?

20. What three general designs of four-way valves are available?

21. What is the difference between a two-position and a three-position four-way valve?

22. Illustrate the various center configurations that are standard with most four-way three-position directional control valves.

23. Explain the difference between a closed-center system and an open-center system.

24. In general, what are the maximum operating pressures for spool-type valves? What are the maximum operating pressures for shear-seal-type valves?

25. How is pressure drop across directional control valves determined?

26. What are the three ways of applying flow control valves?

27. What is meant when a flow control valve is said to be *pressure-compensated*?

28. What is the difference between a bypass type of flow control valve and a restricted inlet type of flow control valve?

29. How are bypass flow control valves used in a circuit?

30. What is the purpose of a deceleration valve?

8

Actuators Provide Flexibility in the Use of Fluid Power

Fluid power can provide either linear or rotary motion through the use of actuators, called *cylinders,* and fluid motors. A tremendous advantage of using fluid power rather than mechanical systems is the fact that it can be applied directly to the work itself. In recent years fluid power has replaced the old-fashioned factory system of shaft and belt drives, where steam engines provided the input power. This was a dangerous hazard to operators and workers and caused many accidents, some with the loss of arms and legs. Fluid power is doing many of the jobs formerly done by mechanical linkage, such as mechanical brakes, chain drives, drive shafts, cams, and other devices.

The flexibility of distributing fluid power through conductors and directly to the actuator does not affect the geometry of the machine. This gives excellent freedom for the design of equipment.

Generally, there is no need for heavy, cumbersome mechanical linkage between the power source and the work. Actuating cylinders and fluid motors can be directly attached to the work, and the power can be piped in to meet any force, speed, or positioning requirement. Fluid power can be transmitted and controlled to provide the correct force and speed for any job from simple to complex.

ACTUATING CYLINDERS

A hydraulic cylinder is a device that converts fluid power into linear mechanical force and motion. It usually consists of a movable element, such as a piston and piston rod, plunger or ram, operating within a cylindrical bore.

The operating principle of the piston and piston rod type is that fluid entering one port drives the movable piston and rod assembly in one direction as shown in Fig. 8-1. Fluid from the opposite side is exhausted back to the reservoir through a directional control valve.

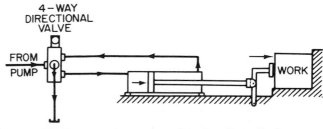

FIGURE 8-1 Function of hydraulic cylinder.

When the directional control valve is shifted to the opposite position, fluid then enters the cylinder port at the rod end, pushing the piston and rod assembly back again as shown in Fig. 8-2. Fluid exhausting from the large area side is directed back to the reservoir again through the four-way valve.

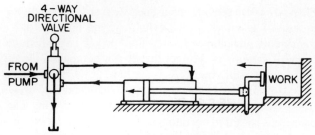

FIGURE 8-2 Function of hydraulic cylinder retract.

CYLINDER VELOCITY

The velocity of a cylinder as it extends or retracts depends on how fast the *oil* is fed into the cylinder and on the *area* of the cylinder. With 10 gpm being delivered to each of the cylinders in Fig. 8-3, the illustration shows the length of travel in 1 minute (231 in.^{3} = 1 gal). Velocity varies directly with gpm. Velocity varies inversely with the area. Mathematically, this may be expressed as

$$V = \frac{gpm}{A}$$

Substituting units into the formula, we have

$$V \text{ (ft/min)} = \frac{gpm}{in.^2}$$

In order to get the correct units on both sides of the equation, convert gallons to cubic inches (231 in.^{3}/gal), and inches to feet (1 ft/12 in.). Now,

$$V \text{ (ft/min)} = \frac{gal/min \times 231 \ in.^3/gal \times 1 \ ft/12 \ in.}{in.^2}$$

Grouping and simplifying, we have

$$V \text{ (ft/min)} = \frac{gal/min}{in.^2} \times \frac{231 \ in.^2\text{-ft}}{12 \ gal} = \frac{gpm}{A} \times 19.25$$

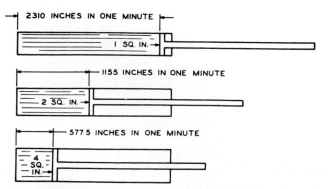

FIGURE 8-3 Length of travel in 1 minute.

The factor 19.25 in.^{2} ft/gal is a constant that can be used for finding the velocity of any cylinder in feet per minute. If feet per second is required, the formula is V (ft/sec) = gpm $\times$ 0.3208/A and is derived in the same manner as the previous formula.

EXAMPLE

Ten gpm is being fed into a cylinder whose area is 5 in.^{2}. What is the velocity of the cylinder in feet per minute?

Solution:

$$V \text{ (ft/min)} = \frac{gpm \times 19.25}{A} = \frac{10 \times 19.25}{5} = 38.5 \text{ ft/min}$$

The velocity in feet per second is

$$V \text{ (ft/sec)} = \frac{gpm \times 0.3208}{A} = \frac{10 \times 0.3208}{5}$$

$$= 0.64 \text{ ft/sec}$$

Being mathematical expressions, these formulas may be expressed in different forms.

Formula:

$$V \text{ (ft/min)} = \frac{gpm \times 19.25}{A}$$

Other forms:

$$gpm = \frac{V \text{ (ft/min)} \times A}{19.25}, \qquad A = \frac{gpm \times 19.25}{V \text{ (ft/min)}}$$

Formula:

$$V \text{ (ft/sec)} = \frac{gpm \times 0.3208}{A}$$

Other forms:

$$gpm = \frac{V \text{ (ft/sec)} \times A}{0.3208}, \qquad A = \frac{gpm \times 0.3208}{V \text{ (ft/sec)}}$$

EXAMPLE

A cylinder 12 in. long with a cross-sectional area of 19.25 in.^{2} contains 231 in.^{3} of hydraulic fluid (see Fig. 8-4).

$$\text{Volume} = \text{area} \times \text{length}$$

$$= 19.25 \times 12$$

$$= 231 \text{ in.}^3$$

$$(231) \text{ in.}^3 = 1 \text{ gal})$$

This indicates that if a hydraulic pump delivers 1 gpm it will take exactly 1 minute to move the piston assembly 12

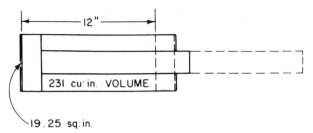

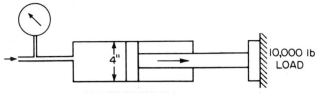

FIGURE 8-4   Diagram for example.

FIGURE 8-6   Diagram for example.

in. If the pump delivery is 2 gpm, it would take only 30 sec. A pump delivering 60 gpm would require only 1 sec to move the piston assembly 12 in.

resistance against the cylinder rod causes fluid pressure to oppose the pump flow as shown in Fig. 8-6.

EXAMPLE

If the resisting force against the piston rod is 10,000 lb, how much pressure will be developed in the system using a 4-in.-diameter cylinder?

$$\text{Speed (ft/min)} = \frac{\text{gpm} \times 19.25}{\text{area}}$$

This formula can be used to calculate cylinder piston speed in feet per minute.

Solution:

$$\text{Pressure} = \frac{\text{force}}{\text{area}}$$

$$\text{Area of cylinder} = 0.7854D^2$$

$$= 0.7854 \times 16$$

$$= 12.56 \text{ in.}^2$$

$$\text{Pressure} = \frac{10,000 \text{ lb}}{12.56 \text{ in.}^2}$$

$$= 796 \text{ psi}$$

EXAMPLE

If the rod diameter in the previous cylinder is 2 in., what will the return speed of the piston assembly be if the pump delivers 10 gpm (see Fig. 8-5)?

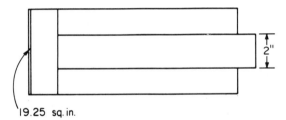

FIGURE 8-5   Diagram for example.

The ram- or plunger-type actuating cylinders have a movable element of the same cross-sectional area as the piston rod (see Fig. 8-7). The ram or plunger cylinder is generally

Solution:

Find the rod area.

$$\text{Area} = 0.7854D^2$$

$$= 0.7854 \times 4$$

$$= 3.1416$$

Subtract the rod area from the cylinder area.

$$19.25 \text{ in.}^2 - 3.14 \text{ in.}^2 = 16.11 \text{ in.}^2$$

$$\text{Speed (ft/min)} = \frac{\text{gpm} \times 19.25}{A}$$

$$= \frac{10 \times 19.25}{16.11}$$

$$= 12.9 \text{ ft/min}$$

The volume of the cylinder and pump delivery determine the speed of a piston assembly, but the area and pressure determine the amount of force being applied. Work or

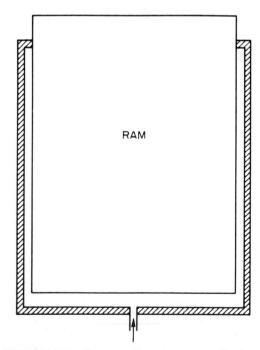

FIGURE 8-7   Ram or plunger-type cylinder.

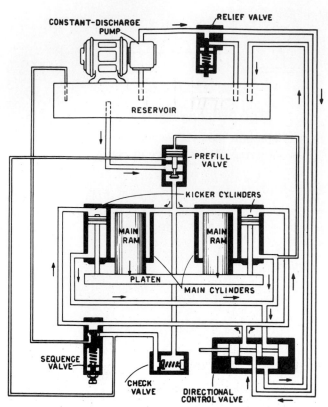

FIGURE 8-8 Press. (Courtesy of Mobil Oil Corporation.)

powered in only one direction. Auxiliary cylinders called "kicker cylinders" are sometimes used to retract the ram. This arrangement is shown in Fig. 8-8. They can also be retracted by gravity or weighted machine members if applied in the inverted position. These designs are available for creating tremendous forces such as those needed in large presses.

Cylinders may also be used for precisely clamping delicate objects with controlled force, as shown in Fig. 8-9. Hydraulic cylinders are manufactured to produce a few ounces of force or thousands of tons. Some metal fabricating presses used for aircraft parts are equipped with large cylinders that produce up to 35,000 tons of force.

In addition to the high force capabilities, cylinders give high power per weight and size, mounting flexibility, and excellent speed control. They are available in a wide range of sizes, strokes, and pressure ranges.

Cylinders may be single-acting, double-acting, or telescoping. Single-acting cylinders (see Fig. 8-10) use fluid power only in one direction. The return stroke uses a gravity load or a spring retracting force. Single-acting cylinders with spring return are sometimes used for clamping operations, small presses, and other such applications where little or no load is subjected on the return stroke (see Fig. 8-11).

Gravity return cylinders are used on mobile equip-

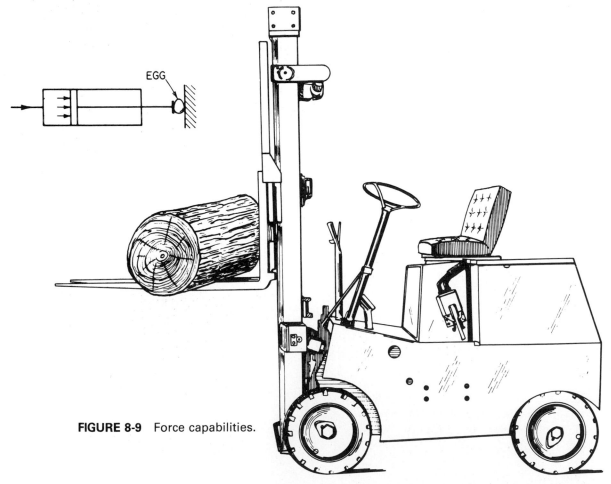

FIGURE 8-9 Force capabilities.

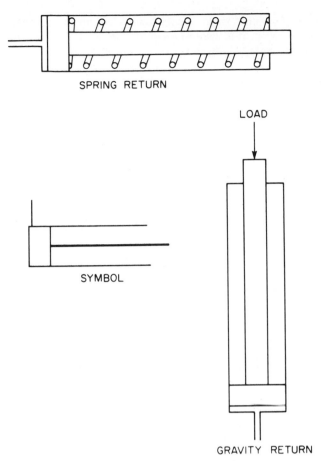

FIGURE 8-10   Single-acting cylinders.

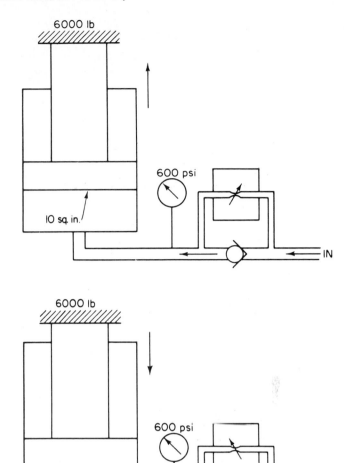

FIGURE   8-12   Vertical   load–single-acting (gravity return).

ment, farm implements, and other machine applications controlling vertically suspended loads. Fluid power entering the cylinder raises the load, and fluid under control leaving the cylinder lowers the load. Figure 8-12 shows that fluid pumped into the cylinder causes it to rise, and fluid metered out of the cylinder allows the load to drop. A pressure slightly higher than 600 psi would be needed to raise the load because of friction and other factors. Fluid metered out through the flow control valve would allow the cylinder to retract.

Double-acting cylinders use fluid power in both directions and may be of either the single- or double-rod types (see Fig. 8-13). Double-acting cylinders of the single-rod type are by far the most popular design. This type generally uses the large-area side to develop the greatest power requirement in a reciprocating circuit.

EXAMPLE

If the cylinder in Fig. 8-14 has a piston area of 4 in.^{2} and is fitted with a piston rod having 2 in.^{2}, what is the difference in force that can be developed between the two sides if the relief valve is set for 2000 psi?

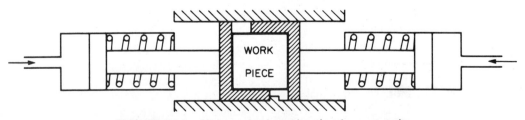

FIGURE 8-11   Clamp–single-acting (spring return).

SINGLE ROD END

DOUBLE ROD END

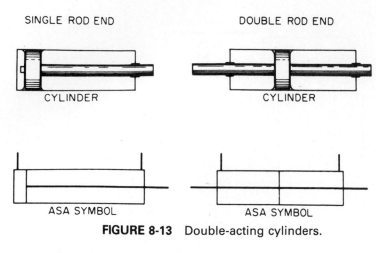

FIGURE 8-13   Double-acting cylinders.

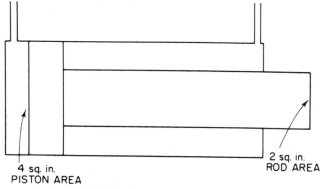

FIGURE 8-14   Example problem.

Solution:

Force extending:

$$\text{Force} = \text{pressure} \times \text{area}$$
$$= 2000 \times 4$$
$$= 8000 \text{ lb}$$

Force retracting:

$$\text{Force} = \text{pressure} \times \text{area}$$
$$= 2000 \times 2 \text{ in.}^2$$
$$= 4000 \text{ lb}$$

Thus, it is apparent that the piston end without the rod has the most effective area and is capable of twice the force. However, it is also true that the piston and rod assembly will travel twice the speed retracting as it will extending.

When one is selecting tubing and component sizes it is very important that the difference in volumes exhausting from this type of cylinder during extending and retracting be taken into consideration. Cylinders with large piston rods return rapidly, exhausting large quantities of fluid that may cause excessive back pressures and heat buildup. This indicates the importance of a separate drain line for sequence

valves or other such components. When cylinders of this type extend, it is also possible to intensify pressure within the rod end if fluid is restricted or metered out of that end.

However, a volume differential cylinder of this type generally conforms to almost all of the important requirements of any circuit. The cylinder's rod size can be varied to conform to side loads, tensional and compressional stresses, or other factors. The cylinder bore can be varied according to the greatest force requirement. There are essentially no limiting factors, because the blind end is solid metal and can be made with tremendous strength.

A cylinder with a single piston and a piston rod extending from each end is a double-acting, double-rod type (see Fig. 8-15). This design of cylinder can provide the same force and speed in both directions. They are generally applied to machines where work is being done in both directions and at equally controlled speeds. This design of cylinder is also used as a metering cylinder where the fluid is directed to another actuator for controlling speed or position.

Telescoping cylinders (see Fig. 8-16) are used where long work strokes are needed. A telescoping cylinder has nested multiple tubular rod segments which provide a long working stroke in a short retracted envelope. It operates on the principle that the rod with the largest area gives a greater force at the least pressure and moves out first; the next largest moves at a slightly higher pressure; the next at a pressure higher yet; and so on for as many stages as the unit may contain. The symbol for a telescoping cylinder is shown in Fig. 8-17.

Other special design types are shown in Fig. 8-18. Where very long strokes are needed, this cylinder design uses a cable which provides a long pull with a relatively short stroke. Fluid entering port B ports extends one or both piston assemblies, which stroke the cable a distance determined by the pulley arrangement. Fluid entering ports A and C retracts the piston assemblies.

Another cylinder design uses three cylinders nested together for greater strength with a long stroke (see Fig. 8-19). Cylinder ports labeled A extend all three piston assemblies, and ports labeled B retract piston assemblies. Positional or tandem designs have two or more cylinders with interconnecting piston assemblies (Fig. 8-20).

Fluid entering, leaving, or holding at any port or combination of ports can actuate piston assemblies to the various positions. Greater force can be exerted when both piston assemblies are moving in contact with the work.

The positional-type cylinder in Fig. 8-21 uses a moving cylinder barrel, one moving piston assembly, and one stationary piston assembly. This design allows a full stroke, which might be used to place the work into a machine, and then a half stroke for repeating a series of work strokes. This technique is sometimes used on welding machines, where a wide stroke is necessary to insert the workpiece into the welder. The short stroke is then used to make a

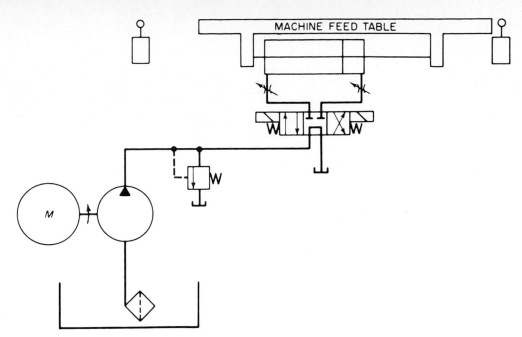

FIGURE 8-15 Double-acting, double-rod-type cylinder.

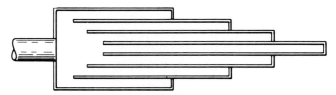

TELESCOPING CYLINDER

FIGURE 8-16 Telescoping cylinder.

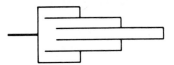

SYMBOL FOR
TELESCOPING CYLINDER

FIGURE 8-17 Symbol for telescoping cylinder.

FIGURE 8-18 Cylinder with cable.

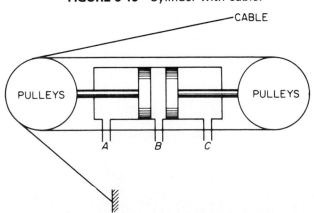

CABLE

PULLEYS   PULLEYS

A   B   C

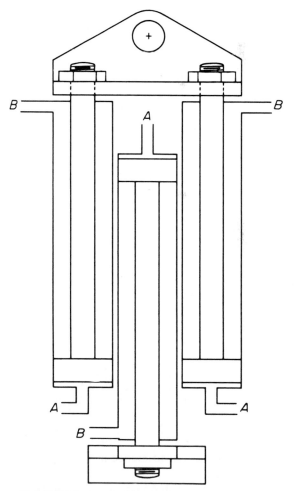

FIGURE 8-19 Three cylinders nested together.

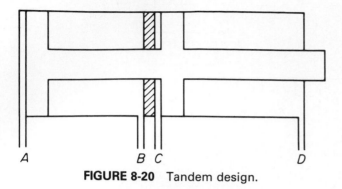

FIGURE 8-20 Tandem design.

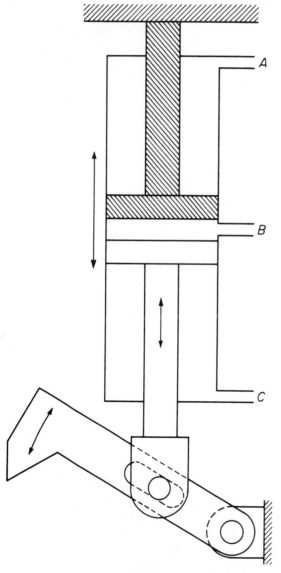

FIGURE 8-21 Positioning-type cylinder; moving barrel and lower piston; fixed upper piston.

number of welds on the work. Figure 8-22 shows popular body styles of cylinders.

Figure 8-23 is a standard cylinder identification coding system which is used by practically all cylinder manufacturers. Specific dimensions can be compared for interchangeability or for ordering replacement cylinders.

International standards groups such as ISO and U.S. groups are standardizing many fluid power devices, one of which is fluid power cylinder characteristics and measuring techniques. The U.S. standards association is the American National Standards Institute (ANSI). The members of the National Fluid Power Association (NFPA) in the United States and other interested engineers and manufacturers have combined their talents to provide these standards for fluid power devices. Figure 8-24 shows recommended bore and piston rod sizes for fluid power cylinders. Figure 8-25 provides information as to typical cylinder mounting methods.

As we refer to forces affecting a fluid power cylinder, we can see in Fig. 8-26 what we are describing. Figure 8-27 illustrates some methods of applying linear motion using cylinders and how to determine resultant loads.

SEALS FOR CYLINDERS

Proper seals are important to prevent leakage and loss of hydraulic power. The sealing material must be compatible with the hydraulic fluid and the operating temperatures. The seals must be applied to prevent seal "breakoff," which contaminates the system. They should also be easy to replace.

Sealing devices are used in cylinders to seal piston and piston rod assembly, end caps to barrel, and for the piston rod sealing gland (see Fig. 8-28). Rod scrapers or wipers should be used to prevent abrasives or dirt from damaging the rod gland seal and getting into the system.

CUSHIONING

For the prevention of shock due to stopping loads at the end of the piston stroke, cushion devices are used. Cushions may be applied at either end or both (see Fig. 8-29a). They operate on the principle that as the cylinder piston approaches the end of the stroke, exhaust fluid is forced to go through an adjustable needle valve which is set to control the escaping fluid at a given rate. This allows the deceleration characteristic to be adjusted for different loads.

When the cylinder piston is actuated, fluid enters the cylinder port and flows through the little check valve so that the entire piston area can be utilized to produce force and motion.

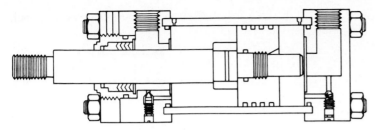

TIE ROD CONSTRUCTION

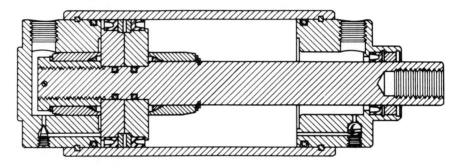

ONE PIECE WITH RETAINER

MILL TYPE CONSTRUCTION

THREADED CONSTRUCTION

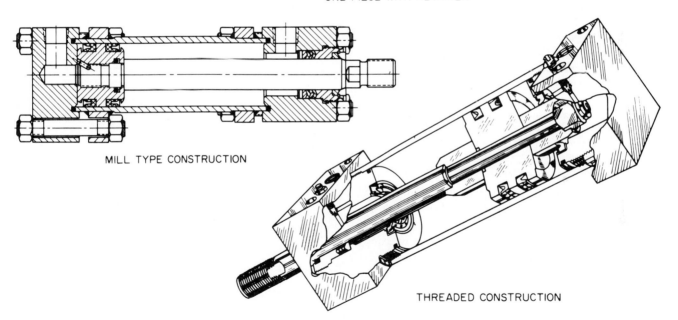

FIGURE 8-22 Cylinder body styles.

FIGURE 8-23 Standard cylinder identification code.

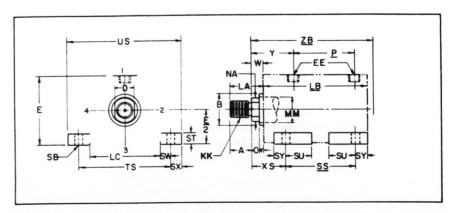

RECOMMENDED BORE AND PISTON ROD SIZE FOR FLUID POWER CYLINDERS
(Nominal)

Bores (in.)		Piston rod diameters (in.)		
1	4½	$\frac{5}{16}$	1¼	4
1⅛	5	$\frac{3}{8}$	1⅜	4½
1¼	6	$\frac{1}{2}$	1½	5
1½	7	$\frac{5}{8}$	1¾	5½
2	8	$\frac{11}{16}$	2	6
2½	10	$\frac{3}{4}$	2¼	7
3	12	$\frac{7}{8}$	2½	8
3¼	14	1	2¾	9
3½	16	1⅛	3	10
4	20		3½	

FIGURE 8-24 Standard bore and rod sizes.

CYLINDER MOUNTING CHART

A. Fixed centerline mounts

Tie rod end
Blind end flange
Rod end flange
Centerline lugs

B. Fixed noncenterline mounts

End lug mounts
Side lug mounts
Integral key mounts
Flush mounts

C. Pivoted centerline mounts

Rod end trunnion
Blind end trunnion
Clevis trunnion
Center trunnion

A. Fixed centerline mounts are used for thrusts that occur linearly or in a centerline with the cylinder. Proper alignment is essential to prevent compound stresses that may cause excessive friction and binding as the piston rod extends. Additional holding strength may be essential with long stroke cylinders.

B. Fixed noncenterline mounts are convenient where exceptionally heavy linear thrusts are encountered. Generally, integral keys or pins are used if excessive hydraulic shock is expected. This helps to relieve shear loads. Since the cylinder has to expand and contract with temperature change, only one end should be keyed or pinned.

C. Pivoted centerline mounts are used to compensate for thrusts occurring in multiple planes or if the attached load travels in a curved path. Ball joints, trunnions, and clevis mounts allow thrusts to be taken up along cylinder centerlines.

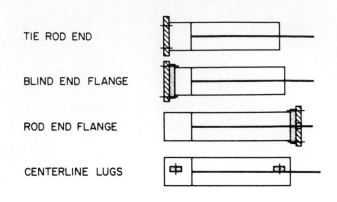

(a) FIXED CENTERLINE MOUNTS

TIE ROD END

BLIND END FLANGE

ROD END FLANGE

CENTERLINE LUGS

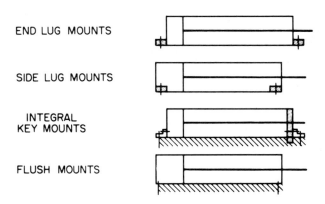

(b) FIXED NON-CENTERLINE MOUNTS

END LUG MOUNTS

SIDE LUG MOUNTS

INTEGRAL KEY MOUNTS

FLUSH MOUNTS

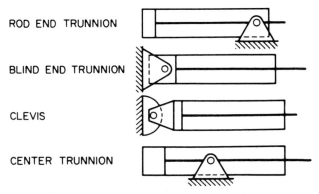

(c) PIVOTED CENTERLINE MOUNTS

ROD END TRUNNION

BLIND END TRUNNION

CLEVIS

CENTER TRUNNION

FIGURE 8-25 Cylinder mounting charts.

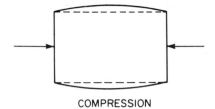

COMPRESSION

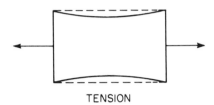

TENSION

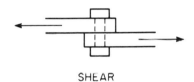

SHEAR

FIGURE 8-26 Forces affecting a cylinder.

M = slugs

W = weight (lb)

S = distance (in.)

t = time (sec)

a = acceleration (in./sec^2)

V = velocity (in./sec)

Decelerating distance:

$$d = V_0 t - \tfrac{1}{2}at^2$$

where V_0 is the initial velocity.

ACCELERATION AND DECELERATION FORCE DETERMINATION

The uniform acceleration force factor chart (see Fig. 8-29b) and the accompanying formulas can be used rapidly to determine the forces required to accelerate and decelerate a cylinder load. To determine these forces, the following factors must be known: total weight to be moved, maximum piston speed, distance available to start or stop the weight (load), direction of movement (i.e., horizontal or vertical), and load friction. By use of the known factors and the g factor from the chart, the force necessary to accelerate or decelerate a cylinder load may be found by solving the formula applicable to a given set of conditions. (See Table 8-1.)

Nomenclature

V = velocity (ft/min)

S = distance (in.)

F = force (lb)

W = weight of load (lb)

g = force factor

f = friction of load on machine ways (lb)

To determine the force factor g from Fig. 8-29b, locate the intersection of the maximum piston velocity line and the line representing the available distance. Project downward to locate g on the horizontal axis.

METHOD OF SOLVING FOR LINEAR ACCELERATION AND DECELERATION

$$F = MA$$

$$S = \tfrac{1}{2}at^2$$

$$V = at$$

Combining the above gives

$$S = \frac{1V^2}{2a}$$

$$a = \frac{V^2}{2S}$$

$$F = \frac{W}{g} \times \frac{V^2}{2S} = \frac{MV^2}{2S}$$

Force F acts over distance S to start and stop mass M to or from velocity V.

$$g = acc^2 \text{ of gravity}$$

$$F = \text{force (lb)}$$

HORIZONTAL CASE

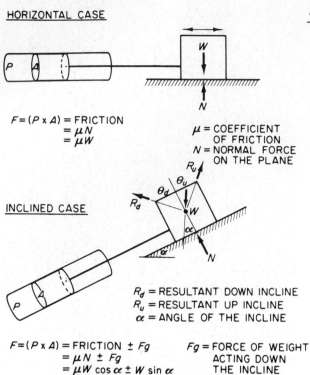

$F = (P \times A) = $ FRICTION
$= \mu N$
$= \mu W$

$\mu = $ COEFFICIENT OF FRICTION
$N = $ NORMAL FORCE ON THE PLANE

INCLINED CASE

$R_d = $ RESULTANT DOWN INCLINE
$R_u = $ RESULTANT UP INCLINE
$\alpha = $ ANGLE OF THE INCLINE

$F = (P \times A) = $ FRICTION $\pm Fg$
$= \mu N \pm Fg$
$= \mu W \cos\alpha \pm W \sin\alpha$
$= \dfrac{W \sin(\theta \pm \alpha)}{\cos\theta}$

$Fg = $ FORCE OF WEIGHT ACTING DOWN THE INCLINE

$\tan\theta = \mu$

OR $\theta = \tan^{-1}\mu$

(a)

VERTICAL CASE

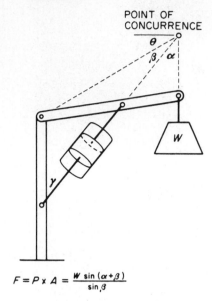

$F = (P \times A) = W$

(b)

JIB CRANE

POINT OF CONCURRENCE

$F = P \times A = \dfrac{W \sin(\alpha + \beta)}{\sin\beta}$

NOTE: AS THE JIB RAISES TO VERTICAL, F APPROACHES W; AS THE ANGLE γ BECOMES LARGER, F APPROACHES ∞

(c)

"LEVER" JIB CRANE

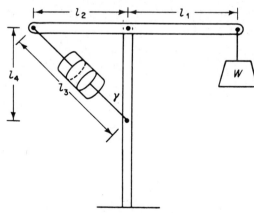

$F = P \times A = \dfrac{l_1}{l_2} W$ SECANT $\gamma = \dfrac{l_1}{l_2} \dfrac{l_3}{l_4} W$

NOTE: AS γ BECOMES SMALLER F APPROACHES $l_1/l_2 W$

AS γ BECOMES LARGER F APPROACHES ∞

(d)

BENT LEVER

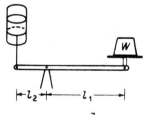

$F = P \times A = \dfrac{l_1}{l_2} W$

(e)

FIRST – CLASS LEVER

$F = P \times A = \dfrac{l_1}{l_2} W$

(f)

FIGURE 8-27 Various methods of applying linear motion using cylinders.

HORIZONTAL RACK AND PINION

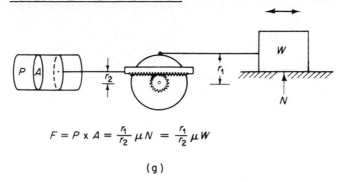

$$F = P \times A = \frac{r_1}{r_2} \mu N = \frac{r_1}{r_2} \mu W$$

(g)

WEDGE

FORCE ACTING THROUGH INCLINE:

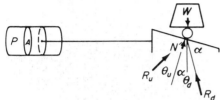

$$F_i = P \times A = \text{TAN}(\theta \pm \alpha) W$$

R_u = RESULTANT UP INCLINE $\qquad$ α = ANGLE OF THE INCLINE

R_d = RESULTANT DOWN INCLINE $\qquad$ θ = $\tan^{-1} \mu_{\text{INCLINE}}$

μ = COEFFICIENT OF FRICTION

FORCE ACTING THROUGH SLIDING SURFACE:

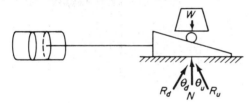

$$F_h = A \times P = \mu \frac{W}{\text{HORIZONTAL}}$$

$$\text{TOTAL } F = P \times A = \left[\tan(\theta \pm \alpha) + \mu_h \right] W$$

(h)

BLOCK AND TACKLE

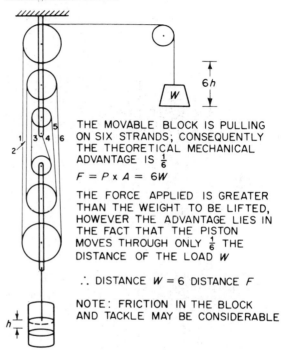

THE MOVABLE BLOCK IS PULLING ON SIX STRANDS; CONSEQUENTLY THE THEORETICAL MECHANICAL ADVANTAGE IS $\frac{1}{6}$

$$F = P \times A = 6W$$

THE FORCE APPLIED IS GREATER THAN THE WEIGHT TO BE LIFTED, HOWEVER THE ADVANTAGE LIES IN THE FACT THAT THE PISTON MOVES THROUGH ONLY $\frac{1}{6}$ THE DISTANCE OF THE LOAD W

$\therefore$ DISTANCE $W = 6$ DISTANCE F

NOTE: FRICTION IN THE BLOCK AND TACKLE MAY BE CONSIDERABLE

(i)

TOGGLE JOINT

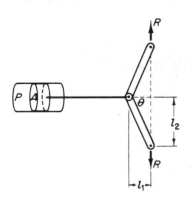

$$F = P \times A = \frac{2R}{\tan \theta} = 2 \frac{l_1}{l_2} R$$

NOTE: AS ANGLE θ BECOMES SMALLER, F APPROACHES ∞; AS ANGLE θ BECOMES LARGER, F APPROACHES ZERO

(J)

SECOND-CLASS LEVER

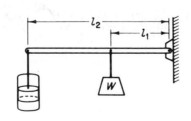

$$F = P \times A = \frac{l_1}{l_2} W$$

(K)

THIRD-CLASS LEVER

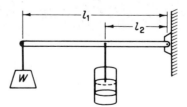

$$F = P \times A = \frac{l_1}{l_2} W$$

(L)

FIGURE 8-27 (continued).

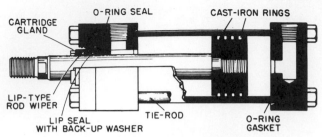

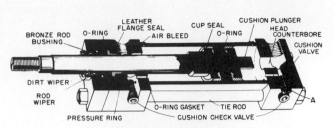

FIGURE 8-28 Cross section of cylinder rod gland; cutaway of a hydraulic cylinder; cross section showing seals. (Courtesy of Mobil Oil Corporation.)

EXAMPLE 1

Horizontal motion of a freely moving 6000-lb load is required with a distance of $\frac{1}{2}$ in. to a maximum speed of 120 ft/min. Formula (1), $F = Wg$, should be used.

$$F = 6000 \text{ lb} \times 1.50 \text{ (from chart)} = 9000 \text{ lb}$$

Assuming a maximum available pump pressure of 1000 psi, a 4-in. bore cylinder should be selected, operating on push stroke at approximately 750 psi pressure at the cylinder to allow for pressure losses from the pump to the cylinder.

Assume the same load to be sliding on ways with a coefficient of friction of 0.15. The resultant friction load would be $6000 \times 0.15 = 900$ lb. Formula (2), $F = Wg + f$, should be used.

$$F = 6000 \text{ lb} \times 1.5 \text{ (from chart)} + 900 = 9900 \text{ lb}$$

Again allowing 750 psi pressure at the cylinder, we find that a 5-in. bore cylinder is indicated.

EXAMPLE 2

Horizontal deceleration of a 6000-lb load is required by using a 1-in.-long cushion in a 5-in. bore cylinder having a 2-in.-diameter piston rod. Cylinder bore area (19.64 in.²) minus the rod area (3.14 in.²) results in a minor area of 16.5 in.² at the head end of cylinder. A 1000-psi pump delivering 750 psi at the cylinder is being used to push the load at 120 ft/min. The friction coefficient is 0.15, or 900 lb.

In this example, the total deceleration force is the sum of the force needed to decelerate the 6000-lb load, and the force required to counteract the thrust produced by the pump.

$$W = \text{load (lb)} = 6000$$

$$S = \text{deceleration distance (in.)} = 1 \text{ in.}$$

$$V = \text{maximum piston speed (ft/min)} = 120$$

$$g = 0.74 \text{ (from chart)}$$

$$f = 900 \text{ lb}$$

Use formula (3), $F = Wg - f$.

$$(F = Wg - f) = (F = 6000 \times 0.74 - 900) = 3540 \text{ lb}$$

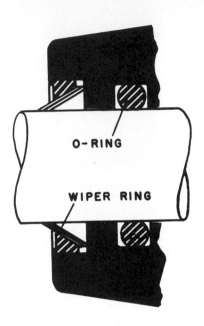

METALLIC ROD WIPER

The pump is delivering 750 psi acting on the 19.64-in.² piston area, producing a force (F_2) of 14,730 lb. This force must be included in our calculations. Thus $F + F_2 = 3540 + 14,730 + 18,270$ lb total force to be decelerated.

The total deceleration force is developed by the fluid trapped between the piston and the head. The fluid pressure is equal to the force (18,270 lb) divided by the minor area (16.5 in.²) equals 1107 psi. This pressure should not exceed the nonshock rating of the cylinder. Table 8-2 is useful for determining flow velocity and pressure data for hydraulic systems.

ROTARY ACTUATORS

Rotary actuators are devices for producing limited reciprocating rotary force and motion. They are used for lifting, lowering, opening, closing, indexing, and transferring movements (see Fig. 8-30).

When the fluid enters on one side it causes the movable element to rotate the shaft in one direction. The rotation is limited, depending on the specific type. Fluid entering the opposite port will reverse the actuator shaft. The single vane

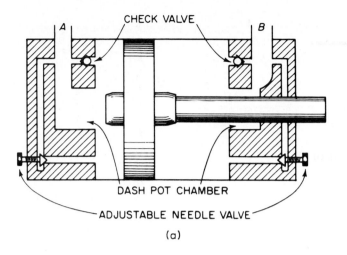

(a)

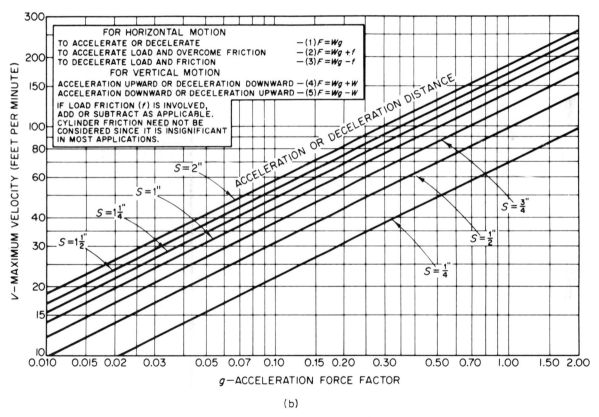

FOR HORIZONTAL MOTION
TO ACCELERATE OR DECELERATE   —(1) $F = Wg$
TO ACCELERATE LOAD AND OVERCOME FRICTION   —(2) $F = Wg + f$
TO DECELERATE LOAD AND FRICTION   —(3) $F = Wg - f$
FOR VERTICAL MOTION
ACCELERATION UPWARD OR DECELERATION DOWNWARD —(4) $F = Wg + W$
ACCELERATION DOWNWARD OR DECELERATION UPWARD —(5) $F = Wg - W$
IF LOAD FRICTION (f) IS INVOLVED,
ADD OR SUBTRACT AS APPLICABLE.
CYLINDER FRICTION NEED NOT BE
CONSIDERED SINCE IT IS INSIGNIFICANT
IN MOST APPLICATIONS.

(b)

FIGURE 8-29   (a) Cushioning; (b) acceleration force factor.

rotates through an arc of 280°. Power in the two-vane design is doubled, but the turning arc is reduced to 100°.

The opposed piston type produces a high turning force at low pressure. As the pistons extend or retract, they rotate the pinion gear, which is in mesh with each rack. It produces the same turning force in both directions.

The piston chain type has two cylinders in parallel.

The large piston is used for powering the chain and drive sprocket, while the smaller cylinder provides a seal. Fluid entering one end operates against both pistons, but the large piston creates the greater force, so it moves, causing rotation of the output shaft. Flow directed to the other port reverses the operation.

Rotary actuators are available for operating pressures

TABLE 8-1
THEORETICAL PUSH AND PULL FORCES FOR HYDRAULIC CYLINDERS

Push Force and Displacement

Cyl. bore size (inches)	Piston area (sq in.)	Cylinder push stroke force in pounds at various pressures										Displacement per inch of stroke (gallons)
		50	80	100	250	500	750	1000	1500	2000	2500	
1	.785	39	65	79	196	392	588	785	1177	1570	1962	.00339
1½	1.767	88	142	177	443	885	1325	1770	2651	3540	4425	.00765
2	3.14	157	251	314	785	1570	2357	3140	4713	6280	7850	.0136
2½	4.91	245	393	491	1228	2455	3682	4910	7364	9820	12275	.0213
3¼	8.30	415	664	830	2075	4150	6225	8300	12450	16600	20750	.0359
4	12.57	628	1006	1257	3143	6285	9428	12570	18856	25140	31425	.0544
5	19.64	982	1571	1964	4910	9820	14730	19640	29460	39280	49100	.0850
6	28.27	1414	2262	2827	7068	14135	21203	28270	42406	56540	70675	.1224
8	50.27	2513	4022	5027	12568	25135	37703	50270	75406	100540	125675	.2176

Deductions for Pull Force and Displacement

Piston rod dia. (inches)	Piston rod area (sq in.)	Piston rod diameter force in pounds at various pressures										Displacement per inch of stroke (gallons)
		To determine cylinder pull force or displacement, deduct the following force or displacement corresponding to rod size, from selected push stroke force or displacement corresponding to bore size in table above										
		50	80	100	250	500	750	1000	1500	2000	2500	
½	.196	10	16	20	49	48	147	196	294	392	490	.00085
⅝	.307	15	25	31	77	154	230	307	461	614	768	.0013
1	.785	39	65	79	196	392	588	785	1177	1570	1962	.0034
1⅜	1.49	75	119	149	373	745	1118	1490	2235	2980	3725	.0065
1¾	2.41	121	193	241	603	1205	1808	2410	3615	4820	6025	.0104
2	3.14	157	251	314	785	1570	2357	3140	4713	6280	7850	.0136
2½	4.91	245	393	491	1228	2455	3682	4910	7364	9820	12275	.0213
3	7.07	354	566	707	1767	3535	3502	7070	10604	14140	17675	.0306
3½	9.62	481	770	962	2405	4810	7215	9620	14430	19240	24050	.0416
4	12.57	628	1006	1257	3143	6285	9428	12570	18856	25140	31425	.0544
5	19.64	982	1571	1964	4910	9820	14730	19640	29460	39280	49100	.0850
5½	23.76	1188	1901	2376	5940	11880	17820	23760	35640	47520	59400	.1028

up to about 5000 psi. Shaft sizes, type of bearings, and cycle frequency are important factors to consider when rotary actuators are selected. They are generally mounted by foot, flange, and end mounts. Cushioning devices are also available in most designs.

FLUID MOTORS

A fluid motor is a device that converts fluid power into mechanical force and motion. It usually provides rotary mechanical motion. Fluid motors and pumps from the same manufacturer use interchangeable parts and look alike in many respects, but the motors work in a manner just op-

posite to the way in which the pumps work. They use the fluid delivered by the pump to provide a rotating force and motion.

There are two general classes of fluid motors, *fixed displacement* and *variable displacement*. In the fixed-displacement class, gear, vane, and piston designs are used. In the variable-displacement class, piston designs are used.

As the term implies, fixed-displacement fluid motors displace a specific amount of fluid for each revolution. Therefore, the speed of any fixed type of fluid motor depends on the displacement per revolution and the amount of fluid supplied to it by the pump.

A variable-displacement piston-type fluid motor is built with a device that can adjust the displacement per revolution.

TABLE 8-2
FLOW VELOCITY AND PRESSURE DATA FOR HYDRAULIC SYSTEMS

The chart below may be used to calculate pressure loss in connecting lines at various flow velocities. The data is useful when one is determining hydraulic cylinder size and port size for applications where cylinder force and speed requirements are known.

S = standard (schedule 40) pipe
H = extra strong (schedule 80) pipe
EH = double extra strong pipe

Tabulations based on a hydraulic oil having a viscosity of 155 SSU at 100° F.—specific gravity of 0.87.
To determine tubing or hose losses, use I.D. closest to tubing or hose I.D.
Pressure drop does not vary with operating pressure. Avoid high pressure losses in low pressure systems. Use largest pipe size practical. Avoid flow velocities greater than 15 ft/sec to reduce hydraulic line shock.

Pressure loss (pounds per square inch per foot length) in pipes at average flow velocity (feet per second) of — *Equivalent straight pipe length (feet) for circuit components**

Nominal size (In.)	Sch.	O.D. (In.)	I.D. (In.)	Wall thick-ness (In.)	I.D.-area (sq. in.)	5 Gal./min.	5 Loss	7 Gal./min.	7 Loss	10 Gal./min.	10 Loss	15 Gal./min.	15 Loss	20 Gal./min.	20 Loss	25 Gal./min.	25 Loss	30 Gal./min.	30 Loss	Tee	Tee	Elbow std.	Elbow sq.	45°
1/8	S	0.405	0.269	0.068	0.057	0.89	1.25	1.24	1.79	1.75	2.60	2.67	3.16	3.56	5.47	4.45	6.20	5.34	7.07	—	—	—	—	—
1/8	H		0.215	0.095	0.036	0.56	1.89	0.78	3.05	1.12	4.26	1.68	5.20	2.24	8.38	2.80	1.10	3.36	12.65	—	—	—	—	—
1/4	S	0.540	0.364	0.088	0.104	1.62	0.67	2.27	1.05	3.24	1.64	4.96	1.92	6.48	2.97	8.10	2.73	9.72	8.73	—	—	—	—	—
1/4	H		0.302	0.119	0.072	1.12	1.11	1.57	1.49	2.24	2.11	3.36	2.84	4.48	4.15	5.60	5.08	6.72	6.30	—	—	—	—	—
3/8	S	0.675	0.493	0.091	0.191	2.98	0.39	4.18	0.57	5.96	0.86	8.94	1.05	11.92	1.69	14.90	4.27	16.88	5.78	2.7	2.7	1.2	0.8	0.6
3/8	H		0.423	0.126	0.140	2.18	0.54	3.06	0.74	4.36	1.10	6.54	1.34	8.72	1.97	10.90	5.19	13.08	7.20	—	—	—	—	—
1/2	S	0.840	0.622	0.109	0.304	4.74	0.24	6.65	0.37	9.48	0.49	14.22	0.68	18.96	2.09	23.70	3.38	28.44	4.28	3.5	3.5	1.5	1.05	0.75
1/2	H		0.546	0.147	0.234	3.65	0.30	5.12	0.45	7.30	0.71	10.95	0.78	14.60	2.47	18.15	3.61	21.90	5.00	2.9	2.9	1.4	0.9	0.68
1/2	EH		0.252	0.294	0.050	0.78	1.54	1.09	2.19	1.56	3.08	2.34	3.65	3.12	6.13	3.90	7.48	4.68	9.55	—	—	—	—	—
3/4	S	1.050	0.824	0.113	0.533	8.32	0.14	11.65	0.22	16.64	0.27	24.96	0.78	33.28	1.47	41.60	2.19	49.92	3.00	4.5	4.5	2.1	1.4	1.0
3/4	H		0.742	0.154	0.432	6.74	0.16	9.45	0.26	13.48	0.37	20.22	0.87	26.96	1.71	33.70	2.48	40.44	3.52	4.0	4.0	1.6	1.2	0.8
3/4	EH		0.434	0.308	0.148	2.31	0.53	3.24	0.67	4.62	1.05	6.93	1.31	9.24	1.94	11.55	5.06	13.86	7.02	—	—	—	—	—
1	S	1.315	1.049	0.133	0.863	13.45	0.10	18.85	0.15	26.90	0.34	40.35	0.57	53.80	1.42	67.25	1.64	80.70	2.24	5.7	5.2	2.6	1.7	1.2
1	H		0.957	0.179	0.719	11.21	0.11	15.70	0.15	22.42	0.24	33.63	0.62	44.84	1.23	56.05	1.84	67.26	2.93	5.2	5.2	2.5	1.6	1.1
1	EH		0.599	0.358	0.282	4.39	0.26	6.16	0.37	8.78	0.53	13.17	0.67	17.56	2.25	21.95	3.29	26.34	3.30	3.0	3.0	1.5	1.0	0.75
1¼	S	1.660	1.380	0.140	1.496	23.35	0.05	31.68	0.08	46.70	0.25	70.05	0.39	93.40	0.78	116.75	1.18	140.10	1.47	7.5	7.5	3.7	2.4	1.6
1¼	H		1.278	0.191	1.280	19.95	0.07	28.06	0.09	39.90	0.26	58.85	0.44	79.80	0.85	99.75	1.27	119.70	1.80	7.0	7.0	3.5	2.1	1.5
1¼	EH		0.896	0.382	0.630	9.83	0.13	13.75	0.16	19.66	0.24	29.49	0.71	39.32	1.35	49.15	2.01	58.98	2.76	4.9	4.9	2.3	1.5	1.05
1½	S	1.900	1.610	0.145	2.036	31.75	0.04	44.49	0.11	63.50	0.19	95.25	0.33	127.00	0.64	158.75	0.96	190.50	1.26	9.0	9.0	4.3	2.8	2.0
1½	H		1.500	0.200	1.767	27.55	0.08	38.62	0.08	55.10	0.21	82.65	0.36	110.20	0.71	137.75	1.06	165.30	1.36	8.2	8.2	4.0	2.6	1.8
1½	EH		1.100	0.400	0.950	14.81	0.09	20.75	0.09	29.62	0.32	44.43	0.51	59.24	1.05	74.05	1.51	88.86	2.14	6.5	6.5	3.0	2.0	1.4
2	S	2.375	2.067	0.154	3.355	52.30	0.04	73.45	0.08	104.60	0.14	156.90	0.24	209.20	0.48	261.50	0.69	313.80	0.85	11.0	11.0	5.5	3.5	2.5
2	H		1.939	0.218	2.953	46.00	0.03	64.60	0.09	92.00	0.15	138.00	0.26	184.00	0.52	230.00	0.73	276.00	0.98	10.8	10.8	5.0	3.4	2.4
2	EH		1.503	0.436	1.773	27.65	0.04	38.78	0.12	55.30	0.21	82.95	0.36	110.60	0.72	138.25	1.34	165.90	1.36	8.2	8.2	4.0	2.6	1.8
2½	S	2.875	2.469	0.203	4.788	74.75	0.03	104.80	0.07	149.50	0.11	224.25	0.20	299.00	0.37	373.75	0.53	448.50	0.72	14.0	14.0	6.5	4.7	3.0
2½	H		2.323	0.276	4.238	66.11	0.04	92.60	0.07	132.22	0.12	198.33	0.21	264.44	0.39	330.55	0.57	396.66	0.87	13.0	13.0	6.1	4.0	2.9
2½	EH		1.771	0.552	2.464	38.45	0.03	53.40	0.10	76.90	0.17	115.35	0.30	153.80	0.59	192.25	0.79	230.70	1.15	10.3	10.3	4.8	3.1	2.2

*Consult valve manufacturer for pressure drops in a particular type of valve and port-to-port flow pattern.

Courtesy of Parker-Hannifin Corp.

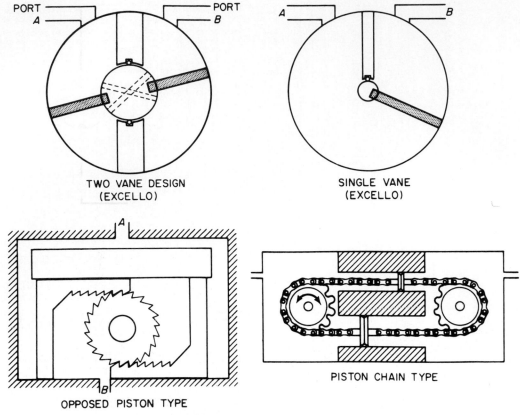

FIGURE 8-30 Rotary actuators.

The speed of these designs can be adjusted over a range from zero to the maximum speed for the size of the unit being used.

Fluid motors are becoming more and more popular for use with modern machinery. Fluid motors can be applied directly to the work, and they provide excellent control for acceleration, operating speed, deceleration, smooth reversals, and positioning. They provide flexibility in design and eliminate much of the bulk and weight of mechanical and electrical power transmissions.

The application of fluid motors in their various combinations with pumping units are termed *hydrostatic transmissions*. There are two classes of applying hydrostatic transmission, the *open circuit* and the *closed circuit*.

The open circuit shown in Fig. 8-31 uses the conventional reservoir and a directional control valve for reversing the direction of motor rotation. The fluid motor speed depends on the capacity of the pump and the displacement of the motor.

The closed-circuit hydrostatic transmission shown in Fig. 8-32a uses a separate replenishing pump to keep the system charged at all times. The pump of Fig. 8-32b is fitted with a charge pump. The charge pump is driven by an extension of the main drive shaft. Only a small reservoir is needed, because most of the fluid is retained and circulates within the system.

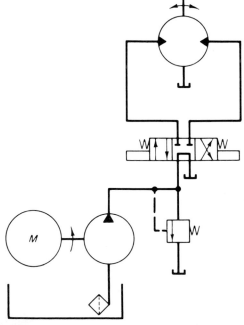

FIGURE 8-31 Open-circuit hydrostatic transmission.

Figure 8-32c illustrates a self-contained hydrostatic transmission with the pump and motor back to back and a

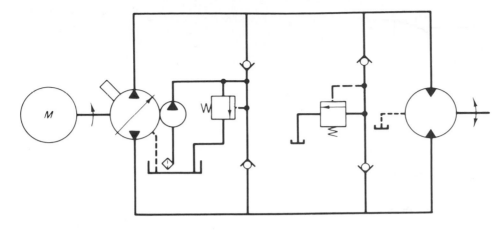

(a)

(b)

(c)

FIGURE 8-32 (a) Closed-circuit hydrostatic transmission. (b) Variable-displacement axial piston pump with integral charge pump. (Courtesy of Commercial Shearing, Inc., Youngstown, Ohio.) (c) Integrated closed-circuit hydrostatic transmission, (Courtesy of Abex Corporation, Denison Division, Columbus, Ohio.)

small reservoir equipped with a filter and heat transfer devices to house the total assembly.

The variable reversible (over-the-center) type of pump is controlled to vary the speed and direction of the fluid motor.

Regardless of the transmission type, fluid motors provide rotational power to do work. The turning force is a measure of torque which is equal to the force multiplied by the radius arm. This means that torque produced by a fluid motor depends on its pressure capabilities and its geometric configuration. Torque produced by a fluid motor is expressed by the following equations:

$$\text{Pounds inch of torque} = \frac{\text{pressure (psi)} \times \text{displacement (in.}^3\text{)/rev.}}{2\pi}$$

$$\text{Pounds foot of torque} = \frac{\text{pressure (psi)} \times \text{displacement (in.}^3\text{)/rev.}}{24\pi}$$

EXAMPLE

How many lb-in. of torque will a gear-type motor produce at 600 psi if its displacement is 1.25 in.^{3} of oil per revolution? (See Fig. 8-33; Fig. 8-34 shows typical performance curves.)

Solution:

$$\text{Torque (lb-in.)} = \frac{600 \text{ psi} \times 1.25 \text{ in.}^3}{6.28}$$

$$= 114.6$$

The relationship between the output horsepower, torque, and speed of a fluid motor is shown by the following:

FLUID AT 600 psi

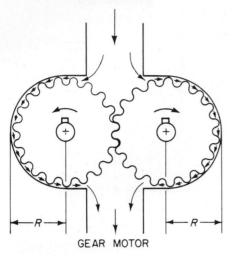

GEAR MOTOR

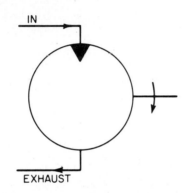

IN

EXHAUST

ASA SYMBOL

FIGURE 8-33 Diagram for example.

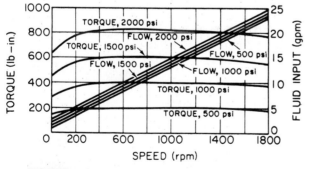

FIGURE 8-34 Typical performance curves: horsepower, speed, torque.

$$\text{Horsepower (hp)} = \frac{\text{torque} \times \text{rpm} \times 2\pi}{33,000 \text{ ft/lb-min}}$$

or

$$\text{hp} = \frac{T \times \text{rpm}}{63,025} \qquad \text{(torque in lb-in.)}$$

or

$$\text{hp} = \frac{T \times \text{rpm}}{5252} \qquad \text{(torque in lb-ft)}$$

EXAMPLE

In the preceding problem, how much horsepower would the gear motor produce if the pump delivers 20 gpm?

Solution:

$$\text{hp} = \frac{T \times \text{rpm}}{63,025}$$

First,

$$\text{Fluid motor speed (rpm)} = \frac{\text{gpm} \times 231 \text{ in.}^3}{\text{disp.(in.}^3)/\text{rev.}}$$

$$= \frac{10 \text{ gpm} \times 231 \text{ in.}^3}{1.2 \text{ in.}^3}$$

$$= \frac{2310}{1.2}$$

$$= 1924$$

Then,

$$\text{hp} = \frac{T \times \text{rpm}}{63,025}$$

$$= \frac{114.6 \times 1924}{63,025}$$

$$= 3.4$$

There are various combinations of pumps and motors used for hydrostatic transmission drives. The type of hydrostatic transmission drive shown in Fig. 8-35 provides fixed ratios of horsepower, torque, and speed.

The pump capacity is selected to provide the maximum output speed needed for a motor whose displacement and operating pressure meet the torque requirements.

Flow controls are generally applied to control speeds less than maximum. The setting of the relief valve in the system controls torque level.

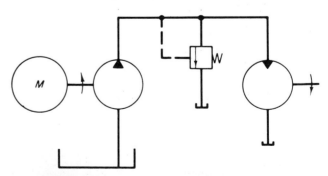

FIGURE 8-35 Constant horsepower, torque, and speed.

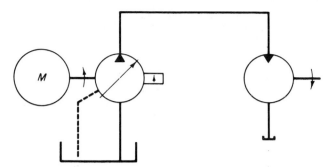

FIGURE 8-36   Variable horsepower and speed, constant torque.

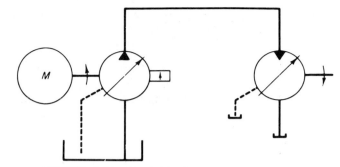

FIGURE 8-38   Variable horsepower, speed, and torque.

Constant torque will result when the type of transmission drive shown in Fig. 8-36 is used. The pump delivery is adjusted to vary the motor speed, but the torque remains practically constant. This combination is excellent for systems that have constant loads.

In the transmission drive shown in Fig. 8-37, the input horsepower remains constant, but the output will vary by adjusting the fluid motor displacement. Speed will depend on the maximum delivery of the pump and the maximum displacement adjustment in the fluid motor. The speed is generally limited to a ratio of 4:1 in this type of transmission.

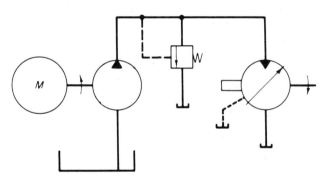

FIGURE 8-37   Constant horsepower, variable speed and torque.

The torque developed during operation depends on the displacement and relief valve settings.

The combination of hydrostatic transmission shown in Fig. 8-38 provides for a wide range of varying loads. The maximum horsepower, torque, and speed depend on the size of the units selected. Infinitely variable speed and torque between zero and maximum provide the most flexibility available in transmission today.

New fields of application are being discovered constantly for hydrostatic transmissions. Farm implements, road machinery, materials handling equipment, numerically controlled machines, high performance aircraft, military uses, and special machinery are only a few of the new fields expanding through the use of fluid power transmissions. Some farm tractors are completely hydraulic, as well as combines and other mobile machinery.

Gear-Type Fluid Motors

Gear-type fluid motors are often used as either pumps or motors. However, since fluid motors are supplied with fluid from the pump and are under pressure while doing work, bearings, type of seals, and drains are sometimes affected.

The type of bearings used is important because of the high operating speeds that are sometimes encountered. It is not uncommon for gear-type motors to operate up to 5000 rpm.

Seals are affected by both operating speed and pressure. Since most fluid motors are reversible, it means that both ports are alternately under pressure or exhaust. Shaft seals are generally limited to a specific operating pressure. Also, proper lubrication and low-pressure drains are essential.

Most reversible types of fluid motors must be externally drained to prevent shaft seals from blowing and to provide adequate lubrication and circulation of fluid for keeping bearings cool.

Gear-type motors are either internal or external gear designs. The external gear design shown in Fig. 8-39 is economical and is available as single or tandem units. A set of matched gears is fitted into a closely machined housing. The gears rotate together as fluid enters the space be-

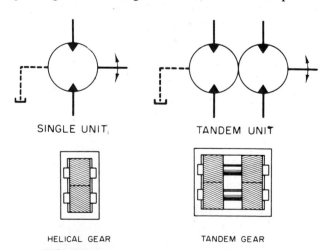

SINGLE UNIT,   TANDEM UNIT

HELICAL GEAR   TANDEM GEAR

FIGURE 8-39   External-gear-type design.

tween the major and minor diameters. Some designs use pressure plates on either side of the rotating gears to prevent leakage as the pressure increases with resistance due to the load. System pressure is directed through internal passages to the outside of the pressure plates or wear plates, holding them closely against the gears as the motor operates. This technique has helped to increase the efficiency of gear units.

The internal gear design shown in Fig. 8-40 consists of a pair of rotating gears, one inside the other. With fluid entering one port and the other open to exhaust, the gear elements rotate. As the teeth disengage, the space between the two elements increases on the pressure side and then decreases on the exhaust side as they gradually engage again. During one-half of one revolution, as the space increases, fluid enters from the pump; during the other one-half of one revolution, as the space decreases, fluid is exhausted from the motor. The internal gear design has high starting torque characteristics and operates at relatively high speeds.

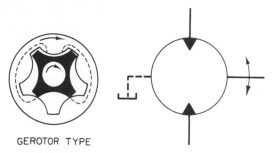

GEROTOR TYPE

FIGURE 8-40 Internal-gear-type motor.

Vane-Type Fluid Motors

Vane-type fluid motors are of the same general design as vane pumps, except that the vanes must be held tightly against the contour of the retaining ring to provide torque. This is generally done with springs applied underneath the vanes to hold them against the intersurface of the retaining ring. Seals, bearings, and drains are important considerations, as they are with all other designs of fluid motors.

Vane motors of the design shown in Fig. 8-41 are balance-type units. Fluid enters from the pump at two points 180° apart. Fluid also exhausts from two points 180° apart. This means that the rotating element of the motor is pressure-balanced, preventing uneven loads on the bearings.

Fluid forced into the vane motor from the pump pushes against the vanes and causes the rotor and shaft assembly to turn. Flow directed to the opposite port reverses the rotor and shaft assembly rotation. The porting arrangements can be located in the retaining ring or in the side plates.

Another type of vane motor uses rolling vanes (see Fig. 8-42). The vanes are attached to a gearing arrangement that operates with the rotating shaft. This maintains the

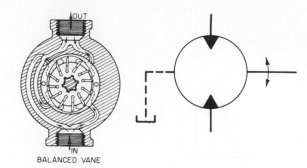

BALANCED VANE

FIGURE 8-41 Balance vane motor.

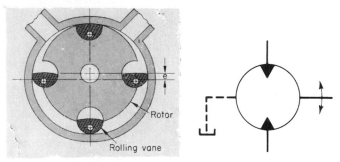

FIGURE 8-42 Roll vane motor.

timing of the rolling vanes as they roll in and out of the rotor. Close tolerances are maintained between the vanes and the retaining ring to provide high efficiencies.

Piston-Type Fluid Motors

Piston designs may be axial piston units or radial piston-type fluid motors. Both designs can be either fixed or variable motors. Their construction features are nearly the same as the pump units.

Axial piston motors use the valve plate method of displacing fluid, and the radial design uses the pintle arrangement of porting. Piston-type fluid motors provide excellent operating characteristics and low leakage. They have high starting torque and can operate at much higher pressures than other types. Output speeds normally range between 0 and 3000 rpm. Operating pressures range from 0 to 5000 psi. In addition to high efficiency, piston units give long life expectancy.

Axial Piston Fluid Motors

Axial piston motors (see Fig. 8-43) operate on the principle that fluid entering a port pushes against the pistons during one-half of one revolution, which causes the cylinder barrel and shaft assembly to turn. During the next one-half of one revolution the same pistons exhaust, while other pistons following in sequences provide continuous operation.

As shown in Fig. 8-44, fluid entering port A from the pump passes through kidney port A in the valve plate and pushes against the piston and shoe assembly, causing it to seek the lower side of the sloping face of the cam plate. As the slope decreases, the piston moves outward in relation to the rotating cylinder barrel. Each sequential piston and shoe assembly imparts a tangential force to the cylinder barrel, causing it to rotate.

When the piston and shoe assembly reaches the other one-half of the revolution, the cam pushes the piston back into the cylinder barrel, exhausting fluid through kidney port B and back to the pump. Each sequential piston and shoe assembly reacts to the cam plate in the same manner to exhaust their fluid.

Piston-type motors usually have an odd number of pistons—five, seven, nine, eleven, or more, depending upon the size of the unit. Generally, the greater number of pistons means more power and a smoother operation. (An odd number of pistons is used for positive starting torque.) Torque characteristics of a typical fluid motor are shown in Fig. 8-45. In the fixed unit the cam plate is stationary, whereas in the variable unit the cam angle is adjustable.

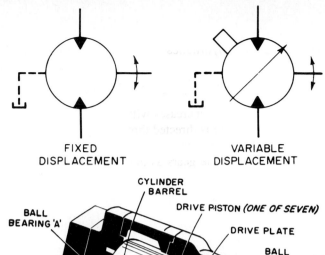

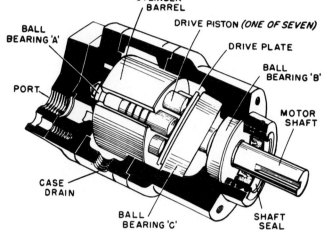

FIGURE 8-43 Axial piston motor. (Courtesy of Mobil Oil Corporation.)

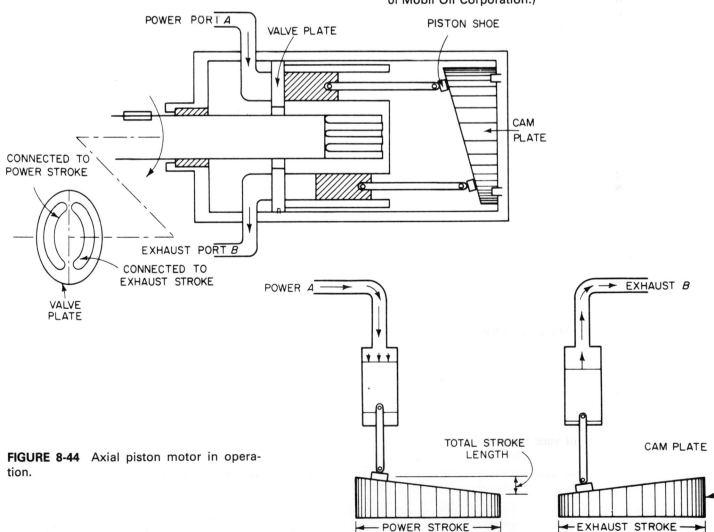

FIGURE 8-44 Axial piston motor in operation.

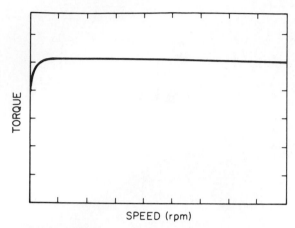

FIGURE 8-45 Example of a torque curve for a piston motor.

FIGURE 8-46 Orbit motor with tapered roller shaft bearings. (Courtesy of Orbmark Division of Regal-Beloit Corporation, Beloit, Wis.)

Radial Piston Fluid Motors

The radial piston design fluid motor has a cylinder barrel with an attached output shaft. The cylinder barrel contains a number of radial bores fitted with precision-ground pistons. The cylinder barrel and shaft assembly rotate within a thrust ring. If the fluid motor is a variable displacement unit, the thrust ring is adjustable. If the unit is fixed displacement, the thrust ring is stationary.

Fluid entering the motor from the pump passes through the pintle arrangement and into the cylinder chamber beneath the pistons. The pistons are forced against the thrust ring, which imparts a tangential force to the cylinder barrel and shaft assembly, causing it to rotate. Mass rotating forces are low with this design, providing excellent performance characteristics for slow speeds with high efficiency.

With new metal alloys and improved techniques in manufacturing, both pumps and motors have improved tremendously, giving higher efficiency than ever before. This high efficiency, coupled with flexibility in design and control, offers many new possibilities for the machines of the future using hydrostatic transmission drives.

Orbit-Type Fluid Motors

Internal gear motors such as that illustrated in Fig. 8-40 are designed with the internal pinion gear shaft eccentric to the centerline of the ring gear. This eccentricity is necessary to accommodate the difference in the number of teeth of the pinion and ring gear. Both pinion and ring gear rotate in the same direction. Carefully machined fluid directing ports permit pressurized fluid to enter one side of the assembly. Fluid from the opposite side of the centerline is directed to a tank or another low-pressure area. The amount of eccentricity, the geometry of the rotor set, and the width of the gear set determine the torque potential as a pressure differ-

ential is created across the motor displacement set. Rpm (revolutions per minute) and torque will respond directly to the available flow, pressure, and resistance to movement of the motor output shaft.

The displacement set of Fig. 8-46 functions differently from that of Fig. 8-40. The outer gear of Fig. 8-46 motor is held stationary. The inner gear orbits in response to carefully directed flow through a commutating valve mechanism to pockets created by the relative position of the teeth of the inner and outer gear. The name *gerotor* is given to this type of gear set. The inner gear has one tooth less than the outer gear. The commutating valve mechanism directs flow in a predetermined pattern to the roots of the outer gear structure.

Because of the eccentricity it is necessary to permit the pinion to orbit within the ring gear. A splined drive shaft acting as a universal joint connects the orbiting pinion gear to the valving mechanism and to the motor output shaft (see Fig. 8-47). Note the movement of the pinion in Fig. 8-48a–c as the pressure differential causes the orbiting motion. The mechanical relationship is such that it requires seven orbits of the pinion to cause one complete rotation of the output shaft. This, of course, is an inherent speed-reduction function. Thus the motor is basically a low-speed, high-torque machine.

The rotor set can have the ring gear machined as shown in Fig. 8-48a–c or structured as shown in Fig. 8-48d. Inserted circular members provide a smooth friction-free surface for the pinion to mate with, as shown in Figs. 8-46 and 8-47. Figure 8-49 illustrates one type of orbit motor structure.

Figure 8-50 illustrates another type of orbit motor, which employs a multiplicity of orbiting motions to provide the commutating valving function, the inherent speed re-

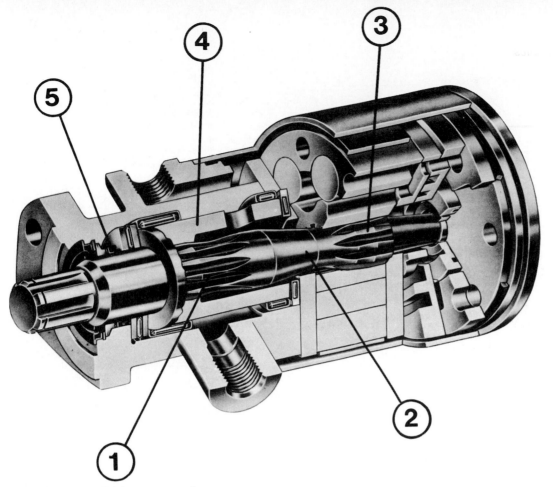

FIGURE 8-47 Orbit motor. 1, Spline drive shaft, output; 2, power transfer; 3, orbiting pinion and spline; 4, shaft and radial bearing structure; 5, thrust bearing and seal assembly. (Courtesy of Ross Gear Division of TRW, Lafayette, Ind.)

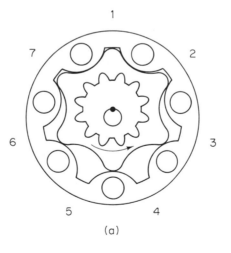

(a)

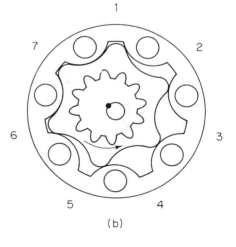

(b)

FIGURE 8-48 Orbiting pattern: (a) start; (b) 1/7 shaft revolution; (c) 1/14 shaft revolution; (d) with insert members for higher efficiency. (Courtesy of Ross Gear Division of TRW, Lafayette, Ind.)

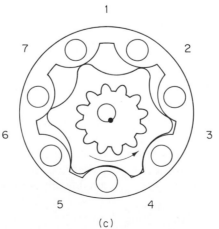

(c)

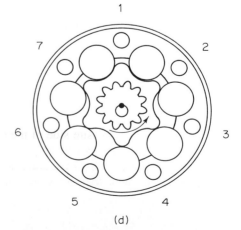

(d)

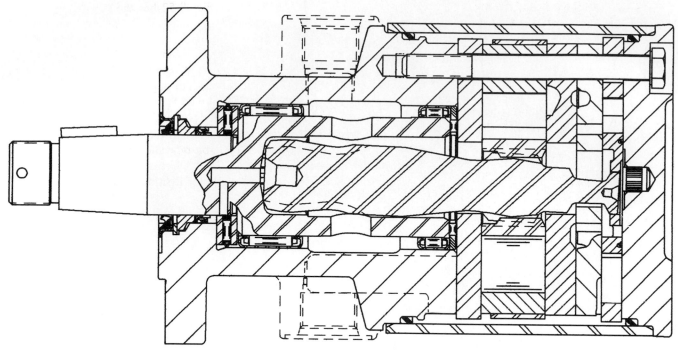

FIGURE 8-49   Cross-sectional view of orbit motor showing drive structure. (Courtesy of Ross Gear Division of TRW, Lafayette, Ind.)

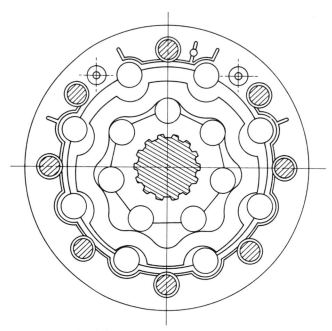

FIGURE 8-50 Internally generated rotor structure. (Courtesy of Nichols Fluid Power Division, Parker Hannifin Corp., Sturtevant, Wis.)

FIGURE 8-51   Valving structure for internally generated rotor type of motor. (Courtesy of Nichols Fluid Power Division, Parker Hannifin Corp., Sturtevant, Wis.)

duction, and the power transfer action. Figure 8-51 illustrates the mechanical structure of the motor of Fig. 8-52.

Orbit motors can be structured as a wheel-type assembly with the motor acting as the hub of the wheel and providing a direct traction-drive function. The motors are often connected directly to a brake assembly to provide a parking brake function.

The orbit-type motor is an ideal device for agricultural applications such as those needed for harvesting, tilling land, and processing foods. Compatibility with adverse surroundings make it an excellent choice in hostile environments. The small relative size makes it possible to apply power in many areas where mechanical or electrical means are difficult or impossible.

The motor of Fig. 8-52 has the connecting ports at the rear of the housing. Many different configurations are commercially available. Thus some designs are adapted for use directly on an axle within a wheel structure for multi-wheel drives and some for direct drive-shaft applications. Some models can have a through shaft.

Many structural modifications are available to reduce physical size, tolerate side loads, operate at high temperature, and to be compatible with hostile environments. The pressure range for orbit-type motors varies with different applications to values above 3000 psi for high-torque units and as low as 250 psi for light-duty applications.

FIGURE 8-52 Orbit-type motor. (Courtesy of Nichols Fluid Power Division, Parker Hannifin Corp., Sturtevant, Wis.)

QUESTIONS

1. What is one of the main advantages in applying actuating cylinders or fluid motors over mechanical devices?

2. What is the function of a hydraulic cylinder in a hydraulic system?

3. What kind of a valve is necessary to operate a double-acting cylinder?

4. How many feet per minute will a cylinder's piston travel if the pump is rated for 12 gpm and the diameter of the cylinder is $2\frac{1}{2}$ in.?

5. How many seconds does it take for a cylinder's piston to stroke 8 in. if the area is 7 in.^{2} and it is operated by a 6-gpm pump?

6. What operating pressure is necesary to produce a force of 12,000 lb if the area of the cylinder is 3.14 in.^{2}?

7. What is the area of a cylinder if the force is 10,500 lb and the operating pressure is 1500 psi?

8. What force will a cylinder produce if it has a bore diameter of $3\frac{1}{2}$ in. and is operating at a pressure of 2200 psi?

9. What precautions are necessary in the selection of conductors when cylinders operate faster on the return stroke than they do on the extending stroke?

10. When is a telescoping cylinder used?

11. Explain the operation of a tandem-type cylinder?

12. Why are wiper rings used on cylinder rods?

13. Explain the function of cushioning in cylinders.

14. Explain the operation of a rotary actuator.

15. What is the main difference between a rotary actuator and a fluid motor?

16. What are the two general classes of fluid motors?

17. How many rpm would a fluid motor make if the displacement of the motor per revolution is 0.85 in.^{3} and it is operated by a 7.5-gpm pump?

18. How much torque would this same motor produce if it is operated at 3000 psi? What horsepower would this same motor produce?

19. What are the four general combinations of hydrostatic transmissions? Explain their output characteristics.

20. How are most reversible types of fluid motors drained?

21. Name the construction types of fluid motors available.

9
System Components and Circuits

The material presented in previous chapters has dealt with basic fundamentals and system components. In addition to the material already covered, there are numerous component parts of a hydraulic system that can be more simply explained in a discussion of circuitry. Designing circuits is an intriguing responsibility. The circuit designer is limited only by the boundaries of his or her own ingenuity and knowledge of the components available.

A circuit is an arrangement of components interconnected to provide a desired form of fluid power. The *circuit* is generally a part of the hydraulic system, whereas *system* refers to the complete assembly of component parts that transmits and controls fluid power. A system may be composed of several circuits.

There are thousands of different components available for designing hydraulic circuits. Many of these components have multiple uses with merely the change of a cover, the use of an adapter, different ways of connecting, the addition of a port, or some other simple procedure.

Never before has a technology been so simple or so complex. It extends from an ordinary hydraulic jack to a space ship control system. Fluid power overlaps practically every branch of engineering, especially electrical, electronic, and mechanical engineering. All of these branches of engineering sciences are teaming up to use the best of each for any given design problem.

With all of this flexibility to feed the designer's imagination and ingenuity, he or she has a big responsibility to perform. Among these factors of responsibility there are three important considerations:

1. The system must be designed to operate safely. Safe operations of a system may include such things as:
    a. Pressure ratings of system components.
    b. Temperature ratings of system components.
    c. Operating speeds where pertinent.
    d. Compatibility between system components.
    e. Environmental conditions.
    f. Interlocks for sequential operations.
    g. Emergency shutdown features.
    h. Power failure locks or safety pins.
    i. Fire hazards.
    j. Protective devices for operators and servicemen.
2. The system must be functional. This may include such things as the following:
    a. System components must meet the required performance specifications.
    b. System components should be selected according to the duty cycle for which they are designed.

c. The life expectancy of the component parts should be essentially the same as the life of the machine.

d. When the system involves electrical or mechanical hookups with fluid power, compatibility for fabrication and control is essential.

e. System components should facilitate good maintenance practices.

f. System components should be selected with a margin of safety to withstand operational hazards, such as hydraulic shock.

g. Accumulators should be discharged when being stored or in a rest condition to avoid danger to personnel or equipment.

3. The system must be efficient. The following are some of the factors to consider:

a. Keep the system as simple as possible, but safe and functional.

b. Use good fabrication techniques.

c. Standardize component parts as much as possible.

d. Maintain clean design with access to important parts that may need repair or adjustment.

e. Consider the availability of replacement parts.

f. Design to keep operational costs to a minimum.

g. Design the system to prevent and remove contamination.

h. Supply good drawings.

The factors mentioned concerning safety, function, and efficiency of a system are only a few of the many important considerations in designing hydraulic systems. One of the most important tools for anyone working in fluid power is a knowledge of the components available and how they operate in a circuit.

The discussion that follows will cover components and circuitry that are representative of the present technology. The various circuits presented will cover component selection factors, methods of calculating specification data, operating information, and graphical diagramming.

LINEAR CIRCUITS

Figure 9-1a shows a simple reciprocating circuit for bending a piece of sheet steel. The system has the following components:

1. Reservoir
2. Strainer
3. Pump inlet line
4. Pump
5. Flexible coupling
6. Electric motor
7. Discharge conductor and connectors
8. Relief valve
9. Directional four-way valve (manually controlled)
10. Double-acting single-rod-end cylinder

How is a circuit problem of this nature approached?

1. What are the specifications for the job?
   a. Force requirement to bend the steel = 1800 lb.
   b. Length of work stroke = 6 in.
   c. Speed of piston and rod assembly extending = 0.50 sec.

2. What size cylinder is needed? Since the force required to do the work is 1800 lb, cylinder area and operating pressure must be selected.

$$F = PA \qquad A = \frac{F}{P} \qquad P = \frac{F}{A}$$

Referring to standard bore and rod sizes for cylinders, we make a selection by reasoning that a larger-diameter cylinder operates at a lower pressure but requires a bigger pump to give the required cylinder speed. We also reason that a small-diameter cylinder must operate at a higher pressure, but a smaller pump is capable of the cylinder speed.

Cylinder selected = 2-in. bore.

Operating pressure:

$$P = \frac{F}{A}$$

$$= \frac{1800}{3.14}$$

$$= 573 \text{ psi}$$

In considering the types of pumps available, both vane and gear pumps are relatively low in cost and give excellent life at this pressure level.

3. What capacity pump is needed? The pump delivery is selected to provide the maximum cylinder speed required by the circuit.

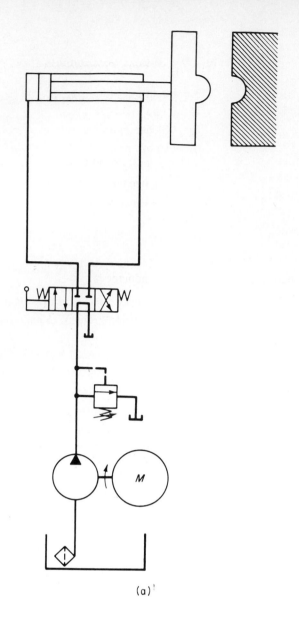

(a)

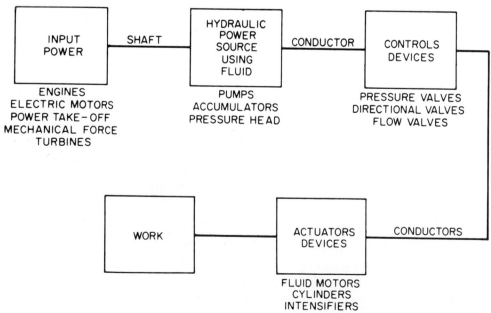

(b)

FIGURE 9-1 Linear circuit: (a) simple reciprocating circuit; (b) block diagram.

gpm =

$$\frac{\text{area of cyl. (in.}^2) \times \text{length of stroke (in.)} \times 60 \text{ sec}}{231 \text{ in.}^3 \times \text{time (sec)}}$$

$$= \frac{3.14 \text{ in.}^2 \times 6 \text{ in.} \times 60}{231 \text{ in.}^3 \times 0.50}$$

$$= 9.8$$

The calculations above indicate that 9.8 gpm is needed to extend the 2-in.-diameter bore cylinder 6 in. in 0.5 sec. It must deliver this amount at a pressure that will reach 573 psi. The next step is to determine the pump to be used by considering the selection factors as described in Chapter 6. It is generally good practice to select a standard pump that provides the nearest specifications to those desired. It is also important to remember that the pump selected usually dictates the choice of other system components. However, if there are unique design characteristics about a system, such as being mobile or stationary, directly or remotely controlled, having limitations in size and weight, or operating at extreme temperatures, then *complete system compatibility is a requirement*.

4. What size electric motor is needed to drive the pump?

$$\begin{aligned} \text{hp} &= \text{gpm} \times \text{psi} \times 0.000583 \\ &= 9.8 \times 573 \times 0.000583 \\ &= 3.27 \end{aligned}$$

This indicates that if the pump and circuit are 100% efficient, then 3.27 hp is needed. A gear or vane-type pump is generally about 85% efficient, and this means that the power losses in the pump would be about 15%. The pump capacity would have to be greater than 9.8 gpm to make up the loss and still satisfy the cylinder speed. It is possible that other losses may also occur, such as fluid friction, mechanical friction, and others.

Electric motors generally have an overload factor, but this should not be used under continuous duty operation. A good selection in this case would be a standard 5-hp electric motor of the enclosed type. (Enclosed designs prevent a fire hazard due to arcing.)

Shaft size and type, direction of rotation, and mounting standards should also be indicated. The type of flexible coupling between the pump and electric motor is merely a matter of choice. The mating shafts should be compatible.

5. What size reservoir should be used? The rule of thumb is $2\frac{1}{2}$ to 3 times the capacity of the pump.

In this case, a standard 30-gpm reservoir would be adequate. If there were a number of large cylinders in the circuit or if fluid were continuously blowing over the relief valve, causing heat, the size must be calculated.

6. What size pump inlet would be needed? Inlet flow velocities normally must be maintained between 2 and 5 ft/sec. The total pump capacity, including 9.8 gpm for the cylinder speed plus about 1.5 gpm to make up for pump slip, equals 11.3 gpm.

The formula for determining the area size of the conductor is

$$\begin{aligned} \text{Area (in.}^2) &= \frac{\text{gpm} \times 0.3208}{v} \\ &= \frac{11.3 \text{ gpm} \times 0.3208}{3 \text{ ft/sec}} \\ &= 1.2 \text{ in.}^2 \end{aligned}$$

The pipe tables indicate that by using standard schedule 40 pipe the inlet size would be $1\frac{1}{4}$-in. pipe. If the system had to use fire-resistant fluid with a high specific gravity or if the circuit operated at a high altitude with less atmosphere, a larger size might be used.

7. What size discharge conductors are needed? The same procedure is used to determine the size of conductors and connectors on the discharge side of the pump. The velocity of flow for the discharge side should be maintained between 7 and 15 ft/sec.

Since the operating pressure is relatively low, either pipe or tubing can be selected. Tubing offers fewer fittings and is generally easier to install. In this case, standard $\frac{3}{4}$-in. pipe or SAE 1010 dead soft steel tubing would be adequate. Trends indicate that tubing is usually preferred because of less potential leakage. The inlet strainer would be selected on the basis of the inlet pipe size.

8. How is the relief valve selected? A number of component manufacturers make only a few standard sizes of valves. A standard relief valve having $\frac{3}{4}$-in. NPT ports would be a logical choice. The system capacity determines the control valve sizes, which are compatible with standard conductor sizes.

A simple relief valve fitted with a spring having a range setting between 0 and 1000 psi would be selected. Usually, the relief valve is set from 10 to 12% higher than necessary to perform the work.

9. What determines the type of directional four-way valve? A manual operator was specified for valve

actuation; however, this might have been electrical, mechanical, pilot, or other. The type of valve actuation depends on the application.

Figure 9-1 indicates that an open center four-way valve is used. Open center systems allow the pump delivery to bypass back to the reservoir when not being used to extend or retract the cylinder. If a closed center valve had been used, the pump delivery would blow across the relief valve and heat would result. The size of the four-way directional control would be the same as the relief valve and have the same port configuration. The procedure used for selecting components for this simple linear circuit generally applies to any circuit. Selecting components for any hydraulic system is usually a compromise, with cost as the major factor. Figure 9-1b provides a block diagram showing the system.

REGENERATIVE CIRCUITS

A regenerative circuit is one in which pressurized fluid discharged from a component is returned to the system to reduce power input requirements (see Fig. 9-2). On single-

rod-end cylinders the discharge from the rod end is often directed to the opposite end to increase rod extension speed. If equal speeds in both directions are required, a 2:1 area ratio is used.

EXAMPLE

A cylinder with a 2-in. area bore and a 1-in. area piston rod is shown in Fig. 9-3. The pump has a volumetric capacity of 10 gpm and the cylinder stroke is exactly 12 in. What speed will the piston and rod assembly travel in each direction? (Calculations to be made in ft/min.)

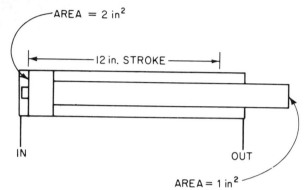

FIGURE 9-3   Cylinder with 2-in.² area bore and 1-in.² area piston rod.

Solution: Using the formula

$$\text{ft/min} = \frac{\text{gpm} \times 19.25}{\text{area (in.}^2)}$$

we have

Forward stroke:

$$\text{ft/min} = \frac{10 \times 19.25}{2} = 96.25$$

Reverse stroke:

$$\text{ft/min} = \frac{10 \times 19.25}{1} = 192.5$$

Thus it is shown that a speed ratio of 2:1 is given when one is using a pump with a constant displacement of 10 gpm. Because the cylinder area or volume is a ratio of 2:1, the piston and rod assembly travels twice as fast, retracting as it does on the forward stroke. By connecting the fluid from the rod end to that of the large area end, equal speeds would result in both directions. This principle is used to provide equal work speeds in both directions, which is desired in many applications, such as presses and bulldozer scraper blades.

It is important to remember that only one-half of the area is effective for producing force as the cylinder's piston and rod assembly extends during regeneration of the fluid. A regenerative circuit does not reduce power input requirements; it trades force for speed.

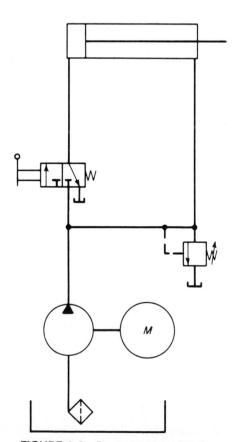

FIGURE 9-2   Regenerative circuit.

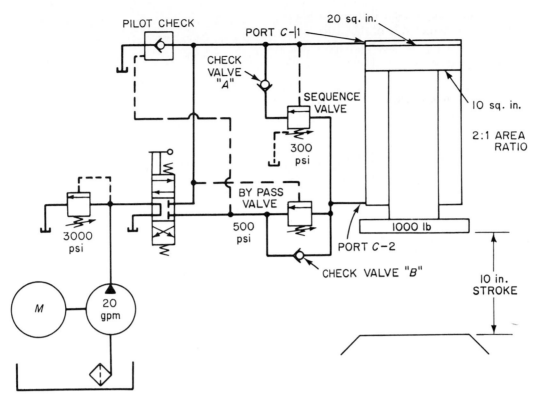

FIGURE 9-4 Press circuit.

The press circuit shown in Fig. 9-4 uses a bypass valve to release fluid on the rod side of the piston during the pressing action. This facilitates the use of the entire area on the large piston end to produce the required force.

With the pump in operation and the manually actuated four-way valve positioned for press down, 20 gpm is delivered through port C-1 and to the large area end of the cylinder. The sequence valve is set to open when the pressure reaches 300 psi. This control also acts as a counterbalance valve to prevent the press from dropping due to gravity.

If friction is neglected, the force relationship may be expressed by using Pascal's principle: force acting to push the press down,

$$F = P \times A = 300 \times 20 = 6000 \text{ lb}$$

plus 1000 lb of weight for press member, which gives a total of 7000 lb of force.

Force resisting the downward motion of the press equals pressure on the rod side due to fluid:

$$\text{Pressure on top} = \frac{F}{A} = \frac{300 \times 20}{10} = 600 \text{ psi}$$

$$\text{Pressure due to weight} = \frac{1000}{10} = 100 \text{ psi}$$

Therefore,

$$\text{Total resisting force} = P \times A$$
$$= 700 \text{ psi} \times 10 \text{ in.}^2$$
$$= 7000 \text{ lb force}$$

This proves that the press is balanced on both sides, which means that it would take a slightly higher pressure than 300 psi on top to move the press downward.

As the press moves downward, the fluid in the rod end of the cylinder is forced out and passes over the sequence valve to join the fluid coming from the pump.

Time extending (sec)
$$= \frac{\text{area (in.}^2) \times \text{stroke (in.)} \times 60 \text{ (sec)}}{231 \text{ (in.}^3) \times \text{gpm}}$$
$$= \frac{20 \text{ in.}^2 \times 10 \text{ in.} \times 60 \text{ sec}}{231 \text{ in.}^3 \times 40 \text{ gpm}} = \frac{300}{231}$$
$$= \frac{300}{231} = 1.29 \text{ sec}$$

As soon as the press comes in contact with the work and resistance causes the pressure to build up to the setting of the bypass valve (500 psi), it opens, relieving the back pressure beneath the piston. Check valve A isolates the high pressure side from the low-pressure side of the system.

With the relief valve set for 3000 psi, the maximum force available for the pressing action would be

$$F = P \times A = 3000 \text{ psi} \times 20 \text{ in.}^2 + 1000 \text{ lb}$$

Total force = 61,000 lb

When the pressing action is completed, the four-way valve is positioned to raise the press. Fluid from the pump is directed through port C-2 and check valve B into the rod end of the cylinder.

$$
\begin{aligned}
\text{Time retracting} &= \frac{\text{area} \times \text{stroke} \times 60}{231 \times \text{gpm}} \\[2mm]
&= \frac{10 \text{ in.}^2 \times 10 \text{ in.} \times 60 \text{ sec}}{231 \text{ in.}^3 \times 20 \text{ gpm}} \\[2mm]
&= \frac{300}{231} = 1.29 \text{ sec}
\end{aligned}
$$

Thus, a 2:1 area ratio cylinder provides equal speeds in both directions. The pilot check valve is used in this case to divide the flow coming out of the large side of the cylinder.

Remember, on the press downstroke 20 gpm from the pump plus 20 gpm from the rod side of the cylinder was used to bring the press down in 1.29 sec. This means that with the equal speeds in both directions 40 gpm must be taken out of the large area side on the up stroke.

If a larger conductor is used between the large area end of the cylinder and the pilot check valve, it allows part of the fluid to pass through a separate line out of the check valve to the reservoir. The rest of the fluid passes through the four-way valve and back to the reservoir. This arrangement allows the use of smaller components in the system, which helps to reduce costs.

ACCUMULATOR CIRCUITS

An accumulator is a container in which fluid is stored under pressure as a source of fluid power. There are two general types of accumulators: the hydropneumatic and the mechanical designs. The hydropneumatic design (see Fig. 9-5) uses a compressed gas which applies force to the stored liquid. Mechanical designs use a weighted member or spring which applies force to the stored liquid (see Fig. 9-6).

The hydropneumatic designs are most commonly used for hydraulic systems and use either air or nitrogen for precharge. This type operates by the principle of *Boyle's law:*

$$P_1 V_1 = P_2 V_2$$

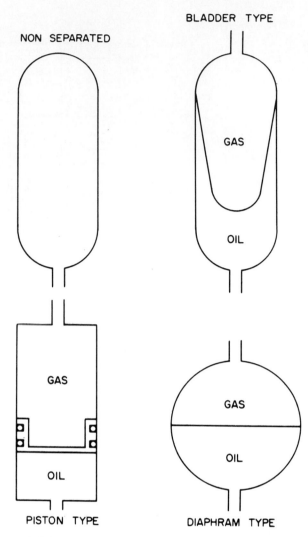

FIGURE 9-5 Design types of hydropneumatic accumulators.

Accumulators are used for a number of different functions. Some of these uses are discussed as follows:

1. *Auxiliary power source* to supplement the pump where the cycle time will allow a charge to be stored for peak requirements (see Fig. 9-7). In an intermittent-duty-type system an accumulator is used as a secondary power supply. This helps to reduce the input horsepower by storing energy during idle times of the machine. The pilot-operated relief or unloading valve opens when the accumulator is fully charged and allows the pump delivery to return to the reservoir at very low pressure. The accumulator maintains its charge, since it is isolated by the check valve. The solenoid actuated on-off type valve is shown as a safety device to unload the accumulator when the pump is shut down.

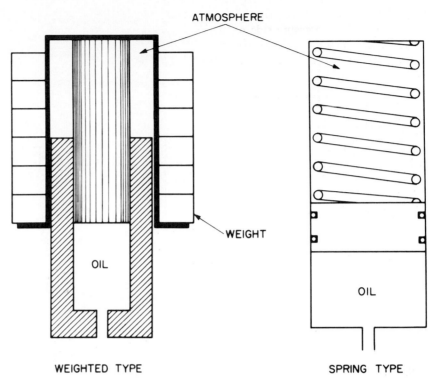

FIGURE 9-6 Mechanical designs of accumulators.

WEIGHTED TYPE   SPRING TYPE

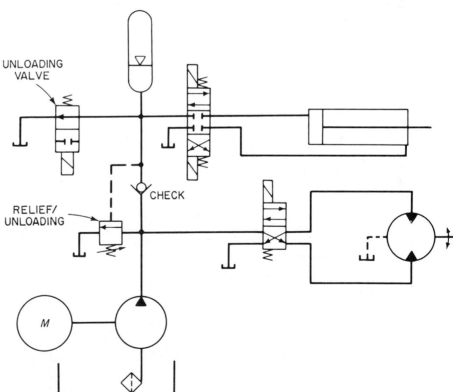

FIGURE 9-7 Auxiliary power source.

2. *Emergency power source* in case of power failure to operate critical circuit functions (see Fig. 9-8). The closed-center-type directional four-way valve allows the accumulator to charge during idle times, and the pump unloads at full charge through the pilot-controlled relief valve. If emergency fluid power is needed, the on-off valve is shifted and fluid enters the system to extend or retract the cylinder through the four-way valve.

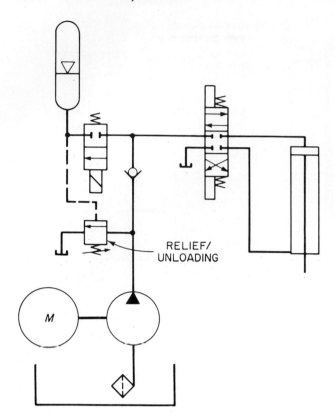

FIGURE 9-8 Emergency power source.

3. *Hydraulic shock absorber* for circuits where sudden impact loads, quick stops, or reversals with heavy loads are a characteristic of the system (see Fig. 9-9). Hydraulic shock loads may be reduced considerably if the deceleration time of the flowing fluid mass can be reduced. The accumulator should be installed as close to the shock source as possible.

Shock can cause hydraulic lines to leak, damage valving, cause pumps to fail and create excessive noise levels. Specialized accumulators can be used to reduce shock in hydraulic systems. Obviously, things to look for are banging pipes when valves shift and leaky piping.

The circuit of Fig. 9-10 provides some typical examples of a circuit that can benefit from the use of a specialized accumulator. The following information must be available:

a. Pipe length in feet between the pump and valve generating the shock $= L = 100$ ft.

b. Internal area of piping in square inches (see Table 9-1) $= A = 1.283$.

c. System operating pressure (psi) $= P_2 = 2500$ psi.

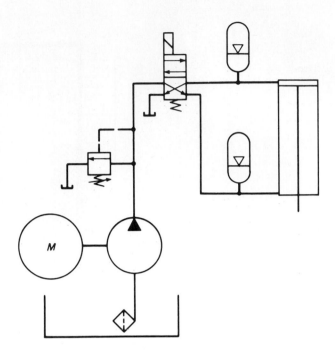

FIGURE 9-9 Hydraulic shock absorber.

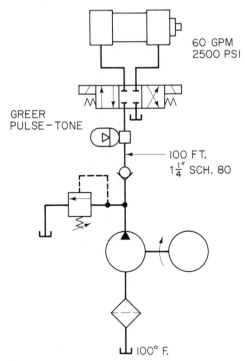

FIGURE 9-10 Specialized shock-absorbing accumulator. (Courtesy of Greer Hydraulics, Commerce, Calif.)

d. Pump flow (gpm) $= Q = 60$ gpm.

e. System operating temperature (°F) $= T = 100$°F.

(*Note:* If T is unknown, use 100°F as a general rule.)

TABLE 9-1
INTERNAL PIPE AND TUBING AREAS

a. Internal Pipe Area (in.^{2})

Pipe size (in.)	Pipe schedule		
	No. 40	*No. 80*	*No. 160*
$1/8$	0.057	0.036	—
$1/4$	0.104	0.072	—
$3/8$	0.191	0.141	—
$1/2$	0.304	0.234	0.171
$3/4$	0.533	0.433	0.271
1	0.864	0.719	0.522
$1 1/4$	1.495	1.283	1.060
$1 1/2$	2.036	1.767	1.410
2	3.356	2.953	2.241
$2 1/2$	4.788	4.238	3.542
3	7.393	6.605	5.416
$3 1/2$	9.886	8.888	—
4	12.73	11.50	9.283

b. Steel Tubing Internal Area

Outside diameter (in.)	*Wall (in.)*	*Internal area (in.^{2})*
$1/4$	0.035	0.255
$5/16$	0.035	0.0460
$3/8$	0.035	0.0731
$1/2$	0.035	0.1452
$1/2$	0.049	0.1269
$5/8$	0.049	0.2181
$5/8$	0.065	0.1924
$3/4$	0.049	0.3339
$3/4$	0.065	0.3019
$3/4$	0.083	0.2679
$7/8$	0.049	0.4742
$7/8$	0.065	0.4359
$7/8$	0.095	0.3685
1	0.083	0.5463
1	0.095	0.5153
1	0.109	0.4803
$1 1/8$	0.083	0.7223
$1 1/8$	0.095	0.6866
$1 1/8$	0.109	0.6461
$1 1/8$	0.120	0.6151
$1 1/4$	0.095	0.8825
$1 1/4$	0.109	0.8365
$1 1/4$	0.120	0.8012
$1 1/2$	0.109	1.291
$1 1/2$	0.120	1.247
2	0.120	2.433
2	0.165	2.190

Source: Courtesy of Greer Hydraulics, Inc., Commerce, Calif.

Using the above, solve for this:

a. *n* from Table 9-2 (average pressure is P_2) = *n* = 1.8.

b. $(L \times A)/144$, which is the total oil volume in the pipe (ft^3) = VA = 0.891.

TABLE 9-2
ADIABATIC EXPONENT FOR NITROGEN GAS, *n*

Average pressure $\dfrac{P_3 + P_2}{2}$	For system temperature of:				
	75° F	*100° F*	*140° F*	*170° F*	*200° F*
100	1.4	1.4	1.4	1.4	1.4
150	1.4	1.4	1.4	1.4	1.4
200	1.4	1.4	1.4	1.4	1.4
250	1.5	1.4	1.4	1.4	1.4
300	1.5	1.5	1.5	1.5	1.5
350	1.5	1.5	1.5	1.5	1.5
400	1.5	1.5	1.5	1.5	1.5
500	1.5	1.5	1.5	1.5	1.5
600	1.5	1.5	1.5	1.5	1.5
700	1.5	1.5	1.5	1.5	1.5
800	1.6	1.5	1.5	1.5	1.5
900	1.6	1.6	1.5	1.5	1.5
1000	1.6	1.6	1.6	1.5	1.5
1250	1.6	1.6	1.6	1.6	1.6
1500	1.7	1.7	1.6	1.6	1.6
2000	1.8	1.7	1.7	1.7	1.6
2500	1.9	1.8	1.8	1.7	1.7
3000	1.9	1.9	1.8	1.8	1.7

Note: In calculating the *n* factor, use next higher average pressure and next lower system temperature (e.g., if the average pressure is 1400 psi and the system temperature is 125° F, *n* should be 1.7; if the system temperature is unknown, use 100° F as a general rule).

Source: Courtesy of Greer Hydraulics, Inc., Commerce, Calif.

c. Weight of fluid (lb/ft^3) (53.04 for oil) = W = 53.04.

d. $Va \times W$, which is the total weight of the fluid in the pipe = W = 47.26.

e. $0.3208Q/A$, which is the flow velocity (ft/sec) in the pipe = FV = 15.

f. *R* from Table 9-3 depending on *n* value = R = 4.16.

Now we can calculate the accumulator size required using this formula:

$$V_1 = \frac{(FV)^2 \times W \times (n - 1) \times R}{P_2} = 6.2 \text{ in.}^3$$

V_1 is the accumulator size required in cubic inches to limit pressure peaks to 10% above P_2. Select

TABLE 9-3
CROSS-REFERENCE TABLE

n	*C*	*R*
1.4	30.5	6.20
1.5	32.3	5.54
1.6	34.4	5.01
1.7	36.8	4.54
1.8	39.5	4.16
1.9	42.7	4.00

Source: Courtesy of Greer Hydraulics, Inc., Commerce, Calif.

TABLE 9-4
3000-psi PULSE TONE

Nominal size	Gas capacity (in.³)
1 quart	60.0
1 gallon	226.0
2¹/₂ gallon	555.0
5 gallon	1095.0

Source: Courtesy of Greer Hydraulics, Inc., Commerce, Calif.

the next larger accumulator for this application (see Table 9-4). Precharge should be 65% of P_2.

4. *Leakage compensation* in circuits necessary to hold loads for long periods of time. The pressure switch may be used as shown in Fig. 9-11 to start and stop the hydraulic pump for maintaining the system pressure level for adequate holding force.

5. *Pressure transfer barrier* operations generally use accumulators to apply a static test or transfer fluid energy to a secondary fluid system. In the example shown in Fig. 9-12, the fire extinguisher tank is pressure-tested to 2000 psi. City water prefills the tank, then the on-off valve A is closed. Hydraulic fluid then enters the lower side of the accumulator, acting against the movable barrier and creating a pressure controlled by the relief valve. The operator would hold the test pressure for the required time and then open valve A and close valve B. This would unload the hydraulic pump back to the reservoir and allow a new fire extinguisher to be installed for test.

CHARGING AND SIZING ACCUMULATORS

Precharge pressures are selected so that almost all of the hydraulic fluid charge is used to do useful work. In most cases, the precharge pressure equals the minimum pressure that will perform the work. The operating principle of the accumulator is based on Boyle's law for gases (see Chapter 11).

Accumulators are preloaded with air or an inert gas, such as nitrogen. Nitrogen is generally used because it is relatively inexpensive and helps to prevent corrosion within the component. As hydraulic fluid is supplied to charge the accumulator, it causes the movable element to compress the gas into a smaller volume to make room for the hydraulic fluid which is being stored. The ANSI standard requires that accumulators working above 200 psi must be charged with dry nitrogen.

Standard sizes of accumulators generally range from 1½ in.³ to 10 gallon capacity and operate at pressures up to 6000 psi. The nonseparator type is available in larger sizes, and specially built units for aerospace use operate at 60,000 psi.

In order to determine the size of an accumulator required, the amount of hydraulic fluid needed must be known. This is determined by the cylinder or fluid motor size and the work speed. Total volume of the accumulator is the sum of the hydraulic fluid stored and the volume of compressed gas necessary to supply the fluid at its working pressure.

The rate of charging or discharging determines whether the process is isothermal or adiabatic. Isothermal expansion occurs when the process is slow enough to allow heat trans-

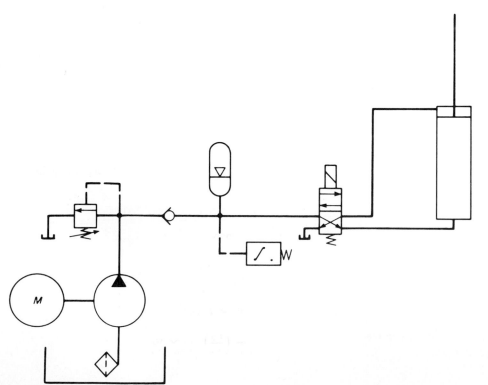

FIGURE 9-11 Leakage compensation.

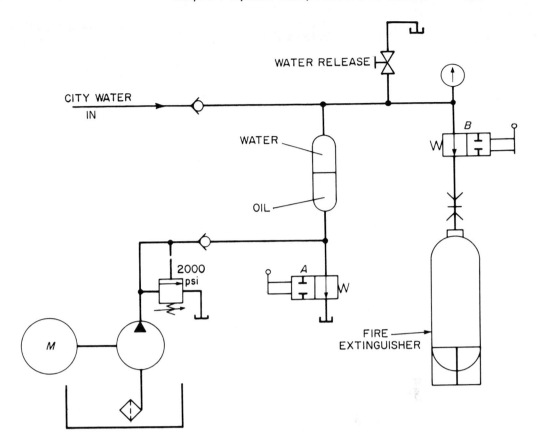

FIGURE 9-12 Pressure transfer barrier.

fer out of the accumulator, which is generally taken as 3 minutes or longer.

The equation for isothermal expansion and compression is

$$P_1V_1 = P_2V_2$$

Adiabatic (without heat) expansion or compression occurs when the process is fast enough to retain all of the heat.

The equation for adiabatic expansion and compression is

$$P_1V_1^{1.4} = P_2V_2^{1.4}$$

EXAMPLE

What size accumulator is necessary to supply 2.5 gpm with an allowable pressure drop from 3000 psi to 2000 psia? The precharge is 1500 psia. (2.5 gpm $\times$ 231 in.^{3} = 577.5 in.^{3}) (*Note:* Absolute pressure must be used when one is working with gases.)

Solution:

Isothermal:

$$P_1V_1 = P_2V_2 = P_3V_3$$

$$P_1 = 1500 \text{ psia} \quad V_1 = ?$$

$$P_2 = 2000 \text{ psia} \quad V_2 = V_3 + 577.5$$

$$P_3 = 3000 \text{ psia} \quad V_3 = V_2 - 577.5$$

$$P_2V_2 = P_3V_3$$

$$2000(V_3 + 577.5) = 3000V_3$$

$$V_3 = 1155 \text{ in.}^3$$

$$V_2 = 1732.5 \text{ in.}^3$$

$$P_1V_1 = P_2V_2$$

$$1500 \ V_1 = 2000(1732.5)$$

$$V_1 = 2306.6$$

$$= \frac{2306.6}{231}$$

$$= 9.9 \text{ gal}$$

Adiabatic:

$$P_1V_1^{1.4} = P_2V_2^{1.4} = P_3V_3^{1.4}$$

$$P_1 = 1500 \text{ psia} \quad V_1 = ?$$

$$P_2 = 2000 \text{ psia} \quad V_2 = V_3 + 577.5$$

$$P_3 = 3000 \text{ psia} \quad V_3 = V_2 - 577.5$$

$$P_1V_1^{1.4} = P_2V_2^{1.4} = P_3V_3^{1.4}$$

$$P_2V_2^{1.4} = P_3V_3^{1.4}$$

$$V_2 = \left(\frac{P_3}{P_2}\right)^{0.714} \times V_3$$

$$= \left(\frac{3000}{2000}\right)^{0.714} (V_2 - 577.5)$$

$$= 2295.5 \text{ in.}^3$$

$$P_1 V_1^{1.4} = P_2 V_2^{1.4}$$

$$V_1 = 2295\left(\frac{2000}{1500}\right)^{0.714}$$

$$= 2820 \text{ in.}^3$$

$$= \frac{2820}{231} = 12.2 \text{ gal}$$

INTENSIFIER CIRCUITS

An intensifier is a device that converts low-pressure fluid power into higher-pressure fluid power. Intensifiers (also called boosters) are used to multiply forces when a great force is needed through a relatively short distance. Hydraulic presses, riveting machines, and spot welders are typical applications.

There are two general types of intensifiers, the air-to-hydraulic and the all-hydraulic. The single-plunger, piston-to-ram intensifier is most commonly found. An automatically reciprocating double-acting design is available for applications needing longer work strokes or for maintaining high pressures for a long period of time (see Fig. 9-13). The air-over-hydraulic intensifier operates on a principle of converting shop air pressure into higher hydraulic pressure (see Fig. 9-14).

A reciprocating intensifier consists of a large double-

rod-end cylinder with the rod ends simulating the operation of two single-acting pumping pistons. Shop air automatically reciprocates the piston assembly through the use of mechanically actuated air pilot valves. The pilot valves supply air to the ends of the directional control valves to alternately shift positions for reciprocation.

As the piston assembly reciprocates by using air, the rod ends act like pumps with an inlet and outlet check valve for each pumping piston. When the intensifier moves toward the *left* side, it causes the inlet check valve to close and the discharge check valve to open and supply high-pressure fluid for the system. At the same time, air pressure over hydraulic fluid in the A tank causes fluid to enter the piston chamber at the right end to replace the moving piston.

At the end of the stroke to the left, a mechanical stem is actuated by the large air piston, which allows the pilot valve to shift the main directional control valve to reverse the intensifier. The procedure repeats until the maximum pressure is reached.

Tank A uses hydraulic fluid pressurized by plant air to give the work cylinder a fast approach when the load is light. Then the intensifier comes into action to do the heavy work. After the work stroke is completed, Tank B supplies hydraulic fluid under pressure to return the work cylinder. The intensifier ratio is determined by

$$\text{Int. ratio} = \frac{\text{intensifier output pressure (psi)}}{\text{intensifier input pressure (psi)}}$$

Intensifiers are available in a great number of ratios, with output pressures to 10,000 psi and even higher.

Because hydraulic fluid exposed to the atmosphere

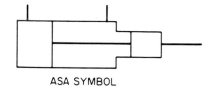

ASA SYMBOL

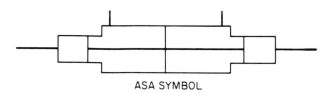

ASA SYMBOL

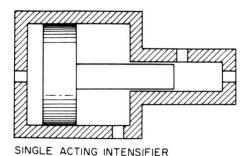

SINGLE ACTING INTENSIFIER

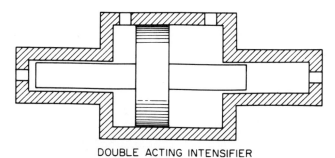

DOUBLE ACTING INTENSIFIER

FIGURE 9-13 Single- and double-acting intensifiers and ANSI symbols.

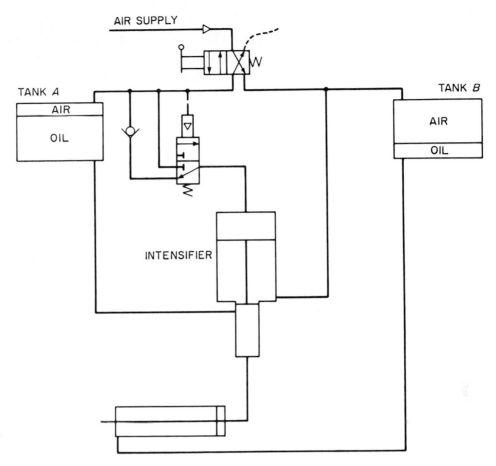

AIR SUPPLY

TANK *A*

AIR

OIL

TANK *B*

AIR

OIL

INTENSIFIER

FIGURE 9-14 Air-over-hydraulic intensifier.

contains entrained air, it compresses. This may cause a delay in the system response between the input flow and the output motion.

The compressibility factor of the hydraulic fluid becomes important when higher working pressures are used. Compressibility of a liquid is the change in volume due to pressure changes. The recommended compression value for intensifiers is about 2% of the total work cylinder volume. When conductors are large or long, all the fluid under compression should be calculated for accuracy.

EXAMPLE

> Required work force = 24,000 lb
>
> Total work cylinder stroke = 12 in.
>
> High pressure work stroke = 4 in.
>
> Available air pressure = 100 psi

What stroke intensifier or booster is required?

Solution:

1. Determine the size of work cylinder needed. (A standard 2.5-in.-bore-diameter cylinder is selected for this application.) This means that a pressure of nearly 5000 psi will be needed for the work portion of the stroke.

$$\text{Pressure} = \frac{\text{force}}{\text{area}} = \frac{24{,}000 \text{ lb}}{4.9 \text{ in.}^2} = 4890 \text{ psi}$$

2. Determine the fluid needed for the high-pressure phase of the work cycle.

$$\text{Volume (in.}^3) = \text{area (in.}^2) \times \text{stroke (in.)}$$

$$= 4.9 \text{ in.}^2 \times 4 \text{ in.}$$

$$= 19.6 \text{ in.}^3$$

3. Calculate the compressibility factor.

$$\text{Total cylinder volume} = 4.9 \text{ in.}^2 \times 12 \text{ in.}$$

$$= 58.8 \text{ in.}^3 \text{ of hydraulic fluid}$$

under 5000 psi

In this case approximately 2% of the total volume is selected for the compressibility factor of the fluid under pressure, or 58.8 in.³. So 0.02%/1.176 in.³ is lost through compressibility of the fluid in the cylinder only.

4. Use the following equation for determining the stroke of intensifier or booster:

Intensifier stroke

$$= \frac{\begin{array}{c}\text{fluid for high-pressure} \qquad \text{volume due} \\ \text{phase of work stroke (in.}^3) \; + \; \text{compress (in.}^3)\end{array}}{\text{area of ram piston (in.}^2)}$$

$$+ \text{ pretravel (in.)}$$

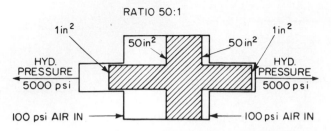

FIGURE 9-15 Intensifier ratio for example.

The pretravel is that portion of the intensifier stroke that is not effective for motion of the work cylinder and will vary with design. (Assume 1.5 in. pretravel.)

Since nearly 25,000 lb of force is needed and 100 psi plant air is available, a 50:1 ratio intensifier is selected (see Fig. 9-15).

$$\text{Intensified hyd. pressure} = \frac{\text{Ratio}}{50:1}$$
$$\text{Intensified hyd. pressure} = \frac{100\text{-psi air} \times 50 \text{ in.}^2}{1 \text{ in.}^2}$$
$$= 5000 \text{ psi}$$

Stroke equation:

$$\text{Intensifier stroke} = \frac{19.6 \text{ in.}^3 + 1.17 \text{ in.}^3}{1 \text{ in.}^2} + 1.5 \text{ in.}$$
$$= 22.27 \text{ in.}$$

Consulting a manufacturer's catalog can determine the exact intensifier or booster that will provide the proper stroke.

The system shown in Fig. 9-16 used on a riveting machine shows the tremendous flexibility in applying hydraulics where force multiplication is advantageous.

Neutral. With both pumps operating, hydraulic fluid flows through the open centered directional control valve and back to the reservoir.

Prefill. Shifting the directional four-way to the A position directs the flow from both pumps through lines 1 and 2 through the prefill passage and into line 3 for extending the work cylinder for a fast approach.

High-Pressure Stroke. When the machine reaches the rivet and resistance due to work causes a pressure increase great enough to open the sequence valve, fluid is directed to the large end of the intensifier through line 4. At the same time, pilot pressure unloads the large pump. The piston assembly of the intensifier advances covering the prefill port, then furnishes intensified fluid pressure to upset the rivet.

Return Stroke. When the rivet is set, the operator shifts the four-way directional valve to the B position. This

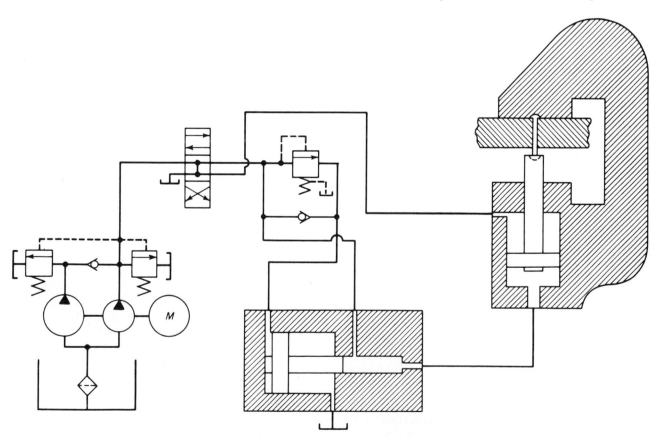

FIGURE 9-16 Force multiplication.

causes a drop in pressure, and the unloading valve closes to direct the flow from both pumps through line 5, returning the work cylinder and the intensifier ram piston assembly. The check valve allows free flow of fluid from the large end of the intensifier back to the reservoir.

HYDROSTATIC TRANSMISSIONS

Hydraulic power transmissions are of two general types, hydrokinetic and hydrostatic. The hydrokinetic type uses the kinetic energy of a high velocity flow of fluid, whereas the hydrostatic power transmission uses high pressures and relatively low velocities.

Detailed Comparison of Hydrodynamic/Hydrostatic Transmissions

There has been confusion in the minds of many engineers over the distinction between a hydrodynamic and hydrostatic transmission. A brief résumé of the points of difference will be given in tabular form below to explain this difference.

Hydrodynamic Transmissions

1. A change in fluid *velocity* yields output energy.
2. Nonpositive displacement units.
3. No output torque available at low input speeds.
4. Approximately 6:1 speed ratio input; output maximum practical.
5. Tend to creep.
6. Nonreversing.
7. Input and output directly coupled.
8. Torque varies over wide range.
9. Practically no internal braking.
10. Speed drops off with load.
11. Only limited control.

Hydrostatic Transmissions

1. A change in fluid *pressure* yields output energy.
2. Positive displacement units.
3. Output torque available at all speeds.
4. Approximately 60:1 speed ratio is practical with hydrostatic (range 50 to 3000 rpm).
5. No creeping tendency.
6. Fully reversible.
7. Input and output can be remote or directly coupled.
8. Substantially constant torque available over full speed range of hydraulic motor.
9. Direct braking available.
10. Substantially constant speed with load variations.
11. Positive control available.

The hydrokinetic type is doing an excellent job on such applications as automatic transmissions for automobiles. However, in the field of fluid power the hydrostatic type provides greater flexibility of application and more positive control. The term *hydrostatic transmission* refers to the use of hydraulic pumps and motors for converting fluid power into mechanical rotary motion.

Both fixed and variable piston pumps and fluid motors have been used on hydraulic drives for many years. Applications that require critical control and reliability, such as steering systems for ships, naval gun turrets, and other uses, have found that hydraulic drives are unmatched for response characteristics and power performance.

During the last two decades, improvements in design and operational performances have opened new fields of usage which hold unlimited possibilities for machines and equipment of the future. Hydraulic drives are now available up to 1000 hp, and the trends indicate even higher horsepower with the advent of higher pressures, which are possible through the availability of better metals and materials.

Because hydrostatic transmissions utilize positive displacement pumps and motors, they offer a number of advantages over other systems:

1. Constant speed can be maintained with varying load requirements.
2. Constant torque can be maintained with varying rotational speeds.
3. A wide range of operating speeds is available up to 10,000 rpm, with some aerospace designs operating much higher.
4. Systems are available for rotation in one direction or for reversible operations.
5. Rotational speeds can be precisely controlled over the full torque range of the motor.
6. Direct braking can be obtained by applying back pressure on the rotating motor.
7. Compact and lightweight units are capable of high power.
8. Both open-circuit and closed-circuit systems are available.
9. High overall efficiencies are available.
10. Hydrostatic transmissions are relatively simple in operation and easy to maintain compared to mechanical transmissions.

Figure 9-17 shows a typical application which uses a hydrostatic transmission drive.

FIGURE 9-17 Hydrostatic transmission application. (Courtesy of Dynex/Rivett Inc., Pewaukee, Wis.)

EXAMPLE

Select the pump and motor for a hydrostatic transmission that has the following requirements:

1. The fluid motor must have a speed range between 0 and 2000 rpm.
2. The maximum torque requirement is 680 lb in.
3. Fluid motor must operate in both directions.

In Fig. 9-18 a typical closed-circuit system is shown in graphical symbols indicating the components necessary for replenishing, filtering, and braking.

(a) Determine the fluid motor displacement necessary for the torque requirement. (Select an operating pressure of 1500 psi.)

(b) Find the maximum pump delivery.

(c) How much horsepower input is needed?

Solution:

(a)
$$T = \frac{\text{psi} \times \text{disp. (in.}^3)/\text{rev.}}{2\pi}$$

$$680 = \frac{1500 \times \text{disp. (in.}^3)/\text{rev.}}{6.28}$$

$$\text{disp. (in.}^3)/\text{rev.} = \frac{6.28 \times 680}{1500}$$

$$= 2.85 \text{ in.}^3 \text{ per revolution}$$

(b) Maximum pump delivery

$$= \frac{\text{maximum motor speed} \times \text{displacement}}{231}$$

$$= \frac{2000 \times 2.85}{231}$$

$$= 24.6 \text{ gpm}$$

If the pump and motor together have a volumetric efficiency of 90%, then 10%, or 2.46 gpm, extra fluid would be needed for making up slip and leakage.

$$\text{Total pump capacity} = 27.06 \text{ gpm}$$

(c)
$$\text{hp} = \text{gpm} \times \text{psi} \times 0.000583$$

$$= 27.06 \times 1500 \times 0.000583$$

$$= 23.66$$

There are many other important factors of consideration which would depend on the specific end use of the system.

1. The duty cycle for the equipment operation may require a heat exchanger.
2. The type of pump control, such as manual, electrical, mechanical, pilot, or servo, may be important for functional machine control.

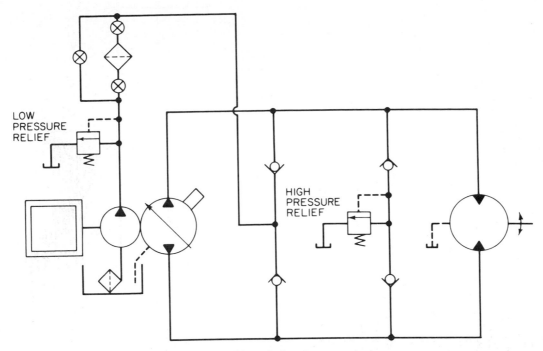

FIGURE 9-18 Closed-circuit transmission.

3. The environmental conditions, such as heat, altitude, contamination, and other physical aspects, may dictate such things as special seals, additional components, or fire-resistant fluid.

4. Cost is a big factor which may suggest a less expensive method of doing the job.

In general, the same method of approach for selecting component types, sizes, ports, and performance data is used for both linear and rotary circuits. The previous circuit selected would be a sophisticated solution to the problem. Reversible or over-the-center pumps are used with the closed-circuit approach. However, an open-circuit system could also resolve the problem with a little less sophistication.

The main difference in the two systems is that in the closed circuit the reversible pump controls the speed in both directions. In the open circuit a directional valve controls the direction of rotation. Fluid in the closed circuit circulates in a closed circuit with only the leakage replenished. The pump in the open circuit uses fluid from the reservoir for the inlet and then exhausts it back to the reservoir through the four-way valve.

Replenishing during the braking period for a heavy load may present a problem in the open-circuit system. Sometimes, heavy loads keep the fluid motor rotating after the four-way valve is shifted to the center position. This means that fluid must be supplied to the motor because, essentially, it is now a pump and could be damaged by lack of fluid.

The method shown in Fig. 9-19 may be a possible solution. Replenishing fluid to the low-pressure side of the circuit can be accomplished by connecting the reservoir fluid to both sides of the motor through a check valve arrangement.

Braking heavy inertia loads might be done as shown, with a counterbalance valve put into operation by an on-off type of valve which directs the flow to the reservoir or through the braking type counterbalance valve.

It is important to remember that shifting four-way valves from one position to the other may give an open center condition or a closed center condition in between the various positions.

A heavy load rotating could feasibly rupture a line if the fluid has no place to go, even for a fraction of a second; therefore, an open center condition is necessary.

It is also important to know how much back pressure a four-way valve can stand if the exhaust fluid going back to the reservoir is restricted. This also makes it imperative to connect drains to a separate line going back to the reservoir when components are specified to be externally drained.

The circuit calculations would be made in the same manner as before. Either method used would still require the same horsepower as before. However, the pressure could be lowered if the displacement of the motor were increased proportionally, which would still provide the same torque as before. This would help resolve problems of cost by perhaps changing to vane equipment instead of piston pump and motor. Figure 9-20 shows one type of hydrostatic transmission using a variable volume reversible pump and a fixed displacement fluid motor.

A third solution might be the use of a fixed delivery type of pump. This would necessitate the use of a flow control valve to vary the speed of the fluid motor.

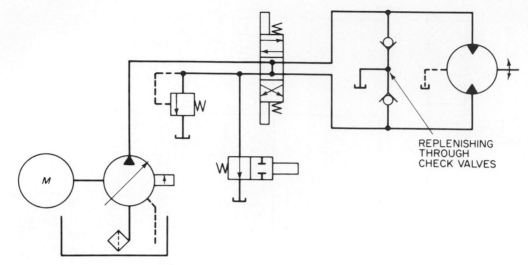

FIGURE 9-19   Open-circuit transmission.

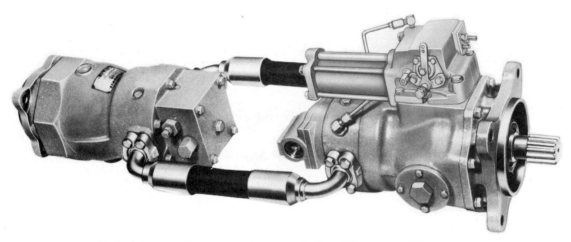

FIGURE 9-20   Closed-circuit transmission. (Courtesy of Dynapower a Unit of General Signal, Watertown, N.Y.)

The foregoing discussion indicates the great amount of flexibility available in fluid power components which can help to solve a problem with any degree of sophistication that may be required. Hydrostatic transmissions hold a tremendous potential for future applications in all fields of usage.

ELECTRICALLY CONTROLLED CIRCUITS

Electrical control devices play a big role in the application of fluid power equipment. Switches, timers, relays, solenoids, and other such devices help to control the stopping, starting, sequencing, speed, positioning, timing, and reversing of actuating cylinders and fluid motors.

Automatic machines such as those used in the machine tool industry rely mainly on the electrical components to control the hydraulic muscles for doing the work. The aircraft and mobile equipment fields have also found that electricity and fluid power work very well together, especially where remote or dual control is essential.

Through the use of a simple push button switch, an operator can put complex machinery into operation that may complete hundreds of machining operations toward the completion of a product such as an automobile or farm implement.

The wiring, starters, relays, and other devices used in an electrical controller for a hydraulic machine are selected according to the operation that must be performed. Push button switches (see Fig. 9-21) are used mainly for starting, stopping, or for providing manual override in case of emergency. Starters are used to protect the electric motor against undervoltage when one is starting hydraulic pumps and bringing them up to operating speed (see Fig. 9-22).

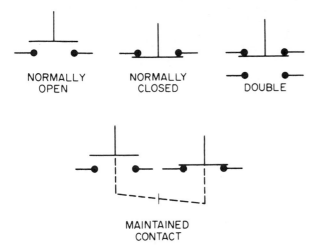

FIGURE 9-21 Push-button switches.

and the contacts that are normally open will close. They are generally spring-loaded to assume their normal position when the coil is deenergized.

Through an arrangement of the contacts in a relay, interlocking can be designed into the controls of a machine to prevent accidental operation of a machine out of the proper work cycle sequence. Limit switches are switches that energize or deenergize coils used for positioning relay contact switches. Solenoids used on hydraulic control valves should be actuated through relay contacts, because they require a high current and need protection. Use of relays permits interlock to avoid accidental actuation of the two solenoids in opposite ends of a valve spool, which would cause burnout of one or both of the solenoids.

Pressure switches utilize the hydraulic system pressure to open or close contact switches. A pressure switch may

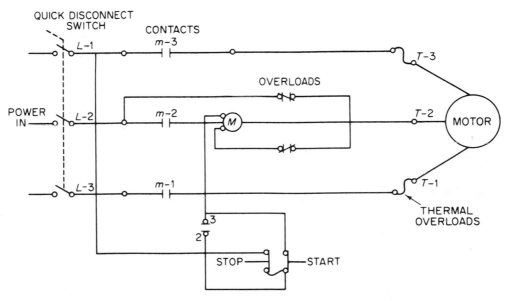

FIGURE 9-22 Starter circuit: three-phase wiring diagram for starting and electric motor drive for a hydraulic pump.

Standard color coding has been adopted by the industry, which requires ac and dc line and load wires to be black or gray; ac control wires should be red, and dc control wires should be blue.

Solenoids (see Fig. 9-23) are used for a push or pull force and consist of a coil and a movable iron core which pushes or pulls due to a magnetic force when the coil is energized. Solenoids are available for forces of a few ounces up to 50 lb. Stroke lengths range from zero to 1 in. Coils capable of producing forces greater than 50 lb are rather large and heavy.

Relays (see Fig. 9-24) are multiple switches which open or close when a coil is energized. A relay may have 15 or 20 contact switches, some being normally closed while others are normally open. In other words, if the coil is energized, the contacts that are normally closed will open

have a high-pressure setting and a low pressure setting. For instance, if it were necessary to start and stop a hydraulic power source to maintain a given pressure, the low side would start the pump and the high side would stop it. They are generally used to initiate some phase of a work cycle, or they may be applied as a safety device (Fig. 9-25).

Time-delay devices are used to adjust the time phase of a working cycle which may provide a dwell or feed condition. A dwell is sometimes needed in such machine operations as drilling, where the drill pauses at the end of the stroke to clean out the hole. Most timers have an adjustment which gives a precise control for specific feed rates or dwells for machining synchronization. Timers as shown are used to provide measured time sequences of a machine's operation (see Fig. 9-26).

The fluid dashpot timing relays are essentially sole-

FIGURE 9-23 Typical solenoids. (Courtesy of Detroit Coil Company, Ferndale, Mich.)

NORMALLY CLOSED   NORMALLY OPEN   OVER LOAD

FIGURE 9-24 General contacts for starters, relays, and so on.

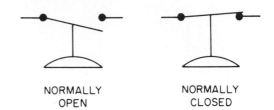

NORMALLY OPEN   NORMALLY CLOSED

FIGURE 9-25 Pressure switch.

COIL ENERGIZED

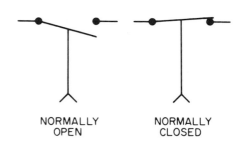

NORMALLY OPEN   NORMALLY CLOSED

COIL DE-ENERGIZED

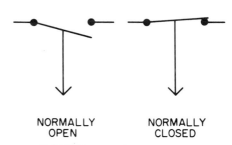

NORMALLY OPEN   NORMALLY CLOSED

FIGURE 9-26 Timer contacts.

noid-operated switches, with the action of the solenoid being delayed by the metering of fluid through an adjustable orifice.

Timers of this type are used where the sequence of operation requires a delay with reasonable accuracy. Because the hydraulic fluid's viscosity varies with temperature change, precise accuracy is not always possible.

Pneumatic timing relays are useful for applications requiring close accuracy. This type is extremely flexible and uses a bellows for controlling the tripping time. They are used for both "on-delay" and "off-delay" operations.

The circuit in Fig. 9-27 shows a three-phase wiring diagram for starting an electric motor to drive a hydraulic pump. When the operator pushes the start button, it completes the circuit from L-1 to L-2, which energizes relay coil M. The relay contacts M-1, M-2, M-3, and M-4 close, completing the circuit to start the electric motor. The contact M-4 around the starter switch provides a holding circuit so the operator can let go of the start button after the motor starts. The quick-disconnect line switch must be closed before the motor can start, and should never be opened with

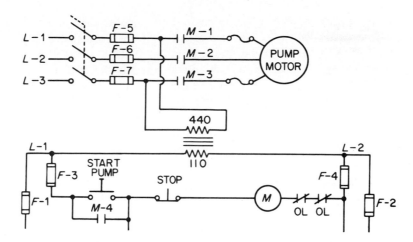

FIGURE 9-27 Electrical circuit pump drive motor.

the motor running under load. Higher-voltage circuits will arc across the switch contacts if opened under load and cause a severe burn or other accident to operating personnel and equipment.

Generally, a transformer is used to reduce the high voltage of the motor circuit to a lower voltage for the control and solenoid circuits. It is also common practice to go through fuses to provide adequate protection for each circuit. The thermal overloads will open if low voltage occurs during startups or if the hydraulic pump is overloaded.

The stop button switch can be operated at any time to stop the electric motor and pump. It is located to have priority control over other devices in the circuit.

TYPICAL APPLICATIONS CIRCUITS

The hydraulic circuit in Fig. 9-28 shows a combination pump unit operated by an electric motor. As the operator pushes the start button, the electric motor starts and the pump rotates. Fluid delivered by both pumps passes through the open-centered four-way valve and back to the reservoir.

When the cycle start button is pushed, solenoid A is energized, and the flow is directed to the cylinder for the rapid advance portion of the cycle with both pumps. Limit switch (LS-1) is a normally closed, held open switch when the cylinder is retracted; hence, it closes at the start of the rapid advance.

FIGURE 9-28 Combination pump unit operated by an electric motor.

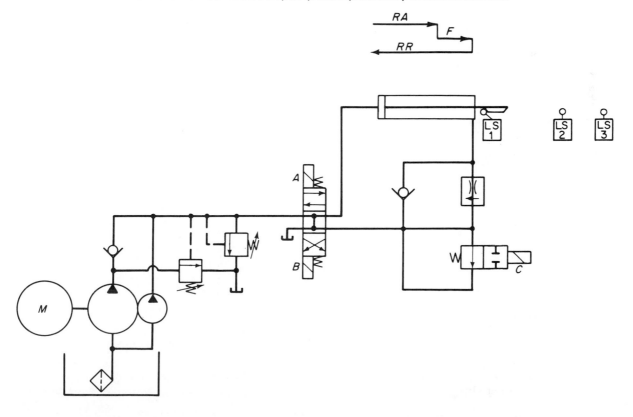

As the cam on the cylinder actuates LS-2, solenoid C is energized, closing the two-way valve. This causes the fluid from the rod end of the cylinder to be metered out through the flow control valve. The machine is now in the feed portion of the cycle, with pilot pressure holding the unloading valve open to unload PF_1 freely back to the reservoir.

At the end of the feed stroke LS-3 is actuated, which must deenergize solenoids A and C, as well as energize solenoid B. As this happens, both the four-way and two-way valves shift to the opposite position. The flow from both pumps is now directed to the rod end of the cylinder, and the machine is in the rapid return stroke.

When the cylinder is fully retracted, the cam actuates LS-1 to open, which deenergizes solenoid B, and the four-way valve is centered by spring force to allow the flow of both pumps to pass back to the reservoir until the cycle is initiated again.

ELECTRICAL CIRCUITS

When the operator pushes the motor start button, the hydraulic pumps operate with the flow going back to the reservoir through the four-way directional valve. Figure 9-29 shows the electrical diagram for this system.

Pushing the cycle start button energizes CR-1 relay, which closes contacts CR1-1, CR1-2, and CR1-3, energizing solenoid A to provide the rapid advance. As the machine

moves forward, LS-1 closes. At the end of the rapid advance LS-2 is actuated to energize relay CR-2, closing contacts CR2-1, CR2-2, and CR2-3, which energizes solenoid C to start the feed stroke.

At the end of the feed stroke LS-3 is actuated, energizing CR-3, closing contact CR3-1; however, because CR3-2 and CR3-3 are normally closed, they will open. Then, as CR3-2 and CR3-3 open, contacts CR3-4 and CR3-5 close. This entire action of the relay causes solenoids A and C to deenergize and solenoid B to energize. The machine rapidly returns, and as LS-1 is held open again, it deenergizes relay CR-3, which deenergizes solenoid B as the contacts in the relay open (CR3-1, CR3-4, and CR3-5) or close (CR3-2 and CR3-3). The machine is now in the neutral or center position.

The normally opened or closed contacts, as indicated in relay CR-3, provide an interlocking safety feature to provide proper machine sequencing and to prevent two solenoids, on the same valve, from being energized at the same time. The forward inch button and the return inch button are used to provide a manual override for machine setups or tool breakage.

ELECTROHYDRAULIC SERVO VALVES

Electrohydraulic servo valves are used for precision control of acceleration, velocity, and position in the hydraulic control of machine tools, spacecraft, radar, submarines, nuclear reactors, heavy-duty robots, and other applications (Fig. 9-30).

A servo valve (Fig. 9-31) is a control actuated by a feedback system which compares the output with the reference signal and makes corrections to reduce the difference. Its function is that of a transducer which transforms an electrical signal into fluid power. Servo valves may be used

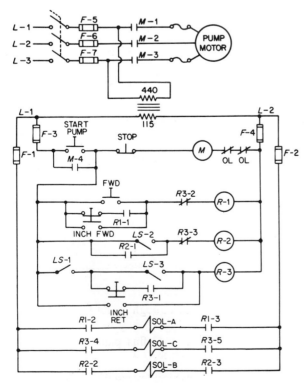

FIGURE 9-29 Circuit.

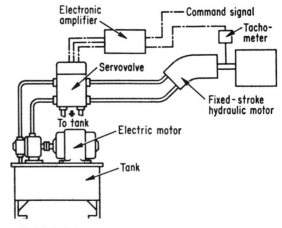

FIGURE 9-30 Complete servo circuit. (Courtesy of Vickers, Incorporated, Troy, Mich.)

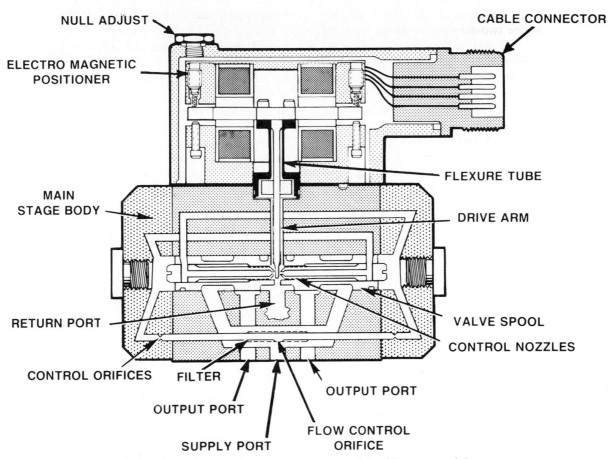

NULL ADJUST

CABLE CONNECTOR

ELECTRO MAGNETIC
POSITIONER

FLEXURE TUBE

MAIN
STAGE BODY

DRIVE ARM

RETURN PORT

VALVE SPOOL

CONTROL NOZZLES

CONTROL ORIFICES   FILTER

OUTPUT PORT

OUTPUT PORT

FLOW CONTROL
ORIFICE

SUPPLY PORT

FIGURE 9-31 Hydraulic servo valve assembly. (Courtesy of Pega-
sus/AMCA International, Troy, Mich.)

to control flow to and from the hydraulic actuator or to
control the volume of a variable pump.

The electrical input signal to the servo valve must be
converted into mechanical motion to actuate the control.
Generally, an electromagnetic torque motor is used to stroke
the mechanical actuator to the valve. The power level of
the electrical input signal to the torque motor is only a watt
or less, but the power level of the output, measured in
hydraulic fluid displacement, may be many thousand times
that of the input. For example, a 15-gpm servo valve at
3000 psi controls 19.6 kW, with as little as 45 mW electrical
input. The resulting power amplification is 35,000:1. The
servo valve provides an output directly proportional to its
input and is a very powerful hydraulic amplifier when interconnected with its associated components in a circuit.

TORQUE MOTORS

In Fig. 9-32, the armature and valve actuator drive arm are
suspended on the flexure tube, which acts as a pivot point.
The top pole piece and lower pole piece are attached to
opposite ends of the permanent magnets. The flux patterns

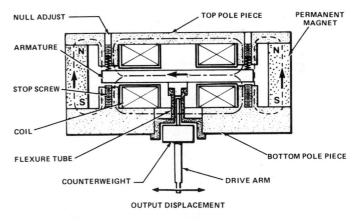

NULL ADJUST

TOP POLE PIECE

PERMANENT
MAGNET

ARMATURE

STOP SCREW

COIL

FLEXURE TUBE

BOTTOM POLE PIECE

COUNTERWEIGHT

DRIVE ARM

OUTPUT DISPLACEMENT

FIGURE 9-32 Torque motor. (Courtesy of
Pegasus/AMCA International, Troy, Mich.)

due to the magnet are shown as dashed lines. This flux
passes from top to bottom through all four of the active air
gaps. The two coils are magnetically in series. If a control
current is applied, the control flux pattern shown as dot-
dash lines will be produced. Note that there are two counter-
rotational paths involved, each having the armature in common. If the flux polarity through the armature is as shown

(from right to left), the control flux will reinforce the permanent flux in the lower left and upper right air gaps and will oppose the permanent flux in the other two air gaps. The unbalanced flux in the air gaps thus produce a rotational force (torque) on the armature, with the left side moving upward and the right side moving downward. This motion also moves the rigidly attached drive arm to the left of the figure. The displacement of the drive arm is proportional to the control current.

The gain of torque motor, drive arm motion versus control current is controlled by several factors. The most significant are: the area of the armature, the armature air gap at null, the number of turns in the coil, the flux of the permanent magnet, and the spring rate of the torque tube. The variations of turns in the coil and magnet charge permits a single-size torque motor to be used for a wide range of stroke and input power. For a typical servo valve up to 10 gpm flow range at 1000 psi, the stroke may vary from 0.005 to 0.015 in. and electrical input from 45 to 1200 mW using the same-size torque motor components.

BOOST SYSTEMS

The valve boost system is a hydraulic amplifier which amplifies the drive arm motion into a high force level to position the four-way spool valve. In the position feedback valve, it is desirable to have the spool follow the position of the torque motor drive arm.

Figure 9-33 shows the symmetrical arrangement of the design with both sides of the valve equal in all respects. Therefore, with the input signal equal to both coils of the torque motor, the drive arm is centered and the main spool is at neutral.

If the current signal increases on coil A, the armature moves to actuate the drive arm against nozzle A. As the flapper arrangement of the drive arm tends to resist the flow out of nozzle A, pressure increases in chamber a. Force due to the increase in pressure causes the main spool to move in direction X, as shown in Fig. 9-34. It will move toward the lower-pressure end b until the flapper is again in the center between the two nozzles, which is shown in Fig. 9-35.

The boost system causes the four-way spool to follow the position output of the torque motor in response to changes in electrical input signals to the torque motor.

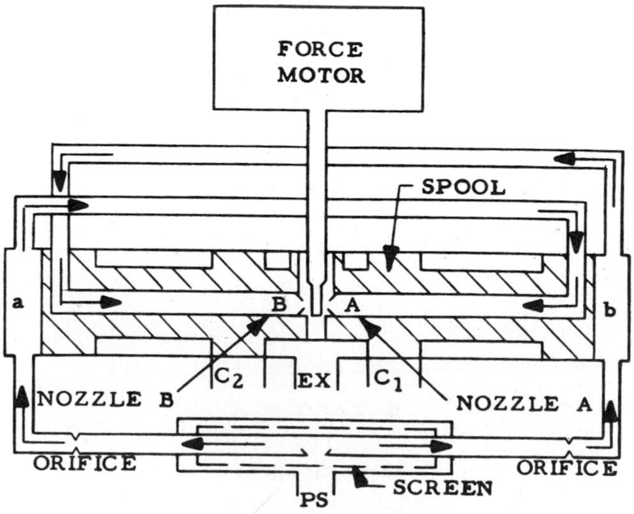

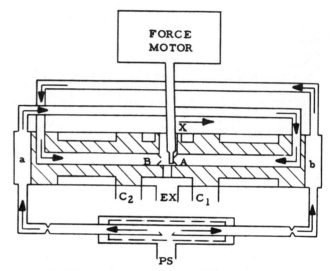

FIGURE 9-34   Force motor. (Courtesy of Pegasus/AMCA International, Troy, Mich.)

FIGURE 9-33   Boost system: center position. (Courtesy of Pegasus/AMCA International, Troy, Mich.)

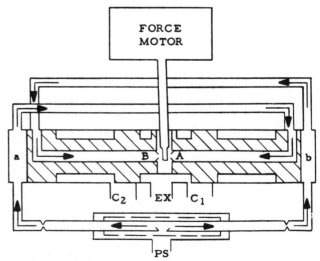

FIGURE 9-35   Force motor. (Courtesy of Pegasus/AMCA International, Troy, Mich.)

FOUR-WAY SPOOL VALVE ACTION

Figures 9-36 through 9-38 show the corresponding motion that would result from a change in electrical input to the torque motor, as described previously. The spool is designed with four metering edges, L-1, L-2, L-3, and L-4, which are approximately on line when the spool is centered. The strength of the electrical signal determines the amount of flow metered to the cylinder and its velocity.

The servo valve design shown in Fig. 9-39a employs a piloted spool with a linkage to provide feedback to the

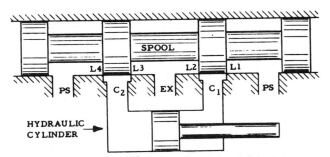

FIGURE 9-36 Four-way spool valve action: center position. (Courtesy of Pegasus/AMCA International, Troy, Mich.)

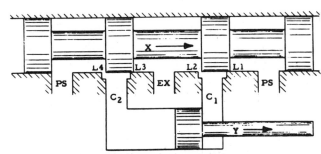

FIGURE 9-37 Four-way spool valve action: cylinder. (Courtesy of Pegasus/AMCA International, Troy, Mich.)

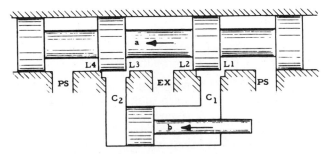

FIGURE 9-38 Four-way valve action: cylinder retracting. (Courtesy of Pegasus/AMCA International, Troy, Mich.)

input signal area. The design shown in Fig. 9-39b uses a swing plate arrangement which is attached to the armature of the torque motor by a push rod. A differential current signal causes the push rod to move in a direction depending on the coil receiving the most current. The swing plate is mounted on flat springs which allow a ±0.015-in. movement from the centered position. The swing plate's underside is contoured and machined flat to match ports. As the push rod moves the plate, hydraulic fluid from the center port will enter the proper actuator port, depending on the direction of the armature. The exhaust or drain connection to the reservoir is provided in the top section of the valve body, which creates a flooded enclosure.

SERVO CONTROL OF VARIABLE VOLUME PUMP

Pump flow, in the system shown in Fig. 9-40, is determined by the position of the slide block race. With the pump rotating, the slide block is positioned eccentric to the cylinder. Pistons move outward on the bottom and inward on top. They receive fluid from the bottom side of the pintle valve and force the fluid into the top side of the pintle valve.

In the view in Fig. 9-41, the slide block is positioned concentric with the cylinder and pump flow to the hydraulic system is zero. Figure 9-42 shows the slide block positioned with eccentricity opposite Fig. 9-40, and flow to the system is reversed.

When the swing-plate type servo valve is deflected downward in Fig. 9-40, fluid from the gear pump enters stroking piston P-1. At the same time, fluid stroking piston P-2 opens to drain, and the slide block moves right.

If the swing plate is deflected upward, the slide block moves to the left. Centering the swing plate stops and holds the slide block. The torque motor armature deflects the swing plate proportionally to current received from the amplifier.

The LVDT (linear variable differential transformer) transmits an electrical voltage proportional to the slide block's position. The slide block receives a remote voltage command from a potentiometer (instrument that controls electrical potential) wired to the amplifier in series with the pump LVDT. The slide block moves until LVDT voltage equals potentiometer voltage. When the electrical differential is zero, the swing plate centers.

A reference transformer provides voltages for exactly full pump flow in each direction. The reference transformer then supplies excitation voltage for command potentiometers, calibrating them for zero to full pump flow over zero to full potentiometer rotation. Figure 9-43 is an excellent example of servo application as shown in the Oilgear steering system for ships.

2. IN NEUTRAL, LARGE
PILOT END IS BLOCKED
AT PILOT VALVE IN
THE STATIC CONDITION.
THIS PRESSURE = 1/2 CONTROL
PRESSURE. (Pc)

3. CONTROL PRESSURE
IS PRESENT HERE AND
AT SMALL END OF
MAIN SPOOL

PILOT STAGE
SLEEVE

SUPPLY PRESSURE

LINKAGE FULCRUM
(VARIABLE)

PILOT SPOOL

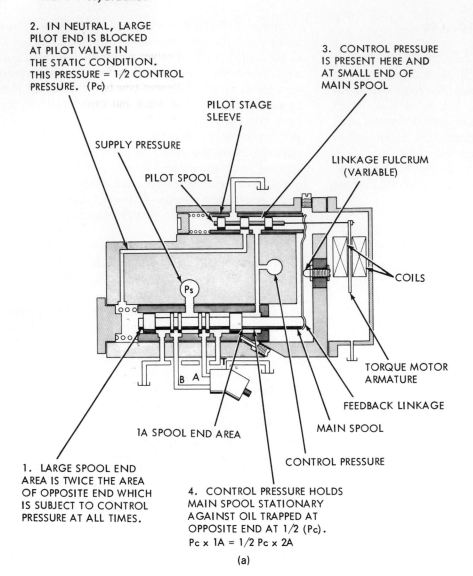

COILS

Ps

TORQUE MOTOR
ARMATURE

FEEDBACK LINKAGE

B A

1A SPOOL END AREA

MAIN SPOOL

CONTROL PRESSURE

1. LARGE SPOOL END
AREA IS TWICE THE AREA
OF OPPOSITE END WHICH
IS SUBJECT TO CONTROL
PRESSURE AT ALL TIMES.

4. CONTROL PRESSURE HOLDS
MAIN SPOOL STATIONARY
AGAINST OIL TRAPPED AT
OPPOSITE END AT 1/2 (Pc).
Pc x 1A = 1/2 Pc x 2A

(a)

ARMATURE
FULCRUM

PUSHROD

SWING PLATE

RETURN
PORT

TORQUE
MOTOR
COILS

SWING
PLATE
MOTION

PERMANENT
MAGNETS

FLAT
SPRING
SWING
PLATE
SUPPORTS

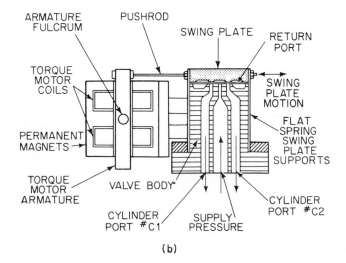

TORQUE
MOTOR
ARMATURE

VALVE BODY

CYLINDER
PORT #C2

CYLINDER
PORT #C1

SUPPLY
PRESSURE

(b)

FIGURE 9-39 (a) Spool-type servo valve with feedback mechanism. (Courtesy of Vickers, Incorporated, Troy, Mich.) (b) Servo valve with swingplate. (Courtesy of The Oilgear Company, Milwaukee, Wis.)

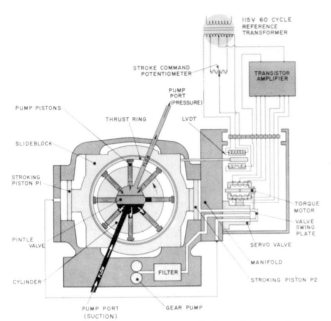

FIGURE 9-40 Servo-controlled variable-volume pump. (Courtesy of The Oilgear Company, Milwaukee, Wis.)

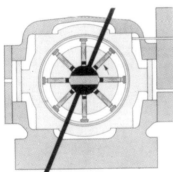

FIGURE 9-41 Slide block concentric with cylinder pump flow equals zero flow. (Courtesy of The Oilgear Company, Milwaukee, Wis.)

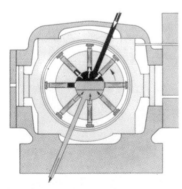

FIGURE 9-42 Slide block: flow reversed. (Courtesy of The Oilgear Company, Milwaukee, Wis.)

LOGIC CIRCUITRY

Servo and proportional valves are useful for control of many pilot-operated devices as well as for the control of spool-type four-way valves. Poppet-type two-way valves are available in a wide range of sizes and capacities. Pressures of 5000 psi and greater can be accommodated. Poppet-type valving structures offer significant design, manufacturing, and usage advantages.

The poppet element is perhaps the simplest of fluid power devices. Yet, with the use of appropriate pilot controls, poppet mechanisms can provide directional, pressure level, and flow rate control with a high degree of accuracy and in an infinitely variable pattern.

We have studied the use of the piloted poppet as a pressure control mechanism (Fig. 9-44). The pilot-operated check valve is a second very important application of the poppet structure as a selectively controlled closure. Check valves are, of course, basic to fluid power systems.

A third use is gaining significant importance in fluid power circuitry. This third use is in the control of direction of flow. The poppet of Fig. 9-45 is designed with three basic area relationships. All three configurations fit into a cartridge-type structure as shown in Fig. 9-46. The manifold can be structured as needed for a specific installation. The cap, too, can be structured to accommodate control devices or merely act as a closure for the cartridge. Pilot lines may be provided in the cap structure.

The first area to be considered is the major diameter of the poppet. This is the largest bore in the envelope in which the poppet is inserted. The envelope may be a shell in which the poppet moves. The shell and poppet is then referred to as a *cartridge*. The sizes of these poppet assemblies are standardized. The major diameter becomes a fixed value as the design is finalized (A_w in Fig. 9-47). It has a relationship with other areas.

The second area considered can be the seal diameter, where the basic closure is encountered (A_a in Fig. 9-48). This area determines the characteristics of two potential control avenues. One is the ratio of this "nose" area to the major diameter of the poppet. The second is the ratio of the major diameter of the poppet minus the nose area or the "shoulder" area. These area relationships can be quite flexible but must be rigidly determined to permit manufacturing economies and a degree of standardization.

In summary we can conclude that

Major area A_w = nose area A_a + shoulder area A_b

or

$$A_w = A_a + A_b$$

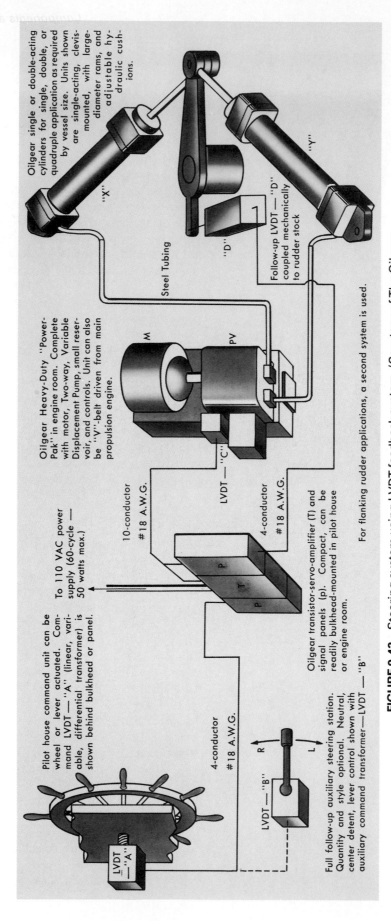

Oilgear single or double-acting cylinders for single, double, or quadruple application as required by vessel size. Units shown are single-acting, clevis-mounted, with large-diameter rams, and adjustable hydraulic cushions.

"X"

"Y"

"D"

Follow-up LVDT—"D" coupled mechanically to rudder stock

Steel Tubing

Oilgear Heavy-Duty "Power-Pak" in engine room. Complete with motor, Two-way, Variable Displacement Pump, small reservoir, and controls. Unit can also be "V"-belt driven from main propulsion engine.

M

PV

LVDT—"C"

10-conductor #18 A.W.G.

4-conductor #18 A.W.G.

To 110 VAC power supply (60-cycle — 50 watts max.)

Pilot house command unit can be wheel or lever actuated. Command LVDT—"A" (linear, variable, differential transformer) is shown behind bulkhead or panel.

Oilgear transistor-servo-amplifier (T) and signal panels (p). Compact, can be readily bulkhead-mounted in pilot house or engine room.

For flanking rudder applications, a second system is used.

P

T

P

4-conductor #18 A.W.G.

LVDT—"B"

R

L

LVDT—"A"

Full follow-up auxiliary steering station. Quantity and style optional. Neutral, center detent, lever control shown with auxiliary command transformer—LVDT —"B"

FIGURE 9-43 Steering system using LVDT feedback system. (Courtesy of The Oilgear Company, Milwaukee, Wis.)

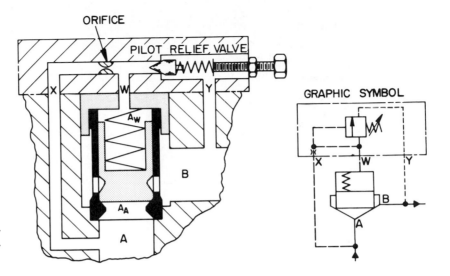

FIGURE 9-44 Logic-type pressure control. (Courtesy of The Oilgear Company, Milwaukee, Wis.)

FIGURE 9-45 Logic cartridge in steel block envelope. (Courtesy of The Oilgear Company, Milwaukee, Wis.)

FIGURE 9-46 Logic cartridge assembly. (Courtesy of The Oilgear Company, Milwaukee, Wis.)

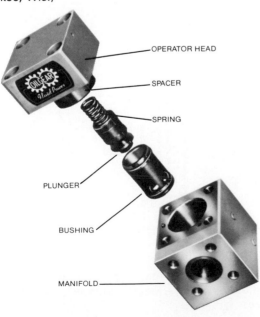

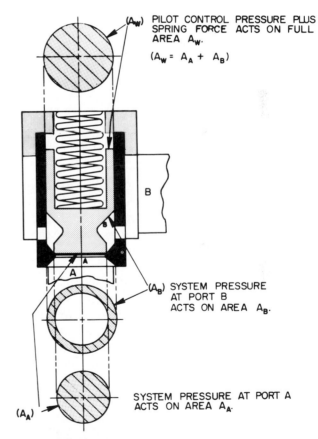

(A_W) PILOT CONTROL PRESSURE PLUS SPRING FORCE ACTS ON FULL AREA A_W.

$(A_W = A_A + A_B)$

(A_B) SYSTEM PRESSURE AT PORT B ACTS ON AREA A_B.

SYSTEM PRESSURE AT PORT A ACTS ON AREA A_A.

(A_A)

FIGURE 9-47 Area ratios. (Courtesy of The Oilgear Company, Milwaukee, Wis.)

Area A_w is always equal to or greater than area A_a or A_b. Thus a series of pressure ratios at A_w, A_a, or A_b can be developed to determine the potential movement or seal characteristics of the poppet assembly. In these relationships we have dealt with a device that can provide a seated seal with virtually no leakage.

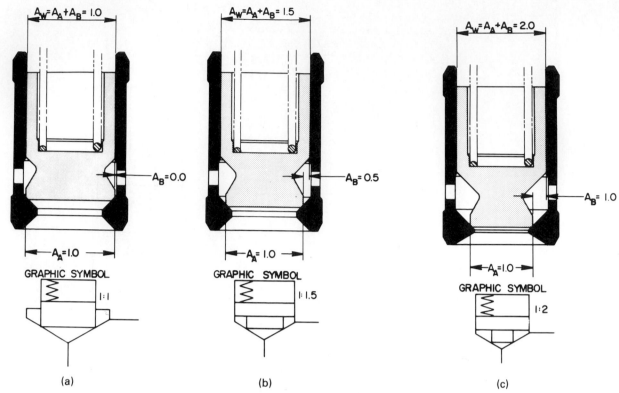

FIGURE 9-48 (a) 1:1 ratio; (b) 1:1.5 ratio (c) 1:2 ratio. (Courtesy of The Oilgear Company, Milwaukee, Wis.)

The assembly of Fig. 9-49 consists of a cylindrical shell with a sliding inner member which can be biased by physical weight or a mechanical spring to a normally open position. Again, consider the major diameter as A_w. The nose diameter here is also equal to A_w and the shoulder diameter is 0. The pilot structure needed for control of the normally open device (if biased to the open position) will be significantly different than that used to control the device shown in Fig. 9-48. Of course, the normally open cartridge will often be used to act as a reducing-valve compensation device for a flow control function. Thus a single poppet assembly can be used for pressure control functions. A single poppet can be used to create a variable passage-way as a flow control function.

A single poppet can also be used as a check device for the simple function of flow in one direction only. Also, a single poppet with selected area relationships can be used to control flow by addition or relaxation of pilot pressure to the major diameter cavity A_w (Fig. 9-50).

Valve plungers are also available with a tapered "cushion nose" (Fig. 9-51). This nose provides gradual opening

FIGURE 9-49 Normally open logic cartridge. (Courtesy of Rexroth Corporation, Bethlehem, Pa.)

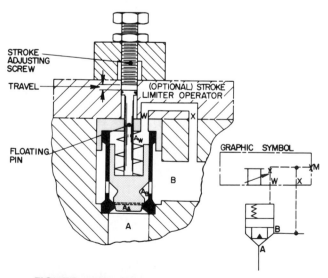

FIGURE 9-50 With stroke limiter option. (Courtesy of The Oilgear Company, Milwaukee, Wis.)

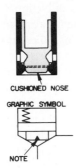

FIGURE 9-51 Cushion nose option for logic cartridges. (Courtesy of The Oilgear Company, Milwaukee, Wis.)

and closing of flow passage to pressure for decompression control and smooth operation of hydraulic system. When used with a stroke-limiting option, it provides a very fine flow adjustment (Fig. 9-50).

Two poppets can be used for a three-way function (Fig. 9-52). Supply flow can be directed to one poppet. Return to tank can be directed or blocked by the second poppet. Thus suitable pilot operation can permit flow to a cylinder, motor, or diversion function. The flow can be blocked at will. The second poppet connected to the reservoir line can be selectively opened to permit flow to the tank to return a cylinder or permit decay of pressure according to the pilot function. Figure 9-53 shows four poppets

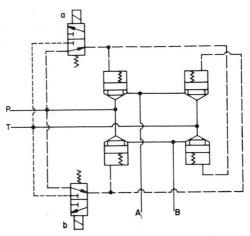

CONVENTIONAL SPOOL TYPE SYMBOL	SOLENOIDS	
	DE-ENERGIZED	ENERGIZED
A.	a & b	
B.		a & b
C.	b	a
D.	a	b

FIGURE 9-53 Four-way function. (Courtesy of The Oilgear Company, Milwaukee, Wis.)

in a circuit for control of a reciprocating cylinder or rotation of a motor.

A normally closed switching element as shown in Fig. 9-54 is vented to close and piloted to open. Pressure at port 1 or 2, acting on areas A1 or A2, plus the bias spring force, tends to close the valve or add to the closing force. Pilot pressure at port 3, acting on area A3, tends to open the valve. Normally closed valves close when the forces affecting the surface areas at port 1 or 2 are greater than the forces affecting the surface area at port 3. Thus, when port 3 is vented, the valve will close and remain closed regardless of the pressure at ports 1 or 2.

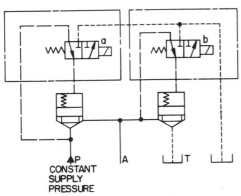

CONVENTIONAL SPOOL TYPE SYMBOL	SOLENOIDS	
	DE-ENERGIZED	ENERGIZED
A.	a & b	
B.	b	a
C.	a	b
D.		a & b

FIGURE 9-52 Three-way function. (Courtesy of The Oilgear Company, Milwaukee, Wis.)

Areas -
Port 1 (A1) = 100%
Port 2 (A2) = 80%
Port 3 (A3) = 180%

Area Ratios
$\dfrac{A3}{A1} = 1.8:1$

$\dfrac{A3}{A2} = 2.25:1$

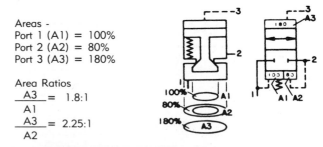

FIGURE 9-54 Normally closed logic cartridge. (Courtesy of Sun Hydraulics Corporation, Sarasota, Fla.)

When pilot pressure is applied at port 3, this pressure, acting on area A3 (minus the bias spring force) tends to open the valve. The valve may be closed by venting the pilot (port 3) to tank, or by introducing a proportionately higher pressure at port 1 or port 2. From the area ratios listed above, it can be seen that

$$\frac{A3}{A1} = 1.8 : 1$$

or 1000 psi at port 3 will open the valve against pressures up to 1800 psi at port 1 (port 2 referenced to tank).

$$\frac{A3}{A2} = 2.5 : 1$$

or 1000 psi at port 3 will open the valve against pressures up to 2250 psi at port 2 (port 1 referenced to tank).

Because logic switching elements are pressure responsive at all three ports, it is essential to consider all aspects of system operation through a complete cycle when designing logic element circuits. If pressure changes at any one port are of sufficient magnitude, they may cause a valve element to switch from a closed to an open condition, or vice versa. All possible pressure changes in the complete circuit cycle must be considered to assure a safe, functional system design.

The selection of a pilot signal source is a major consideration in logic element circuit design. In practice, logic elements are piloted from either an external source or self-piloted from one or both work ports. Logic switching elements are available which offer the user the option of internal self-piloting from either work port, or from both work ports through an integral shuttle valve. Integral self-piloting in the cartridge will often simplify manifold design and reduce or eliminate the drilling, cross drilling, and plugging of many external pilot connections. In both normally open and normally closed configurations, four piloting configurations are available for flexibility in circuit design (Fig. 9-55).

Selection of an appropriate internal piloting source can simplify both the control circuit design and the manifold construction. Since internal piloting requires only on-off control, simple two-way solenoid valves will often meet the control requirements. A more elaborate control circuit, in conjunction with an external pilot source, will allow a single logic element cartridge to perform several different functions during a machine cycle.

In summary, four logic switching elements are required to duplicate the functions of a spool-type four-way directional valve, one element for each port—P, T, A, and B (Fig. 9-56). By controlling the opening and closing of each element, 12 different flow paths can be obtained. Using logic switching elements, it is possible to shift from one flow condition to any other without going through any intermediate (or "center") valve conditions or function. Be-

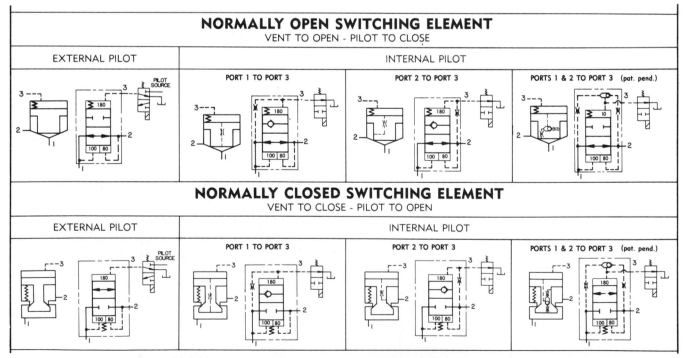

FIGURE 9-55 Comparison of normally open to normally closed logic cartridges. (Courtesy of Sun Hydraulics Corporation, Sarasota, Fla.)

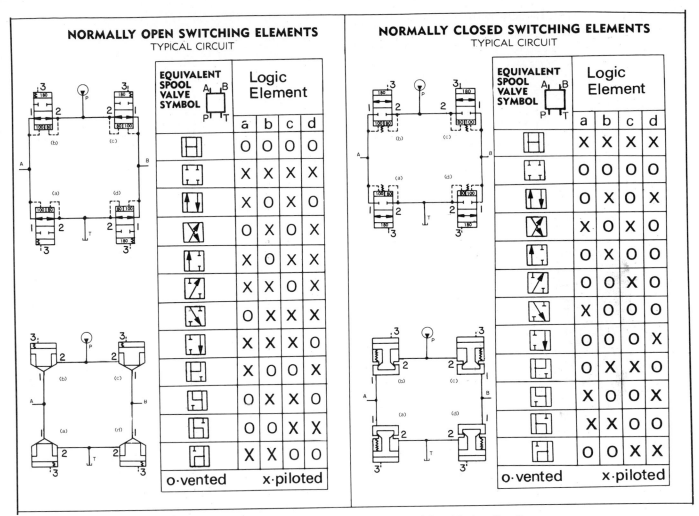

FIGURE 9-56　Typical logic circuits. (Courtesy of Sun Hydraulics Corporation, Sarasota, Fla.)

cause the opening of the elements can be precisely controlled, the logic circuit can anticipate the requirements of the system, allowing flow to shift direction smoothly. By contrast, most spool-type directional valves have only three operating positions and must always pass through a center condition when changing from one flow path to another. The opening order and shifting conditions are rigidly fixed by the spool configuration and overlap.

Typical logic element four-way directional valve circuits illustrating normally open and normally closed elements are described in Fig. 9-56 using both ANSI-style and generic symbols. Although it is not always necessary to show the area ratios when designing with ANSI-style symbols, it is helpful to do so when analyzing the final circuit to determine the effect of pressure changes on each element as a machine goes through a complete cycle.

Logic circuits using these cartridge-type elements often find their greatest use in sizes to pass 100 gpm nominal capacity and larger (Fig. 9-57). This is not to imply that this pattern will not change in the future. As designers become acquainted with the infinite flexibility (Fig. 9-58) of control of this type of logic circuit, we may find them used increasingly in systems of smaller physical size.

In summary, consider the advantages:

1. Poppets can be used in housing which has conventional high-pressure connections, cartridge form, or can be machined into any appropriate structure (Fig. 9-59).

2. Physical location is of minor concern (Fig. 9-60). Spacers can be used to provide alignment. Location in a manifold is at the designer's option (Fig. 9-61).

3. Cartridges can be sized and structured for flow and pressure needs at point of usage. As an example, a relatively small cartridge may be used on the rod end of a cylinder with an oversized rod and a very large cartridge can be installed on the head end where major flow and pressure can be anticipated.

TYPICAL PRESSURE DROPS with 550 SUS Fluid at 100°F.

PRESSURE DROP VS. FLOW
FOR OILGEAR CARTRIDGE VALVES
(W/O SPRING)

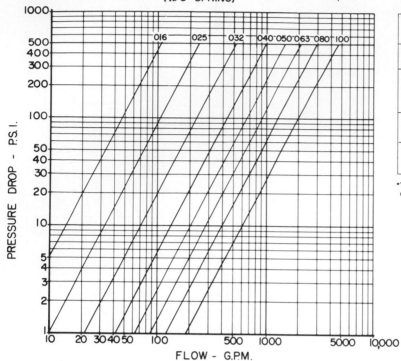

Rated Pressure	Standard	5000 psi
Ports A & B	Special	9100 psi
Pilot Pressure* Ports W, X, Y, Z		5000 psi
Hydraulic Fluids		Mineral Oil
		Fire Resistance
		High Water Base
Viscosity Range*	Minimum	32 SUS
	Maximum	2500 SUS
Temperature Range*	Minimum	0° F
	Maximum	180° F

*These values may be further limited by the pilot valve operatoi or other system components.

Conversions
PSI ÷ 14.5 = bar
GPM x 3.79 = liters/min
Inches x 25.4 = mm
Ounces x 28.35 = gram
Inches² x 645.16 = mm²
Inches³ x 16.39 = milliliters
Feet/sec. x 0.305 = m/sec.
(°F−32) x 0.55 = °C

Curves are based on 550 SUS fluid with a specific gravity (SG) of 0.88. Different viscosity effects can be predicted from

$$\Delta P = \Delta P_{(Curve)} \left[\frac{\text{Actual Viscosity}}{\text{550 SUS}} \right]^{0.1}.$$ Different specific gravity effects can be predicted from $\Delta P = \Delta P_{(Curve)} \left(\frac{\text{used SG}}{0.88} \right).$

FIGURE 9-57 Typical pressure drops. (Courtesy of The Oilgear Company, Milwaukee, Wis.)

FIGURE 9-58 Comparison of typical logic cartridges. (Courtesy of The Oilgear Company, Milwaukee, Wis.)

	SIZE 016		SIZE 025		SIZE 032		SIZE 040	
	TYPE A without cushion nose	TYPE B with cushion nose	TYPE A without cushion nose	TYPE B with cushion nose	TYPE A without cushion nose	TYPE B with cushion nose	TYPE A without cushion nose	TYPE B with cushion nose
Plunger Stroke—inches	0.250	0.410	0.250	0.470	0.320	0.540	0.480	0.760
Pilot Area (A_W)—inches²	0.395	0.395	0.760	0.760	1.491	1.491	2.149	2.149
Pilot Volume to close—inches³	0.099	0.162	0.190	0.357	0.477	0.805	1.032	1.633
Weight of Plunger, ounces	1.2	1.5	2.6	3.0	6.6	7.6	12.0	13.5
Weight of Cartridge Assembly—ounces	7.7	7.9	14.6	14.8	33.0	33.8	60.5	61.0

	SIZE 050		SIZE 063		SIZE 080		SIZE 100	
	TYPE A without cushion nose	TYPE B with cushion nose	TYPE A without cushion nose	TYPE B with cushion nose	TYPE A without cushion nose	TYPE B with cushion nose	TYPE A without cushion nose	TYPE B with cushion nose
Plunger Stroke—inches	0.560	0.870	0.680	1.070	0.950	1.340	1.486	1.956
Pilot Area (A_W)—inches²	3.291	3.291	5.966	5.966	8.793	8.793	13.423	13.423
Pilot Volume to close—inches³	1.843	2.863	4.057	6.384	8.353	11.783	19.947	26.255
Weight of Plunger, ounces	24	27	54	63	105	120	197	250
Weight of Cartridge Assembly—ounces	106	107	229	232	450	457	991	1028

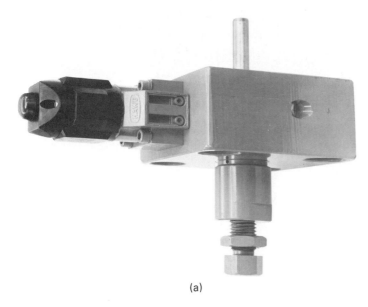

(a)

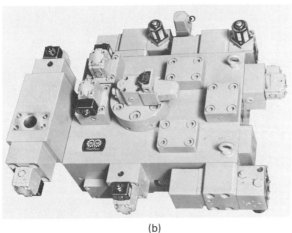

(b)

(c)

FIGURE 9-59 (a) Stop cap; (b) logic valve assembly; (c) subassembly. (Courtesy of The Oilgear Company, Milwaukee, Wis.)

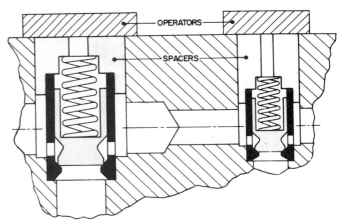

FIGURE 9-60 Sizing manifold structures for alignment. (Courtesy of The Oilgear Company, Milwaukee, Wis.)

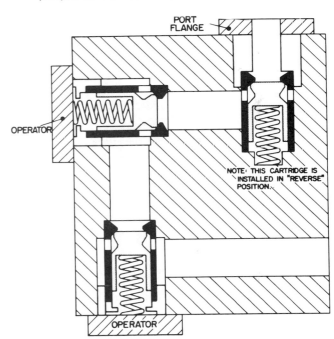

FIGURE 9-61 Logic cartridges can be installed in most convenient position. (Courtesy of The Oilgear Company, Milwaukee, Wis.)

4. The physical size of logic cartridges is usually much smaller than that of spool-type devices for equal flow and pressure.

5. Spool-type flow control capabilities are fixed in a pattern that requires major design changes to alter characteristics. Logic cartridges are available in many sizes and configurations which can be assembled into the control circuit, and the control pattern is easily changed by modifying the pilot control action or changing cartridge format in standardized bores.

6. Logic cartridges can be virtually leakproof at critical seal points. Spool valves must have lubri-

cation and clearance flows to compensate for temperature variables.

Figure 9-62 illustrates a cartridge structure that can be used to create an orifice. It includes a pilot valve (5) with a proportional solenoid (6) and an LVDT (4) connected to the piloted member (3) to provide a signal as to the physical position of the piloted orifice creating member.

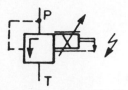

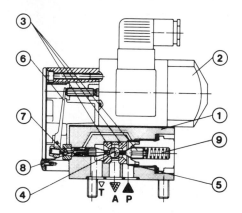

FIGURE 9-63 Symbol for pressure control valve with proportional solenoid and LVDT feedback. (Courtesy of Rexroth Corporation, Bethlehem, Pa.)

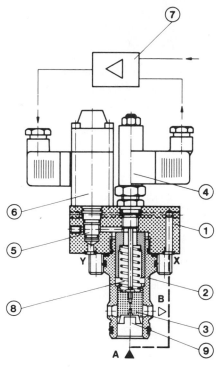

FIGURE 9-62 Logic cartridge assembly with proportional solenoid and LVDT feedback unit. (Courtesy of Rexroth Corporation, Bethlehem, Pa.)

The quantity of pilot fluid needed to pilot logic devices is so small that usual physical sizes do not require special pilot valves. A proportional solenoid-controlled valve such as that indicated by the symbol of Fig. 9-63 can control the pressure level in the cavity at the major diameter and infinitely control the movement of the logic poppet member as it reacts to forces at the shoulder, nose, or combined nose and shoulder areas.

Some circuits can function properly with the use of a spool-type pilot valve. The pilot valve shown in Fig. 9-64 uses a poppet-type structure which is mechanically moved by a digital solenoid (2). A mechanical advantage (6) is provided for solenoid 2 to move ball 4 by pin 8. Spring 9 is used to bias the ball when the solenoid is deenergized. Hardened seals and plunger 3 are fitted into cavity 5.

The choice of the type and sophistication of the control can be an option for the circuit designer. His or her analysis

FIGURE 9-64 Poppet-type solenoid-operated three-way valve. (Courtesy of Rexroth Corporation, Bethlehem, Pa.)

of the control needs for the power transmission function can be converted into economical, efficient industrial machinery.

CHECKING SAFETY IN DESIGNING CIRCUITS

There is no compromise for safety when fluid power circuits are being designed. However, if errors slip through the design stage of a circuit, trouble may occur, causing additional expense or an unsafe condition.

The following checklist of design questions may help to provide safe, efficient, and dependable fluid-powered equipment.

Question 1. Is the input horsepower adequate?

Remarks. Prime movers, such as electric motors, can be overloaded for only short intervals during peak loads. However, if peak pressures are to be held for long periods of time, the motor must be large enough to supply the proper horsepower.

Overload periods also occur when relief valves delay in opening or when pressure-compensated pumps fail to stroke immediately. Power input must be sufficient to meet these conditions.

The retraction of high-speed cylinders and rams may cause high flow rates at back pressures much too high for the input horsepower to handle.

The duty cycle of fluid power equipment must be considered to match horsepower requirements with conditions of safety, efficiency, and dependability.

Question 2. Does the system need a heat exchanger?

Remarks. Generally, it is recommended that hydraulic systems operate between 120 and 140°F. Ambient operating temperatures, flow characteristics within a system, the cycle of operation, and other factors should be recognized to assure adequate heat control.

To ensure proper circulation through a heat exchanger, a separate circulating pump is generally recommended. The electrical hookup should be wired to operate both the cooler circulating pump and the main hydraulic pump simultaneously. Circulating pumps may also be used as a source of pilot pressure or for filtration systems.

Question 3. Should the pump be unloaded between cycles?

Remarks. In the open-circuit type of system the pump can be unloaded at close to zero pressure by the following methods:

1. An open center four-way directional control valve
2. Venting a pilot-operated relief valve with an on-off type of solenoid-actuated valve
3. Pressure-operated unloading valve or two-way directional valve

Closed-circuit systems may give problems of local overheating if the same fluid is circulated constantly for long periods.

A bleed-off and replenishing arrangement may be required. Generally, the heat exchanger works directly from the reservoir and would be ineffective to correct local heat problems. Heat exchangers may be applied elsewhere in the system to take care of local heat problems.

Question 4. Is it possible to start hydraulic pumps regardless of the position of actuators or fluid motors?

Remarks. When pumps are unloaded with all solenoids deenergized and controls in neutral, the pump can be started without moving actuators. If manual, mechanical, or pilot-operated valves are used, they should be checked before startups. It is also important to check out all cylinders or motors that may be of the constant-pressure type or may have trapped pressure energy on one side. (Constant-pressure actuators move by creating a pressure drop on one side or the other.)

Question 5. What checks are essential with vertical presses and other such machines that may drift downward?

Remarks. Limit switches may be held open with the ram or cylinder in the up position. A check should be made to ensure that the proper solenoids are automatically energized to pressurize the system to return the actuators to the up position.

When variable pumps are used, they should be adjusted and controlled to hold the actuator in the up position when operating. Safety locks and pilot check valves are also used to prevent free falling of vertically suspended loads.

Question 6. Is independent movement of any actuator possible regardless of the position of other actuators?

Remarks. When bypass valves are used to move actuators for setups, they must be checked before the circuit is put into service. It may also be important to use an interlocking system to protect personnel from injury and prevent damage to the equipment.

Inching actuators in forward or reverse is sometimes essential for adjusting fixtures and clamps. When accumulators are used as a power source when the pump is off, fluid can be used directly through needle adjusting valves.

Question 7. Should all actuators stop immediately with the stop button?

Remarks. For safety reasons, it is good policy to arrange all controls to prevent motion in case of emergency, such as tool breakage or power failure.

Question 8. Are all conductors and connectors properly sized for operating conditions throughout the entire cycle?

Remarks. If the lines are too small, excessive back pressure will result, causing friction and heat. Any controls operated by pilot pressure may also be affected. Leaks and vibrations may occur from excessive fluid velocity. Some parts of the system may be subjected to mechanical binding and hydraulic unbalance.

When larger lines and valves than necessary are used, the result is higher component costs. In addition, more fluid is needed, which may affect the circuit response characteristics.

Question 9. When large double-acting cylinders are being used, are the lines and components connected to the large area side adequate to handle exhausting fluid?

Remarks. Because the return speed is much faster than the down speed with a given fluid delivery, a greater flow rate will exhaust from the large area side and must be considered.

Back pressure is sometimes utilized to control fluid delivery from the pumps to the small-area side, thus controlling the speed of the actuator.

When the volume exhausting from the large end exceeds approximately four to five times the discharge volume of the pump, a separate unloading or dump valve is recommended.

Question 10. Have drain lines been installed properly on pilot-operated valves?

Remarks. Spring chambers exposed to the entry of fluid must be properly drained when the control valve operates.

Sequence valves, pressure-reducing valves, and other devices, such as pressure switches, must have low-pressure drains because trapped fluid will not allow parts to move.

Question 11. Are all parts of the system protected against local intensification?

Remarks. Hydraulic fluid trapped in the small-area side of a double-acting cylinder may be subject to intensification when fluid is being applied to the large end. The delayed action of control valves is also a factor. Safety fuses should be applied where intensification is likely to occur.

Question 12. Have all ports been indicated properly for each component of the system?

Remarks. Check valves installed backwards and other wrong connections may create serious problems. Factory specialists have been known to travel halfway around the world to turn a check valve around or change a port connection.

Question 13. Are vent connections installed for bleeding off trapped air from vulnerable spots in the hydraulic system?

Remarks. Air trapped in large cylinders can have an accumulator effect which may cause actuators to move without the pump. When air cools in large cylinders during a shutdown, it can cause a negative pressure, which may also cause a machine to jump when a fluid lock is released or the pump is started. Systems using the pump and reservoir below the actuators require frequent venting to help air out of the system.

Question 14. Is the pump protected against over-pressures at all times?

Remarks. The pump is the source of fluid power and should have priority protection in its discharge side with a relief valve or remote unloading control.

Question 15. Will the fluid under pressure create a decompression problem that may cause shock and damage?

Remarks. Large actuators containing fluid at relatively high pressures may need a valve for decompression. Hydraulic fluids compress at the rate of about one-half of 1% for every 1000 psi.

Question 16. Are cycle sequences positive?

Remarks. Sequencing is accomplished with pressure controls, timers or position devices. Sequencing with a pressure control or pressure switch may be subjected to hydraulic shock loads that would create erratic machine operation. Electrically and mechanically actuated directional controls provide a more positive sequencing, but may be more costly.

Question 17. What visual check devices, such as pressure gauges, are needed to assure proper operation or assist in troubleshooting?

Remarks. Properly calibrated gauges should be installed to indicate operating pressure levels or inlet conditions of the pump. Temperature gauges and flow meters may also be important in some applications. All such devices should be protected against hydraulic shock loads with snubbers or check valves.

Question 18. What provisions have been made to facilitate inspection and repair?

Remarks. Working space is important to inspect or remove components for repair. Temperature gauges, pressure gauges, sight glasses on reservoirs, indicators on filters, and other inspection devices should be visible during the equipment operation without getting too close to the moving parts.

Question 19. Under what conditions must the system operate?

Remarks. This question covers all the factors, such as ambient temperature, contamination in the atmosphere, fire hazards, and such other conditions that may affect circuit design. When fire-resistant fluids are essential, this affects the components' seals that are used, and may have a direct bearing on the life of the equipment selected.

Question 20. If accumulators are used in the circuit, have means been provided to unload pressure energy during shutoffs or for maintenance?

Remarks. When the pump has been shut off, the circuit can still operate with line accumulators. It is the responsibility of the circuit designer to provide an unloading device to discharge line accumulators before any machine is inspected or worked on. If this is impossible, accumulators must be isolated automatically from the rest of the system.

Question 21. Does the nature of the work create excessive vibration during operation?

Remarks. Flexible hoses and swivel joints should be used where necessary to connect two parts of a system where motion of one occurs relative to the other. Rigid conductors should be clamped down, and inline components must be self-supported where vibration occurs.

QUESTIONS

1. What are the three main design considerations that are the responsibility of the system designer?

2. What is the most important tool for anyone working in fluid power?

3. Draw a simple reciprocating circuit and name the necessary components of the system.

4. Explain the operating principle of a regenerative-type circuit.

5. Name five different applications for accumulators.

6. Why are accumulators generally preloaded with an inert gas?

7. Illustrate a simple intensifier and explain how it multiplies force.

8. Explain the effect of compressibility when higher pressures are encountered in a hydraulic system.

9. What are the differences between a hydrokinetic type of transmission and a hydrostatic transmission?

10. Draw the graphical diagram for a typical closed-circuit system. Draw the graphical diagram of a typical open-circuit system. Compare the important differences between the closed circuit and the open circuit.

11. Name the common electrical devices that are used to control hydraulic systems.

12. Why are separate circuits used for the electric controls and for solenoids?

13. What is a servo valve, and what is its function in a hydraulic system?

14. Explain the operation of a torque motor.

15. What is the purpose of a feedback in a servo system?

16. What is the difference between a single-stage and a two-stage servo valve?

17. What are some common applications of servo valves?

18. Name several safety checks necessary when one is designing circuits, and explain why they are important.

19. What is a logic circuit? What are the basic valve structures employed with logic circuits?

20. Is a pilot-operated relief valve a logic device?

10
Fluid Power Maintenance and Safety

GENERAL MAINTENANCE REQUIREMENTS

The first requirement in a completely satisfactory maintenance program for fluid power is good equipment properly installed.

The second requirement is properly trained personnel with a thorough knowledge of the equipment's operations, and the ability to inspect, troubleshoot, and remedy problems.

The third requisite of a good maintenance program is the establishment of preventive maintenance. This includes keeping good records, continuous inspection of equipment, and heading off difficulties through system cleanliness and making minor repairs before they become major repairs with costly downtime.

THE TOOLS OF GOOD MAINTENANCE

The most important tools of the fluid power mechanic are equipment manuals, blueprints, data sheets, parts lists, and other such information.

Another important requirement is an adequate testing facility to bench-test pumps, controls, and actuators that have undergone repair. It may be dangerous to test repaired components directly on high-production-type machinery. Highly qualified personnel would be needed for maintenance of this procedure.

A third requirement is adequate tools, such as accurate pressure gauges, flow meters, clean working space, hand tools, hoists, tote carts, and other such devices essential to a mechanic.

Another important factor to prevent delay and downtime is a good inventory of spare parts, such as seals, bearings, repair kits, extra hydraulic fluid, filter cartridges, and, where important, spare components.

RECORDS

Good records concerning hydraulic machinery are a must for preventive maintenance. These records may include the following:

1. The equipment record. This is generally a card providing model number, ratings, manufacturer or source, location, etc.

2. A repair card which tells repair history, running cost, and other such data.

3. An inspection checklist. This is a list of points or things to be checked on fluid power machinery, and when the checks should be made, etc.

4. A maintenance schedule of inspections is a day-by-day listing of duties such as changing fluid or filters, making test samples, greasing, cleaning, etc.

5. An inventory control. This is often combined with the repair history and cost. It is important to keep track of spare parts, extra fluid, seals, repair kits, or spare units where essential.

GENERAL SAFETY RULES FOR THE OPERATOR OF FLUID-POWERED MACHINES

1. Understand the function and operating principle of all the component parts of the machine.

2. Understand the function and operation of the machine's control system.

3. Report any change in operating characteristics, such as abnormal gauge readings, unusual sounds, faulty or erratic performance, and leakage from components.

4. Wear proper clothing that will not catch in rotating or moving parts.

5. Wear safety glasses to protect eyes against flying objects or fluid spray from ruptured lines.

6. Observe the operating rules for the machine, such as safety guards, two-hand control, automatic pull-aways, and other important devices.

7. Know how to shut down the machine and to prepare it for safe inspection or maintenance.

8. Know how to check out controls for restarts after an emergency shutdown.

9. Do not make hasty emergency repairs that may endanger life or damage the equipment.

10. Do not operate controls in a reckless manner that may cause hydraulic shock and damage costly hydraulic components.

11. If the machine is equipped with an accumulator, be sure its pressure energy is released when preparing for shutdowns.

12. Know which valves or controls move the machine parts forward and reverse, in case of an accident.

13. Log any occurrence or observation that may be a clue that preventive maintenance is needed.

14. Contribute information to the safety and well-being of others.

GENERAL SAFETY RULES FOR THE FLUID POWER MECHANIC

1. Before working on a machine circuit, check the drawings to determine all implications with other parts of the system or other machine operations that may be interconnected.

2. Open and lock out any electrical circuits to the electric motor, controls, and related equipment.

3. For central hydraulic systems, isolate the circuit by closing, wiring, and tagging any connecting valves.

4. When engines are used for input power, disconnect the pump from the drive shaft or use other adequate means to prevent accidental starting of the pump.

5. Shore up or block cylinders and machine parts that may drop because of gravity.

6. Bleed fluid to relieve any pressure from the system by cracking fittings with a rag over the joints to prevent velocity spray of fluid until the pressure is zero.

7. Plug all tube ends and ports of components exposed to the atmosphere with sealing caps or plugs to prevent contamination.

8. Before removing heavy components from a circuit, be sure of the weight and whether or not a hoist is required.

9. Make sure all component parts are marked or tagged during disassembly to prevent such items as valve spools from being installed backwards. (A spool installed the wrong way around may allow a press to accidentally fall during a startup.)

10. When disassembly of a component is made, be careful not to unload a spring force that may cause parts to fly.

11. Do not use harmful or toxic cleaning fluids. A good solvent and some of the machine's fluid is generally recommended for washing parts to assemble components.

12. Be sure all replaced parts, such as seals and other items, are of the proper materials for compatibility with the fluid in the system.

13. Do not over-torque housing bolts or other fasteners when important adjustments for clearances may affect internal moving parts.

14. When applying tube or pipe fittings with NPT threads, be careful not to crack the housing of the component by over-torquing the fitting into the port.

15. Maintain cleanliness during maintenance, and refill all components and lines with clean system fluid before replacing them in the circuit to keep out air and contamination.

16. Check and reset all pressure controls with an accurately calibrated pressure gauge.

17. Observe extreme caution when starting up equipment for the first time or after a repair outing.

18. Always use the proper tools for the job.

19. Never use damaging blows or forces when disassembling or assembling fluid power equipment.

20. If conductors or connectors are replaced, be sure they have the proper dimension, pressure rating, threads, ports, and safety factor at operating temperature.

HOW TO TROUBLESHOOT HYDRAULIC CIRCUITS

Fluid power systems depend on flow from the hydraulic pump to provide linear or rotary motion against resistance, which results from the work. Thus, two major measurements are important for troubleshooting hydraulic circuits; these are flow and pressure. Temperature is also a factor that affects flow and must be considered.

Flow, pressure, and temperature are affected by external and internal leakages. Pressure drops due to any leakage will cause friction and heat. External leakage is easy to see, and a good maintenance program can eliminate such leaks, which are generally due to loose connections, bad seals, pin holes in castings, or worn parts. Internal leakage can also be detected through the use of pressure gauges, thermometers, and flow meters.

The operating pressure of a system can provide a good indication if leakage problems occur within a circuit. Less load can be accomplished, and if the leak is too bad the actuator will stall under loaded conditions. Adjusting the relief valve will not remedy the problem.

Gauges used to check the pressure in a hydraulic circuit are generally of the bourdon tube or spring-loaded piston types. The bourdon tube type shown in Fig. 10-1 consists mainly of a socket, tube, and tip. Fluid is communicated through the fitting in the socket to the bourdon tube. As pressure builds up, the internal forces tend to cause the tube to straighten or uncoil. The tip is linked to a movable element which translates the distance moved to the gauge pointer which indicates pressure. A specific ratio between the distance moved by the tip and the rotating distance of the pointer is calibrated for accuracy.

The spring-loaded piston design uses a piston which is opposed by spring force on one side and fluid pressure on the other (see Fig. 10-2). As fluid moves the tightly sealed piston to compress the spring, an indicator rod moves with the piston. The indicator rod motion is translated to a calibrated scale on the gauge body, which gives a pressure reading.

Vacuum and compound gauges are used for determining lower pressures or absolute pressures (see Fig. 10-3). Vacuum gauges are generally of the bourdon tube type and are calibrated in inches of mercury (see Table 10-1). Compound gauges are used to measure in both absolute and gauge pressure (see Fig. 10-4).

Hydraulic system components should be equipped with gauge connections so that pressure checks can be made during a complete cycle of operations. Results can be compared to those when the machine functioned properly.

Flow meters are useful for checking flow rates from the pump, to a control, from a control, to an actuator, from an actuator, and to the reservoir. Complete "hydraulic testers" (see Fig. 10-5) are available for troubleshooting flow

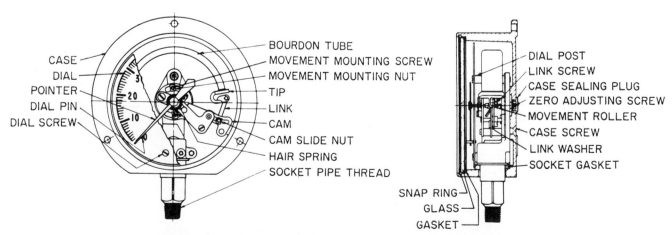

FIGURE 10-1 Bourdon tube gauge. (Courtesy of Helicoid Gauge Division, American Chain and Cable Co.)

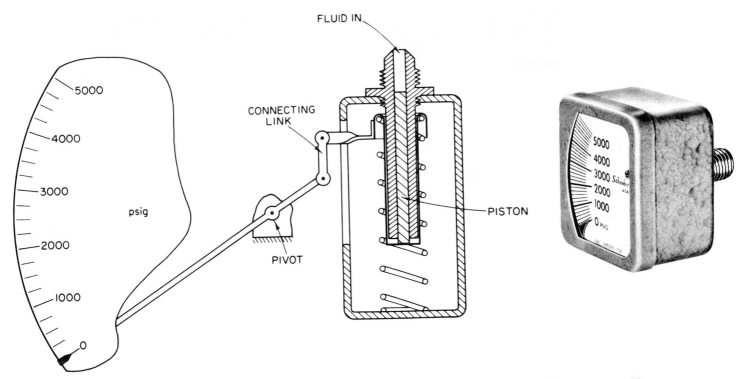

FIGURE 10-2 Pressure gauge. (Courtesy of Schrader Bellows, Akron, Ohio.)

FIGURE 10-3 Vacuum gauge. (Courtesy of Marsh Instrument.)

TABLE 10-1
ABSOLUTE AND GAUGE PRESSURE

Absolute pressure (in. Hg)	U.S. gauge pressure (psi)
1.0	0.48
2.04	1.00
4.1	2.0
6.1	3.0
8.2	4.0
10.2	5.0
12.2	6.0
14.3	7.0
16.3	8.0
18.4	9.0
20.4	10.0
22.4	11.0
24.5	12.0
26.5	13.0
28.6	14.0
29.92	14.7

rate, pressure, and temperature. All the components are built into a convenient-to-carry box, which provides portability. Testers of this kind can be used to troubleshoot a circuit at any point of the circuit by connecting it between desired components and cycling the machine.

When one is troubleshooting individual components, a test stand can be used which is equipped with all the necessary flow-raters, gauges, and other instruments (see Figs. 10-6 and 10-7).

In the field, a stopwatch may be useful to check the cycling time of cylinders. Since the stroke length and bore of a cylinder are known, the cycling time provides information to determine the flow rate to the cylinder. A check of the pump capacity gives comparison data to determine losses.

$$\text{Time (sec)} = \frac{\text{area of cy.} \times \text{stroke (in.)} \times 60}{231 \times \text{gpm}}$$

A hand tachometer is useful in determining the speed of a fluid motor (see Fig. 10-8).

FIGURE 10-4 Compound gauge. (Courtesy of Marsh Instrument.)

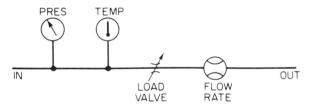

FIGURE 10-5 General test stand setup for controls.

If the pump capacity and the displacement of the fluid motor are known, losses can be analyzed by obtaining the fluid motor speed.

$$\text{Motor speed} = \frac{\text{gpm} \times 231}{\text{displacement/rev.}}$$

Fluid motors make excellent flow meters, and a spare motor may be used in place of a hydraulic tester with the following components as shown in Fig. 10-9: (Temperature can be obtained by using a glass-tube-type thermometer)

Care should be taken to avoid exposing low-pressure seals to high-pressure fluid. External drains may have to be

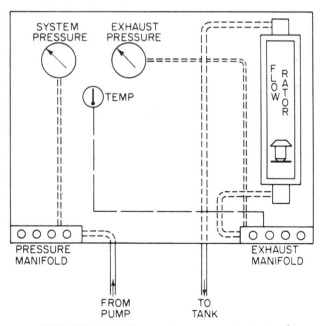

FIGURE 10-6 Flow and pressure test stand.

calculated separately by collecting drain fluid into a graduated container for a period of time.

Pump test stands (see Fig. 10-10) are generally equipped with a variable-speed drive to operate the test pump over a range of speeds (rpm). It should also be equipped with an accurately calibrated flow-rater, various pressure gauges, and a loading device.

A pump testing station of the design shown in Fig. 10-11 provides for a test over normal operating speeds (rpm) and at any desired pressure level within the range of the

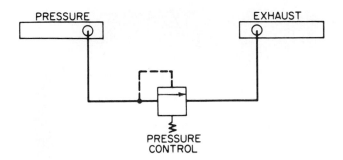

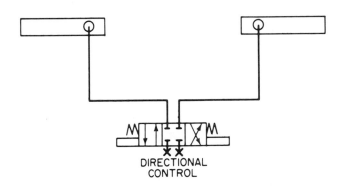

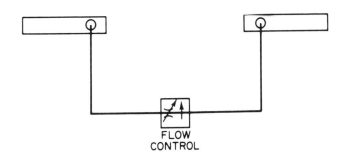

FIGURE 10-8   Hand tachometer.

Test circuit for pressure control	Test circuit for directional control	Test circuit for flow control
Set pressure control for desired operating pressure and flow. Connect to test stand flow rator. Gradually increase the test stand pressure until the control opens. Leakage below the setting of the control will be indicated by the test stand flow rator.	Leakage from a directional control can be found by hooking test stand supply to the valve's pressure port. The valve "out" port is hooked to the exhaust, and the two cylinder ports are blocked. Operate valve and observe leakage.	Connect flow control to test stand supply manifold and exhaust; then operate control between zero and full range to determine its operation.

FIGURE 10-7   Leakage test.

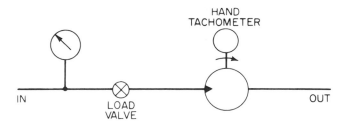

FIGURE 10-9   Flow meter using fluid motor.

test unit. Operating curves can be computed and then compared to those when the pump was factory-tested.

Another easy-to-build test unit is a hydraulic dynamometer arrangement as shown in Fig. 10-12. A standard power unit being used, fluid is directed to operate a fluid motor which, in turn, drives a test pump. The test pump inlet uses a length of non-collapsible hose which is put into a container of oil. The discharge from the pump is connected

FIGURE 10-10 Pump test stand. (Courtesy of Kenosha Technical Institute.)

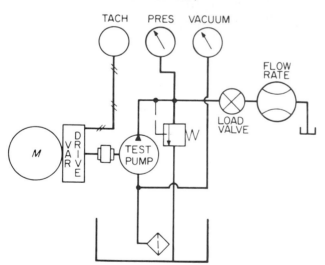

FIGURE 10-12 Dynamometer. (Courtesy of Kenosha Technical Institute.)

TACH   PRES   VACUUM

FLOW RATE

LOAD VALVE

M   VAR DRIVE   TEST PUMP   W

FIGURE 10-11 Pump test stand with variable-speed drive, flow-rater, and so on.

through a "hydraulic tester" and then exhausts back to the same container.

The load valve, pressure gauge, and flow-rater can be used to measure the output of the pump. The speed (rpm) of the fluid motor driving the test pump can be determined by a tachometer. This arrangement could also be used to test the fluid motor that drives the pump. The pump, in this case, would be loaded through the hydraulic tester to de-

termine the torque or horsepower capabilities of the fluid motor.

One of the most important factors to remember in the troubleshooting of hydraulic systems is that pressure is a result of resistance, and if there is no resistance the pressure will be near zero.

TROUBLESHOOTING CHECKLIST

Lack of full pressure in the system:

1. Air in the system
2. Pump starved by clogged inlet strainer

3. Loose inlet connection to pump

4. Relief valve not properly seated

5. Leak across directional control valve ports

6. Inadequate input power

7. Leak in conductor or connector

8. Defective seals on cylinder piston assembly or rod gland

9. Loose or cracked piston rod gland seal

10. Defective pressure gauge

11. Cylinder crack or scored

12. Check valve not seated

13. Defective or worn pump

14. Defective or worn fluid motor

15. Scored valve plate in pump or motor

16. Wrong hydraulic fluid

No pressure develops in the system:

1. Input power not coupled properly

2. Pump not turning in the right direction

3. Pump housing bolts not properly torqued

4. Not enough hydraulic fluid in the reservoir

5. Pressure gauge snubber valve closed, or defective gauge

6. Ruptured conductor

7. No resistance in the system

8. Air breather in the reservoir clogged

9. Defective relief valve

10. Inlet to pump blocked

Actuating cylinder or fluid motor fails to develop full speed:

1. Air in system

2. Wrong viscosity fluid

3. Loss of one pump in a dual pump system caused by unloading valve being set too low

4. Pump operating too slowly

5. A check valve not seating

6. Relief valve defective or set too low

7. Directional valve not shifting far enough, or defective

8. Defective seals in cylinder piston assembly

9. Defective fluid motor

10. Fluid too hot and thin, causing excessive internal leakage

11. Trapped fluid under back pressure or intensification

12. Mechanical binding in linkage

13. Inlet to pump blocked or too small

14. Excessive friction in conductors, or hose kinked

15. Air breather in reservoir blocked

Erratic motion of an actuating cylinder:

1. Air in system

2. Not enough fluid in the system

3. Cylinder rod gland not adjusted properly

4. Mechanical bind

5. Pump capacity too small

6. Defective seals in cylinder piston assembly or fluid motor

7. Unloading or relief valves defective or out of adjustment

8. Air breather in reservoir blocked

9. Excessive back pressure on externally drained control valves

10. Pump inlet leaks

11. Defective pump

12. A defective check valve

13. Flow control valve defective or compensator sticking

14. Slight back pressure needed to stabilize actuating cylinder or fluid motor due to the nature of the work

15. Broken spring in a control valve

Actuating cylinder or fluid motor fails to move:

1. Directional control valve fails to shift

2. Sequence valve not externally drained

3. Solenoid burned out, or limit switch not actuated

4. Relay contacts dirty

5. Broken spring in control valve

6. Control sticks open or closed

7. Pump control defective

8. System hooked up wrong

9. Check valve in backwards

10. Hydraulic lock prevents control from operating

11. Mechanical obstruction
12. Insufficient power

Excessive noise in operation:

1. Air entering the pump inlet line
2. Entrapped air
3. Misalignment of pump and drive unit
4. Conductors vibrating
5. Sudden release of fluid under pressure
6. Turbulent flow characteristics
7. Vibration of system components
8. Relief valve chattering when set too close to load pressure
9. Defective pump or fluid motor

Electric motor driving pump keeps kicking out:

1. Too small to carry load
2. Carrying too high a system pressure
3. Electric motor hooked up wrong or operating on low voltage
4. Misalignment of electric motor and pump
5. Lack of circulating air to cool electric motor

Excessive heat, causing high temperature of hydraulic fluid:

1. Components or conductors and connectors too small for the fluid being delivered to the system
2. Wrong grade of fluid, or dirty fluid
3. Not enough fluid in the system, or reservoir too small
4. Pump operating too fast
5. Small high-pressure leaks within components
6. Dirt-holding relief valve partially opened
7. Defective pump
8. System overloaded
9. Cooling system defective or inadequate
10. Air breather too small or blocked
11. Entrained air in the hydraulic fluid

Air in system:

1. Defective seals
2. Leaks at joints
3. Loose inlet to pump

4. Not enough fluid in the system
5. Loose pump casing
6. Inlet to pump too long
7. Replenishing pump defective
8. Improper maintenance procedures
9. Permeation through accumulator bladders or diaphragms

ANALYSIS AND TROUBLESHOOTING CIRCUITS

Problem 1.   Speed Control Circuit (see Figs. 10-13 through 10-15)

Directions:

1. Trace the flow path of the fluid for each phase of the machining cycle and color lines to indicate pressure, drain, etc. (use the standard color code).
2. List the name of each component in the system and describe its function.
3. Draw the graphical diagram shown in the old symbols, using the new ANSI symbols.

Problem 2.   Regenerative Circuit (see Fig. 10-16)

Directions:

1. Explain the principle of a regenerative circuit.
2. Draw this circuit in ANSI symbols.
3. Color in all lines, using the standard color code.

Problem 3.   Accumulator Circuit (see Fig. 10-17)

Directions:

1. Explain the function of the accumulator circuit.
2. Color in all lines, using the standard color code.
3. Redraw the circuit in ANSI symbols and include some safety means of unloading the accumulator when the pump is stopped.

Problem 4.   Intensifier Circuit (with Error) (see Fig. 10-18)

Directions:

1. Trace the flow path of the fluid through the circuit, and color in all flow lines with standard code.

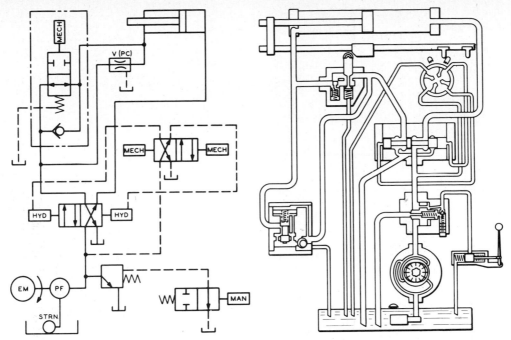

FIGURE 10-13 Circuit problem 1.

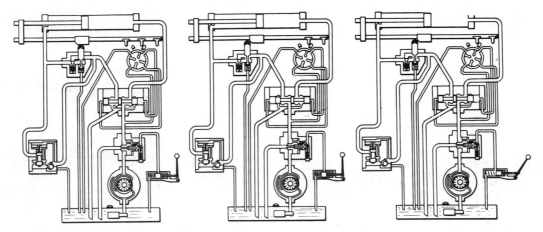

FIGURE 10-14 Circuit problem 2.

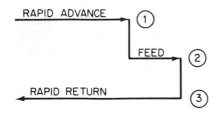

RAPID ADVANCE ——→ ①

FEED ——→ ②

RAPID RETURN ←—— ③

FIGURE 10-15 Desired functions.

FIGURE 10-16 Circuit problem 2 (regenerative circuit).

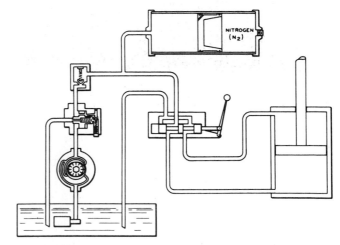

FIGURE 10-17 Circuit problem 3 (accumulator circuit).

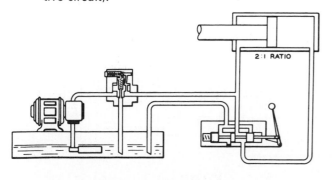

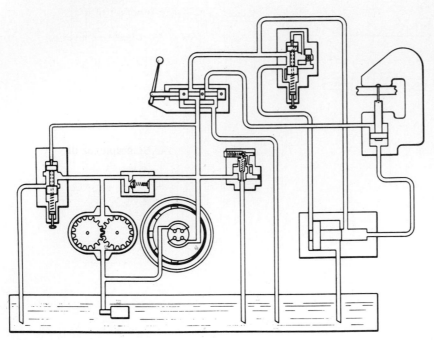

FIGURE 10-18 Circuit problem 4 (intensifier circuit with error).

2. Explain the function of an intensifier.

3. "Troubleshoot" the sequencing phase of the cycle and describe the change needed.

4. Draw ANSI graphical diagram.

Problem 5. Sequencing Circuit (with Error) (see Fig. 10-19)

Directions:

1. Trace the flow path of the circuit to provide the following work sequence:

FIGURE 10-19 Circuit problem 5 (sequencing circuit with error).

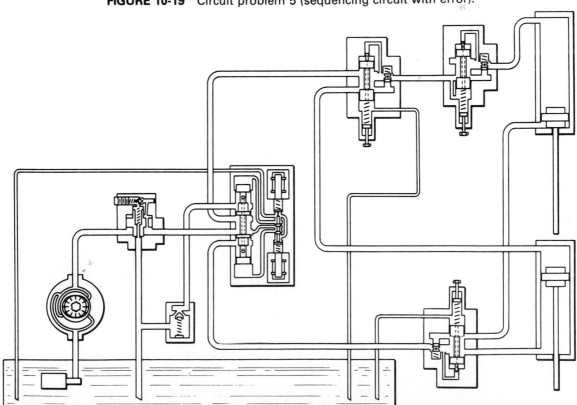

a. Top cylinder extends →

b. Bottom cylinder extends →

c. Bottom cylinder retracts ←

d. Top cylinder retracts ←

2. Redraw the circuit, eliminating any components not needed and making all necessary corrections for proper circuit operation.

Problem 6. Press Circuit (see Fig. 10-20)

Directions:

1. Trace the flow path of the fluid through the system and color in all lines, using standard colors.

2. Describe the function of a prefill valve.

3. Describe the function of kicker cylinders.

4. Draw an ANSI graphical diagram for the circuit.

Problem 7. Closed-Circuit System (see Fig. 10-21)

Directions:

1. Identify and list each of the component parts of the system.

2. Draw the ANSI graphical diagram of the complete system.

3. Color code all active pressure, exhaust, inlet, and drain lines of the circuit.

4. Describe the power and torque output characteristics of the system.

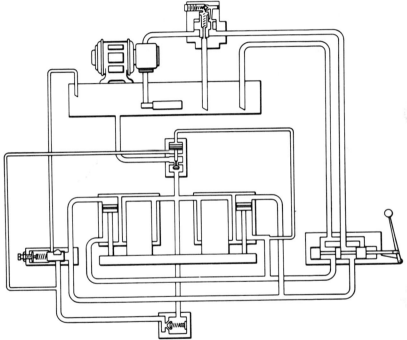

FIGURE 10-20 Circuit problem 6 (press circuit).

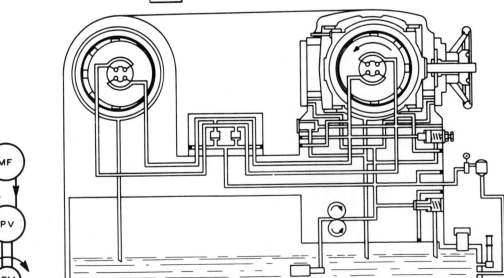

FIGURE 10-21 Circuit problem 7 (closed circuit system).

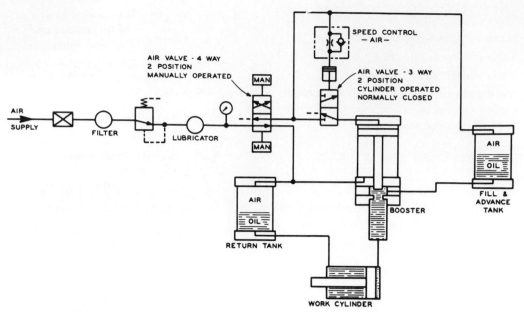

FIGURE 10-22 Circuit problem 8 (combination air and hydraulic circuit).

Problem 8. Combination Air and Hydraulic Circuit (see Fig. 10-22)

Directions:

1. Name and describe the function of each component in the system.

2. Redraw the circuit, using the latest ANSI graphical symbols.

3. Name the advantages of using air.

QUESTIONS

1. What records should be maintained on hydraulic machinery for a good maintenance program?

2. Read and discuss the general safety rules for the operator of fluid-powered machines.

3. Read and discuss the general safety rules for the fluid power mechanic.

4. Explain the importance of the three following statements when one is troubleshooting hydraulic circuits:
   a. Pressure in a hydraulic system is due to resistance.
   b. A pump pumps hydraulic fluid and not pressure.
   c. Atmospheric pressure forces the hydraulic fluid into the inlet side of the pump.

5. Which is more difficult to find and eliminate in a system, external leakage or internal leakage?

6. What is a compound pressure gauge?

7. What measurements can be found with a hydraulic tester?

8. Explain how a fluid motor can be used as a flow-rater.

9. Explain how to build a simple test hydraulic dynamometer and how it can be used as a maintenance tool.

10. Read and discuss the troubleshooter's checklist.

11

Basic Principles of Pneumatics

The term *pneumatics* as treated in this chapter pertains to *power or control signal transmitted and controlled through the use of a pressurized gas.* Pneumatics is the branch of physics that deals with the physical properties of air and other gases. The art of applying gaseous fluid power is expanding in both power and control systems (see Figs. 11-1 and 11-2.)

Most manufacturing and processing plants maintain compressed air for a variety of operations. Because air is convenient and universally available, new uses for it are constantly being discovered. A number of important basic principles for the application of pneumatic power and control systems are discussed in this chapter.

PROPERTIES OF AIR

Air is a mechanical mixture of gases containing, by volume, approximately 78% nitrogen, 21% oxygen, and about 1% other gases, including argon and carbon dioxide. Water vapor, called humidity, is present and varies in amounts almost constantly from hour to hour. Life on earth depends on air for survival, and we harness its forces to do useful work.

Air is compressible and has weight. The weight of air is easily shown in the laboratory by first weighing an airtight container from which all the air has been removed, and then weighing it again after allowing it to fill with air. Because air is heavier than some other gases, balloons inflated with the lighter gases can float in air.

The earth is surrounded by atmosphere and at *any point* exerts a pressure, because of the column of air above *that point*. The reference point is sea level, where air exerts a pressure of 14.7 psi (Table 11-1). At 1000 ft above sea level the atmospheric pressure is about 14.2 psi. This indicates that the air pressure varies about $\frac{1}{2}$ lb for each 1000 ft of altitude. In addition, the air pressure varies constantly, depending upon atmospheric conditions. Weather forecasters use these changes to predict weather conditions.

Atmospheric pressure of 14.7 at sea level is used as a standard for making air circuit computations. The accepted weight for air is 0.08071 lb/ft³ at 14.7 psi and at 32°F. (Dry air weighs about 0.0746 lb/ft³ at 72°F.) Atmospheric pressure is equal in every direction at any given point of contact. This can easily be demonstrated in the laboratory by exhausting all the air from the inside of an airtight container with thin walls and watching it collapse. When the air is removed from the inside, reducing atmospheric pressure, the outer pressure due to the atmosphere creates a force against the walls and the container is crushed. If the sides of this container were 144 in.², the force of the atmosphere would be more than a ton on each side. This shows the tremendous forces available by utilizing atmospheric pressure (Fig. 11-3).

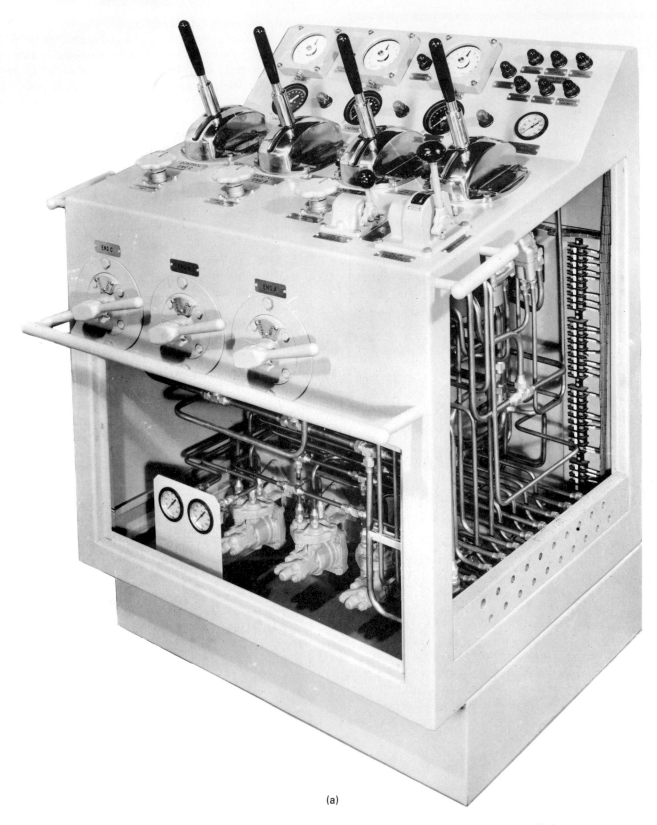

(a)

FIGURE 11-1 (a) Pneumatic controls have contributed much to the safety, efficiency, and economy of diesel-powered craft. Remote panel-mounted devices provide fingertip control, assuring split-second response for any desired maneuver or speed. Interlocking logic control prevents damage to engines and assures proper sequence of operations. Marine console unit. (Courtesy of WABCO.) (b) Vacuum is created by

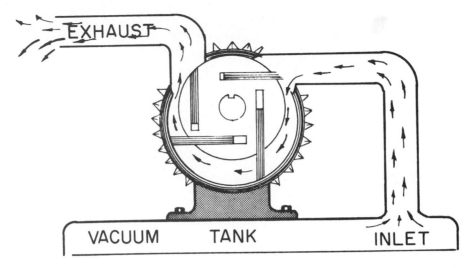

(b)

the rotary motion of the off-center rotor, whose four vanes each "bite" off a slice of air at the inlet side, carry it down and around the bottom of the pump, and then compress it so that it goes out the exhaust. Centrifugal force keeps the vanes in contact with the inside wall of the rotary pump. Notice how the vanes slide out to the greater length at the bottom and are pushed in for shorter length when they get to the top. Each "bite" past the inlet removes more air from the reserve tank, creating the desired vacuum for the milking machine. (Courtesy of Jamesway Manufacturing Co., Inc.)

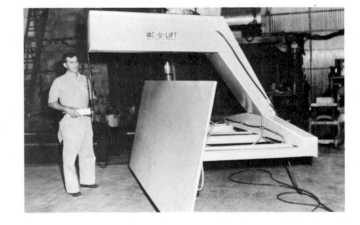

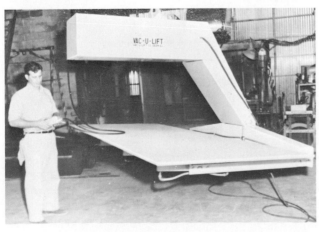

FIGURE 11-2   Vac-U-Lift 180° turnover unit. (Courtesy of Vac-U-Lift Company, Division of Lear Siegler, Inc.)

Table 11-1 Atmospheric pressure at altitudes above sea level

Altitude in feet	Absolute pressure of atmosphere in lb per sq in.	Mercury gauge reading in inches
Sea level	14.7	0
1,000	14.2	1.0
2,000	13.7	2.1
3,000	13.2	3.1
4,000	12.7	4.1
5,000	12.2	5.0
6,000	11.7	6.0
7,000	11.3	6.9
8,000	10.9	7.7
9,000	10.5	8.6
10,000	10.1	9.4
15,000	8.29	13.0
20,000	6.75	16.3
30,000	4.36	21.1
40,000	2.72	24.5
50,000	1.70	26.5
60,000	1.05	27.79
70,000	0.65	28.60
80,000	0.40	29.10
90,000	0.25	29.41
100,000	0.15	29.60
110,000	0.10	29.72
120,000	0.065	29.792
130,000	0.045	29.833
140,000	0.030	29.865
150,000	0.020	29.885
160,000	0.014	29.897
170,000	0.010	29.906
180,000	0.007	29.912
190,000	0.005	29.917
200,000	0.003	29.920

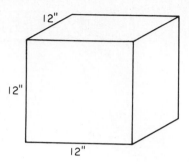

FIGURE 11-3 Can crushed by atmospheric pressure.

$$\text{Force} = \text{pressure} \times \text{area}$$
$$= 14.7 \times 144$$
$$= 2116.8 \text{ lb (one side)}$$

(One pound of air occupies 12.39 ft³) (Fig. 11-4)

$$= \frac{1 \text{ ft}^3}{0.08071 \text{ lb}} = 12.39 \text{ ft}^3/\text{lb}$$

1 ft³ of air at 32°F weighs 0.08071 lb. (Dry air at 72°F weighs 0.0746 lb. per cu ft)

Atmospheric pressure is measured with a barometer, as shown in Fig. 11-5. The barometer consists of a straight glass tube over 30 in. long, having the upper end sealed. It is completely filled with mercury, and the open end is submerged into an open vessel of mercury. The level of the

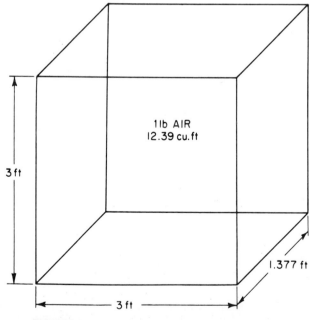

FIGURE 11-4 One pound of air occupies 12.39 ft³.

mercury in the tube will be about 30 in. above the level of the mercury in the open vessel. This is because the weight of the mercury column in the tube is in balance with the 14.7 psi of the atmosphere. The atmospheric pressure acting on the mercury in the open vessel supports the column of mercury in the tube because there is no air pressure in the sealed portion of the tube above the mercury (perfect vacuum).

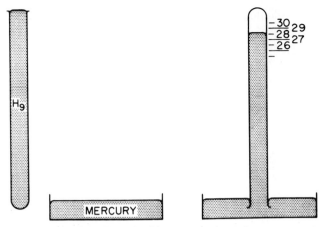

FIGURE 11-5 Mercury barometer.

Pressure readings above zero (perfect vacuum) are called absolute pressures. It could also be stated that absolute pressure is atmospheric pressure plus gauge pressure. Gauge pressure is pressure above atmospheric pressure as measured by the standard bourdon or spring-type gauge.

$$\text{Absolute pressure} = \text{gauge pressure} + \text{atmospheric pressure}$$

$$\text{Gauge pressure} = \text{absolute pressure} - \text{atmospheric pressure}$$

Absolute pressures may be determined by adding atmospheric pressure at any given elevation to gauge pressure.

The height of the mercury column remains at approximately 29.92 in. at 14.7 psi. Any changes in atmospheric conditions will cause the column of mercury to vary in height.

For accurate readings with mercury gauges, tubes should be clean and dry, using pure distilled mercury. After the tubes have been filled, the mercury should be boiled in the tube to expel all moisture.

Other devices for measuring air pressure include the water monometer, the mercury U-tube monometer, and the McLeod gauge, shown in Fig. 11-6.

EXAMPLE

If the gauge of an air tank reads 80 psi, then at sea level the absolute pressure would be 80 + 14.7 = 94.7 psi.

EXAMPLE

Suppose that a vacuum gauge shows 5 in. of mercury and absolute pressure reading is required. Barometric readings in inches of mercury vacuum can be converted to psi by multiplying by the factor of 0.491 psi/in. Hg. To find absolute pressure:

$$5 \times 0.491 - 2.455 \text{ psi}$$

$$14.7 - 2.455 = 12.245 \text{ psi absolute pressure}$$

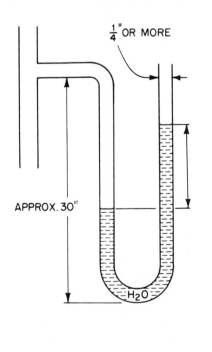

WATER MANOMETER

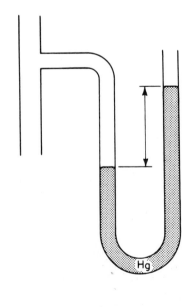

MERCURY U-TUBE

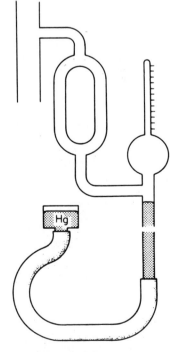

McLEOD GAGE

FIGURE 11-6 Devices for measuring air pressure.

The term *vacuum* is usually applied to any degree of pressure less than atmospheric pressure. If the pressure in an airtight container is 10 psi absolute pressure, it is a partial vacuum. If the pressure is reduced further until it is zero, it would be full vacuum or perfect vacuum. A high vacuum means a very low absolute pressure. Vacuum is measured with a mercury monometer gauge and is usually expressed in inches of mercury.

Bourdon tube gauges are also used for measuring vacuum. A compound bourdon tube gauge reads both vacuum and gauge pressure.

Fluid power in vacuum form (see Fig. 11-7) is being used for many jobs in industry, such as chucking, bottle filling, instrument testing, forming, lifting, feeding, printing machines, milking machines, and farming operations, plus others.

FIGURE 11-7 Vacuum pumps (from top left): four-curved-wing type; two straight-wing type; four-straight-wing type; fan-cooled.

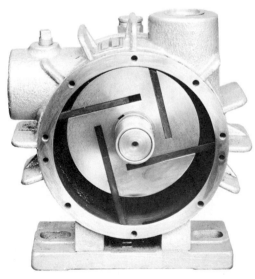

FIGURE 11-7 (*continued*) Vacuum pumps (from top): radiator-cooled; rotary vane; vacuum and pressure.

Figure 11-8 shows the effect of vacuum. With the vacuum pump in operation a degree of vacuum will build up in the container, but it will not disturb the dirt. A static condition of vacuum exists, and there is no motion in the air.

If an orifice is put into the end of the container, the vacuum will cause air to enter through the orifice (Fig. 11-9).

The velocity of the air through the container increases as the vacuum increases. The high velocity disturbs the dirt, causing it to flow toward the pump. Figure 11-10 shows

FIGURE 11-8 Vacuum does not affect dirt in bottle.

Table 11-2   Holding force created by vacuum on sucker

Vacuum inches of mercury	*Diameter of sucker*					
	1 in.	1½ in.	2 in.	3 in.	4 in.	5 in.
5	1.9	4.4	7.8	17.7	31.4	49
10	3.8	8.7	15.4	34.6	62.0	96
15	5.8	13.1	23.3	52.2	93.0	145
20	7.6	17.3	30.8	69.0	123.0	192
25	9.6	21.8	38.6	87.0	154.0	241

that vacuum alone cannot move material, but it will cause air velocity that can move material. When a sheet is being picked up with vacuum, the pickup force is proportional to the hole area (Table 11-2). Figure 11-11 and Fig. 11-12 show a practical application using vacuum power.

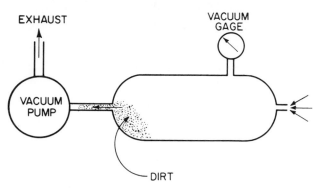

FIGURE 11-9 Effect of air entering vacuum tank.

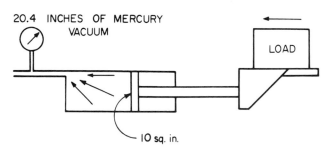

FIGURE 11-11 Vacuum sucker design.

EXAMPLE 1

If a vacuum of 20.4 in. Hg is used with a sucker having an area of 2 in.², what is the lifting force? (See Fig. 11-12.)

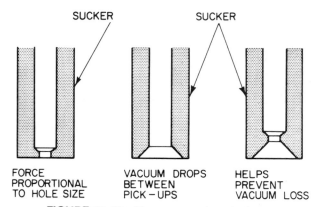

FIGURE 11-12 Example: vacuum cylinder.

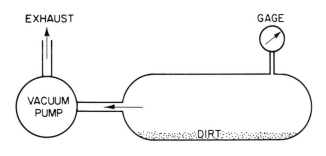

FIGURE 11-10 Effect of vacuum on dirt in tank.

Solution:

Lifting pressure (psi) $= \dfrac{\text{vacuum (in. Hg)}}{2.04}$

$= \dfrac{20.4}{2.04}$ (2.04 = reciprocal of 0.491)

$= 10$ psi

Lifting force = pressure × area

$$= 10 \times 2$$

$$= 20 \text{ lb}$$

EXAMPLE 2

How much pull will a piston exert if the area is 10 in.² and the vacuum is 20.4 in Hg?

Solution:

$$\text{Lifting pressure (psi)} = \frac{\text{vacuum (in. Hg)}}{2.04}$$

$$= \frac{20.4}{2.04}$$

$$= 10 \text{ psi}$$

$$\text{Pulling force} = \text{pressure} \times \text{area}$$

$$= 10 \text{ psi} \times 10 \text{ in.}^2$$

$$= 100 \text{ lb}$$

Thus it can be seen that absolute pressures below zero gauge pressures can be used to perform useful work.

MEASURING AIRFLOW

Impact Tube

The impact tube is an instrument for measuring low-pressure airflow through pipes or ducts. The air impinging on the tube opening at point *A* causes the water in the U tube to react to the impact pressure, and the static pressure is applied at point *B* (see Fig. 11-13). The instrument indicates the difference between the two pressures, which is the velocity pressure. The static pressure reading is subtracted from the impact pressure reading to get the velocity pressure.

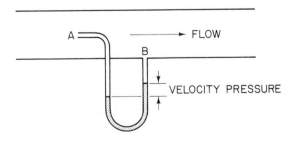

FIGURE 11-13 Impact tube.

The velocity in lineal feet per minute can be obtained by the following equation:

$$\text{LFM} = 4005\sqrt{\text{velocity pressure}}$$

To convert inches of water to pounds per square inch:

$$\text{Inches of water} = 27.686 \times \text{psig}$$

$$\text{Psig} = \frac{\text{inches of water}}{27.686}$$

Pitot Tube

The pitot tube (see Fig. 11-14) is based on the same principle as the impact tube. The tube is placed inside of the static pressure tube, which eliminates the need of two holes in the pipe or duct. The difference would again indicate the velocity pressure for calculating LFM.

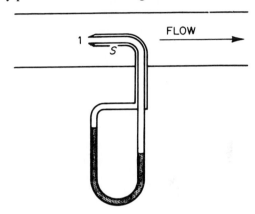

FIGURE 11-14 Pitot tube.

Flow Meter

The principle of the flow meter is shown in Fig. 11-15. The instrument measures pressure drop across the fixed orifice, which has a known coefficient. The pressure upstream is

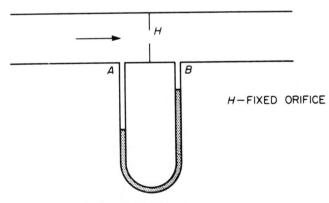

FIGURE 11-15 Flow meter.

picked up at *A* and the pressure downstream registers at *B*. The U tube gives the difference in pressures across the orifice.

The number of cubic feet per minute is then found by the following formula:

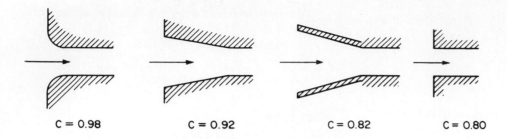

C = 0.98   C = 0.92   C = 0.82   C = 0.80

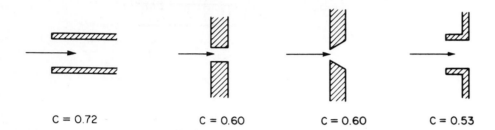

C = 0.72   C = 0.60   C = 0.60   C = 0.53

FIGURE 11-16 Various orifices and their coefficients.

$$\text{cfm} = \frac{C}{60}\sqrt{I \times P}$$

where C = coefficient of orifice

I = inches of water (1 in. of water = 0.0361 psi)

P = absolute pressure on the upstream side

cfm = cubic feet of free air per minute

Figure 11-16 shows typical orifices and their coefficients.

Gasometer

A gasometer provides accurate measure of free air leaving a pipe or compressor. It consists of an inverted tank which is counterbalanced in a tank of water so that it floats freely to any position (see Fig. 11-17).

The air is directed by a pipe to the inside, where it causes the inverted tank to rise. A calibrated scale, based on the volume per inch of rise, is attached to the apparatus. The number of cubic feet per minute is obtained by timing the rising tank with a stop watch.

FREE AIR AND STANDARD AIR

Free air is air at normal atmospheric conditions. Because the atmosphere is subject to change of pressure and temperature, free air characteristics may change from day to

FIGURE 11-17 Gasometer.

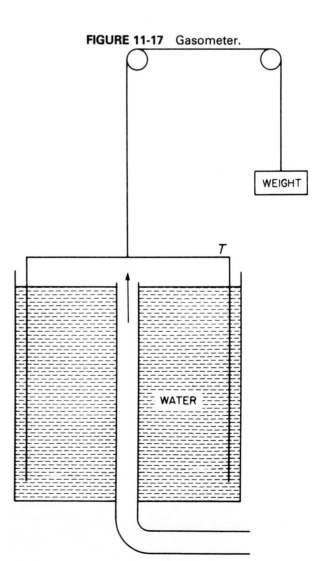

WEIGHT

T

WATER

day. The term *"standard air"* is derived from an average atmospheric condition with the air at a temperature of 68°F, atmospheric pressure at 14.7 psi, and a relative humidity of 36%. (Relative humidity is the percentage of saturation of air with water vapor.)

COMPRESSED AIR AND THE GAS LAWS

When air is being compressed for use in pneumatic systems, pressure, temperature, and volume have an effect on the operation. Thus, it is important to know the effect of these three variables in applying pneumatic power and control. Air resembles a perfect gas, because it follows closely the laws of a perfect gas in expansion, contraction, and absorbing and releasing heat.

Boyle's law states: "When the temperature of a confined gas remains constant, the volume varies inversely as its absolute pressure." This can be stated mathematically as a simple formula:

$$\frac{\text{Initial absolute pressure}}{\text{Final absolute pressure}} = \frac{\text{final volume}}{\text{initial volume}}$$

$$\left(\frac{P_1}{P_2} = \frac{V_2}{V_1}\right)$$

or as

Initial absolute pressure $\times$ initial volume = final absolute pressure $\times$ final volume

$$(P_1 \times V_1 = P_2 \times V_2)$$

(psia = pounds per square inch absolute)

(psig = pounds per square inch gauge)

EXAMPLE 1

An air tank contains 4 ft³ of air at 20 psi. The tank is charged with a liquid until the air occupies only one-half its original volume. What is the final pressure in the tank?

Solution:

$$P_1V_1 = P_2V_2$$

Transpose the formula to proper form.

$$P_2 = \frac{P_1V_i}{V_2}$$

$$P_1 = 20 \text{ psi} + 14.7 = 34.7$$

$$V_1 = 4$$

$$V_2 = 2$$

$$P_2 = \text{unknown}$$

$$P_2 = \frac{34.7 \times 4}{2}$$

$$P_2 = 69.4 \text{ psia}$$

(Remember, if gauge pressure is desired from absolute pressure, subtract 14.7 from the absolute pressure.)

EXAMPLE 2

How many cubic feet of free air per minute are required to operate an air cylinder (single-acting) 100 cycles per minute if the volume of the cylinder is 1.5 ft³ and the operating gauge pressure is 60 psi? (See Fig. 11-18.)

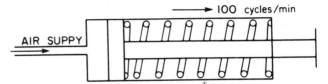

FIGURE 11-18   Diagram for Example 2.

Solution:

$$\frac{\text{Free air}}{\text{cfm}} = \frac{\text{vol. of cylinder}}{\text{ft}^3} \times \frac{\text{cycles}}{\text{min}}$$

$$\times \frac{\text{gauge pressure} + 14.7}{14.7}$$

Thus it is easy to see that the basic transposed formula $V_2 = (V_1 \times P_1)/P_2$ is flexible in use, and units of time may be added (Fig. 11-19).

$$\frac{\text{Free air (ft}^3)}{\text{Time}} = \text{vol. of cylinder (ft}^3)$$

$$\times \text{ no. of cycles/min}$$

$$\times \frac{\text{gauge psi} + 14.7}{14.7}$$

$$V_1 = 1.5 \text{ ft}^3$$

$$P_1 = 60 \text{ psi} + 14.7 = 74.7 \text{ psia}$$

$$P_2 = \text{zero gauge or } 14.7 \text{ atm pressure}$$

$$V_2 = \text{unknown}$$

$$= \frac{1.5 \times 74.7}{14.7}$$

$$= 6.2 \text{ ft}^3/\text{min of free air}$$

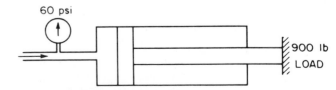

FIGURE 11-19   Diagram for Example 2.

Many of the basic laws pertaining to liquids also apply to gases. For instance, the cylinder in Example 2 must be large enough to overcome the force of the work. Pascal's law would apply as follows: Suppose that the force on the cylinder due to the work load is 900 lb. What piston area would be required?

Given:

$$Force = 900 \text{ lb}$$

$$Pressure = 60 \text{ psi}$$

$$Area = unknown$$

Thus:

$$Area \text{ (in.}^2) = \frac{force \text{ (lb)}}{pressure \text{ (in.}^2)}$$

$$= \frac{900 \text{ lb force}}{60 \text{ psi}}$$

$$= 15 \text{ in.}^2$$

EXAMPLE 1

How many cubic feet of free air are required to raise the receiver from zero gauge to final pressure? Suppose that the receiver is 12 in. in diameter and 24 in. long and that 80 psi is the final pressure (Fig. 11-20).

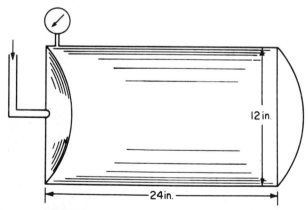

FIGURE 11-20   Diagram for Example 1.

Solution:

$$Volume \text{ of receiver} = area \times length$$

First, find the area:

$$Area = 0.7854D^2$$

$$= 0.7854 \times 12 \times 12$$

$$= 113 \text{ in.}^2$$

$$volume \text{ in cubic inches} = 113 \text{ in.}^2 \times 24 \text{ in.}$$

$$V = 2712 \text{ in.}^3$$

To change cubic inches to cubic feet we must multiply by 1 ft³/1728 in.³

$$Vol. \text{ (ft}^3) = \frac{2712}{1728} = 1.57 \text{ ft}^3$$

$$= 1.57 \text{ ft}^3$$

Applying Boyle's law

$$P_1 = 14.7 \qquad\qquad P_1V_1 = P_2V_2$$

$$P_2 = 80 + 14.7 \qquad\qquad V_2 = \frac{P_1V_1}{P_2}$$

$$V_1 = 1.57 \times X$$

$$X = free \text{ air unknown}$$

$$V_2 = 1.57 \qquad\qquad V_2 = \frac{1.57 \text{ ft}^3 \times 80 \text{ psi}}{14.7 \text{ psi}}$$

Substituting gives

$$(14.7)(1.57 + X) = (80 + 14.7)(1.57)$$

$$1.57 + X = \frac{(80 + 14.7)(1.57)}{14.7}$$

$$X = \frac{(80 + 14.7)(1.57)}{14.7} - 1.57$$

$$= \frac{(94.7)(1.57) - 1.57}{14.7} = 8.54 \text{ ft}^3$$

Example 2

Suppose that the air in the receiver which is discussed in Example 1 were used until the pressure dropped from 80 psig to 30 psig. How much free air would be required to bring the pressure back to 80 psig?

Solution: We now have the following relationship:

Ft³ of free air required to raise receiver from 30 psig to 80 psig

$$= \frac{vol. \text{ of receiver}}{ft^3} \times \frac{final \text{ psig} - initial \text{ psig}}{14.7}$$

$$X \text{ (ft}^3) = 1.57 \text{ cu ft} \times \frac{80 \text{ psig} - 30 \text{ psig}}{14.7}$$

$$= 1.57 \times \frac{50}{14.7}$$

$$= 5.34 \text{ ft}^3$$

Charles' Law.   When the pressure of confined gas remains constant, the volume of the gas is directly proportional to its absolute temperature. This can be stated mathematically with a simple formula:

$$\frac{Initial \text{ volume}}{Final \text{ volume}} = \frac{initial \text{ absolute temperature}}{final \text{ absolute temperature}}$$

or

$$\frac{\text{Initial absolute temperature}}{\text{Initial volume}} = \frac{\text{final absolute temperature}}{\text{final volume}}$$

or

$$\frac{V_1}{V_2} = \frac{T_1}{T_2} \quad \text{or} \quad V_2 T_1 = T_2 V_1$$

EXAMPLE

A tank holding 300 ft³ of air at atmospheric pressure and 60°F is heated to 120°F. What is the volume of air that escapes from the tank?

Solution:

$$V_2 = \frac{V_1 T_2}{T_1}$$

$T_1 = 60 + 460 = 520°F$ initial absolute temperature

$T_2 = 120 + 460 = 580°F$ final absolute temperature

(Absolute temperature in degrees Rankine = F° + 460)

$$V_2 \ (\text{ft}^3) = \frac{300 \times 580}{520}$$

$$= 334.6$$

$$\text{Vol. escaping} = 334.6 - 300$$

$$= 34.6 \ \text{ft}^3$$

Guy-Lussac found that if the volume of a gas remains constant, the pressure exerted by the confined gas is directly proportional to the absolute temperature. This law is expressed as follows:

$$\frac{\text{Initial pressure}}{\text{Final pressure}} = \frac{\text{initial absolute temperature}}{\text{final pressure}}$$

or

$$\frac{\text{Initial absolute temperature}}{\text{Initial pressure}} = \frac{\text{final absolute temperature}}{\text{final pressure}}$$

or

$$\frac{P_1}{P_2} = \frac{T_1}{T_2} \quad \text{or} \quad \frac{T_1}{P_1} = \frac{T_2}{P_2}$$

or

$$P_2 T_1 = T_2 P_1$$

EXAMPLE

A *closed* tank holding 300 ft³ of air at 14.7 psi and 60°F is heated to 120°F. What will be the final pressure? (Remember to use absolute temperatures.)

Solution:

$$T_1 = 460 + 60 = 520°F$$

$$T_2 = 460 + 120 = 580°F$$

$$P_2 = \frac{P_1 \times T_2}{T_1}$$

$$= \frac{14.7 \times 580}{520}$$

$$= 16.2 \ \text{psi}$$

The gas laws may be combined with the following results:
Final pressure:

$$P_2 = \frac{P_1 V_1 T_2}{V_2 T_1}$$

Final volume:

$$V_2 = \frac{P_1 V_1 T_2}{P_2 T_1}$$

Final absolute temperature:

$$T_2 = \frac{P_2 V_2 T_1}{P_1 V_1}$$

EXAMPLE 1

50 ft³ of air at atmospheric pressure and 40°F is compressed to 10 ft³ at a temperature of 160°F. What is the final pressure?

Solution:

$$P_2 = \frac{P_1 V_1 T_2}{V_2 T_1}$$

$$P_1 = 14.7$$

$$V_1 = 50 \ \text{ft}^3$$

$$T_2 = 160 + 460 = 620°F \ \text{absolute temperature}$$

$$T_1 = 40 + 460 = 500°F \ \text{absolute temperature}$$

$$V_2 = 10 \ \text{ft}^3$$

$$P_2 = \frac{14.7 \times 50 \times 620}{10 \times 500}$$

$$= 91.14 \ \text{psi}$$

EXAMPLE 2

60 ft³ of air at 30 psig and 50°F is compressed to 80 psig and 110°F. What is the final volume?

Solution:

$$V_2 = \frac{P_1 V_1 T_2}{P_2 T_1}$$

$P_1 = 30 + 14.7 = 44.7$ psia

$V_1 = 60$ ft³

$T_2 = 110 + 460 = 570°F$ absolute temperature

$P_2 = 80 + 14.7 = 94.7$ psia

$T_1 = 50 + 460 = 510°F$ absolute temperature

$$V_2 = \frac{44.7 \times 60 \times 570}{94.7 \times 510}$$

$$= 31.6 \text{ ft}^3$$

EXAMPLE 3

The air space above the oil in a closed hydraulic reservoir contains 160 in.³ of air at 14.7 and 60°F. Return oil coming back from the system compresses the air into a space of 60 in.³ What is the final temperature of the air if the final pressure is 30 psi?

Solution:

$$T_2 = \frac{P_2 V_2 T_1}{P_1 V_1}$$

$P_2 = 30 + 14.7 = 44.7$

$V_2 = 60$ in.³

$T_1 = 60°F + 460 = 520°F$ absolute temperature

$P_1 = 14.7$

$V_1 = 160$ in.³

$$T_2 = \frac{44.7 \times 60 \times 520}{14.7 \times 160}$$

$$= 593 - 460$$

$$= 133° \text{ F}$$

The basic laws and principles discussed in Section One are also applicable to pneumatics since both gases and liquids are classified as fluids. Air and water set the standards and all other fluids are compared with them for such things as specific gravity, density, and other physical properties.

QUESTIONS

1. What is the most common instrument for measuring the pressure of the atmosphere?

2. Write a comparison of the physical characteristics of gases with liquids.

3. What is absolute pressure?

4. What is gauge pressure?

5. Explain how vacuum can be used to perform work.

6. If the gauge on a receiver tank reads 96 psi, what would the absolute pressure be at sea level?

7. Express 8 in. Hg vacuum in psia.

8. How much pull in pounds would a vacuum cylinder lift provide if the piston area is 12 in.² with a vacuum pump operating at 26 in. Hg?

9. If 1 psi supports a column of water 27.686 in. high, what length of tube would be necessary to measure atmospheric pressure at sea level?

10. Draw the diagram of a gasometer.

11. What is free air?

12. Explain the importance of Boyle's law.

13. A closed air tank contains 16 ft³ of air at 14.7 psi; oil is forced into the tank until the air occupies only one-fourth its original volume. What is the final pressure?

14. How many cubic feet of free air is required to raise the receiver from zero gauge to 80 psi if the receiver is 20 in. in diameter and 36 in. long?

15. Explain the importance of Charles' law.

16. An accumulator that holds 864 in.³ of N_2 at 60°F is heated to 140°F. What is the final pressure?

17. Sixty-four cubic feet of air at atmospheric pressure and 50°F is compressed to 12 ft³ at 154°F. What is the final pressure?

12
How Air Is Compressed

A compressor is a machine for compressing air or other gases from an initial intake pressure, usually atmospheric pressure, to a higher pressure. It may also be stated that a compressor increases pressure exerted by a gas by reducing the volume. The two general classifications of compressors are positive displacement and nonpositive displacement.

Compressors used for pneumatic power are generally reciprocating piston or rotary designs which are positive displacement units. Reciprocating compressors have one or more pistons which reciprocate within a cylindrical bore. In a single piston unit air is compressed from its initial inlet pressure to the desired final pressure in one compressive step. This is known as a single-stage air compressor (see Fig. 12-1).

A compressor having two or more compressive steps in which the discharge from each supplies the next in series is a multistage compressor (see Fig. 12-1).

In a single-stage unit the size, stroke, and speed of the reciprocating piston determines the volume displacement, generally expressed in cubic feet per minute (cfm). For multistage compressors, the piston displacement of the first stage is commonly used as that of the entire machine.

Rotary compressors are available in both single-stage and multistage configurations. Air is moved via the dis-

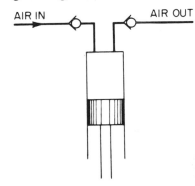

SINGLE-STAGE

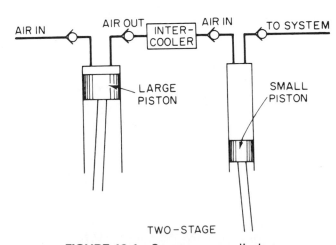

TWO-STAGE

FIGURE 12-1 Compressor cylinders.

placement structure from inlet to outlet. The number of stages determines the pressure available at the outlet. The vane design contains a number of rectangular vanes fitted in a slotted rotor. The rotor is off center to the housing, and as the shaft turns, centrifugal force causes the vanes to move out against the contour of the retaining ring. Air is trapped between the vanes during one-half of one revolution as the space between the rotor and ring increases. Air is pushed out during the other one-half of one revolution as the space decreases (see Fig. 12-2).

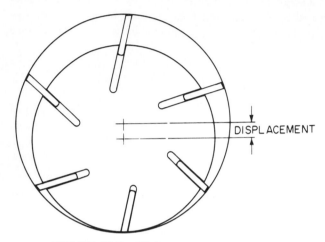

FIGURE 12-2   Sliding vane compressor.

The vanes in a rotary compressor are generally made of an asbestos cloth impregnated with a phenolic resin. The vanes take on a very high polish during operation and provide excellent wear characteristics. When air is compressed, it gains heat as the molecules of the gas become closer together and bounce off each other faster and faster. Because of the difficulty in removing heat with a single-stage unit, there is an economical limit to the pressure level that can be attained in one compressive step. Generally, this limit is between 100 and 150 psig. Multistage compressors are used when high pressures are required, because better cooling between stages can effectively increase the efficiency and reduce the input power requirements (Fig. 12-3).

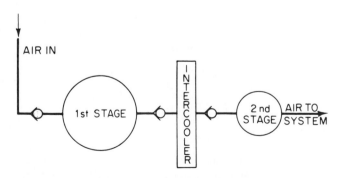

FIGURE 12-3   Rotary compressor.

Type	Capacity (psig)
Single-stage compressor	100–150
Two-stage compressor	150–500
Three-stage compressor	500–2500
Four-stage compressor	2500 up

ADIABATIC AND ISOTHERMAL COMPRESSION

Theoretically, air may be compressed with all the heat of compression retained (adiabatically) or by removing all the heat of compression (isothermally). However, in practical application air compressors give performance between these two theoretical concepts.

Students should keep in mind that the relationship of these two concepts are limits within which the actual compression always occurs. This relationship may be expressed as the ratio of specific heat at constant pressure to the specific heat at constant volume.

Specific heat is expressed by the number of Btu (British thermal units) necessary to raise the temperature of 1 pound of air (or any substance) 1 degree Fahrenheit. Through experiments, the specific heat of air has been determined as 0.2375 Btu/lb. per °F. (Specific heat for water = 1 or 0.2375 btu.)

Isothermal compression occurs when air undergoes a change in pressure or volume at constant temperature, which means that the specific heat equals 1 or 0.2375 Btu. Adiabatic compression occurs when air undergoes a change in pressure or volume while retaining heat of compression, which means a change in specific heat. This change or difference has been determined by Regnault's experiments to be 0.0686 Btu/lb, per °F which gives a difference (0.2375 − 0.0686) = 0.1689 Btu/lb, per °F. The 0.0686 Btu represents energy lost when air is compressed adiabatically.

The ratio of difference between the two concepts may now be expressed as

$$\frac{0.2375}{0.1689} = 1.41$$

(Boyle's law modified)　　　　(Boyle's law)
Adiabatic compression　　Isothermal compression
$P_1V_1^{1.4} = P_2V_2^{1.4}$　　　　$P_1V_1 = P_2V_2$

This concept could also be illustrated with an indicator diagram, which is a graphic representation of the relationship of adiabatic to isothermal compression (see Fig. 12-4). ADE represents the additional power required for adiabatic compression that is greater and above the power required for isothermal compression. In fact, about one-third

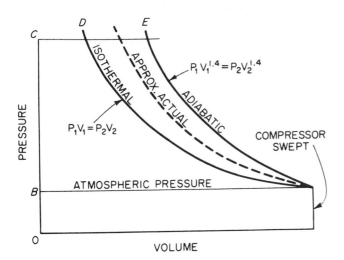

FIGURE 12-4 Relationship of adiabatic to isothermal compression.

more power is required to compress air to 100 psia adiabatically than isothermally, which indicates the importance of cooling during compression.

Figure 12-5 shows a typical idealized indicator card for a single-stage piston-type compressor. At point 1 the piston is at the end of its stroke and a certain percentage of

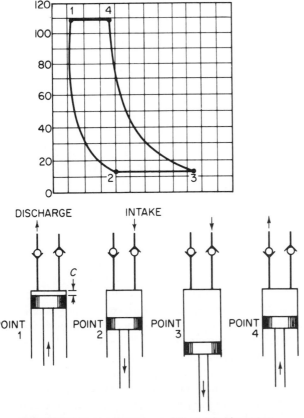

FIGURE 12-5 Indicator card for a single-stage piston-type compressor.

air at discharge pressure is trapped in the clearance volume. The clearance volume is the volume of air in the cylinder over and above the volume swept by the piston. The clearance volume includes air left between the piston and the head and in and under the check valves used for displacement.

When the piston begins its return stroke, the clearance air expands until the pressure in the cylinder is slightly lower than the pressure in the intake line. This occurs at point 2, where the inlet valves open, and the cylinder takes in air between 2 and 3 on the diagram.

At point 3 the piston is at the opposite end of its stroke and it now returns, compressing the air from point 3 to 4. When it reaches point 4, the air is compressed to a point slightly higher than the pressure in the receiver, and the outlet check valve opens, allowing discharge to the system. When the air is discharged from the cylinder, the piston is again back to point 1 and the cycle repeats.

Intercoolers normally cool air by the use of the same cooling medium as the compressor itself (air or water). Air-cooled intercoolers may consist of finned tubes through which air is forced, or it may be of the radiator type. Cooling air is blown over the outer surface of the tubes to effect the transfer of heat caused by compression.

A water-cooled intercooler consists of a nest of tubes through which water is passed to effect the cooling. The tubes are enclosed within a shell, and the air being cooled passes over the outside of the tubes. Baffles may be used to control the paths of both the air being cooled and the coolant for maximum efficiency.

The power savings effected by the staging technique of positive displacement compressors depends upon factors such as the ratio between the suction and discharge pressures, the cooling medium, and the effectiveness of the intercooler.

The capacity of positive-displacement compressors is determined by the actual amount of fluid that is delivered through the discharge valves in 1 minute. This is referred to as the *actual delivered capacity*. The piston displacement volume for 1 minute divided into the actual delivered capacity of the compressor is called the *volumetric efficiency* and is usually expressed as a percentage.

If the fluid were incompressible, it might be supposed that a compressor's capacity would be nearly the same as the piston displacement. However, since air expands and contracts in volume with changes in pressure and temperature, the actual capacity is less than piston displacement. This is mainly because of the clearance between piston and cylinder head at the end of each stroke and because of the valve opening in the cylinder.

At the end of each compression stroke as shown in Fig. 12-6, the clearance volume is filled with compressed air that has not been delivered through the discharge valves.

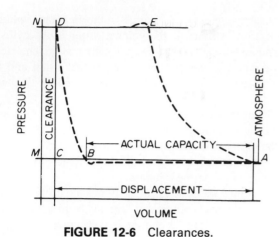

FIGURE 12-6 Clearances.

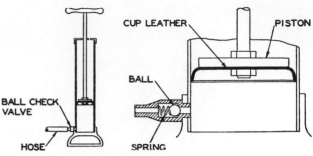

FIGURE 12-7 Hand air pump.

On the intake stroke this air will reexpand in the cylinder until the pressure is below the suction pressure to allow new air into the cylinder through the inlet valve. The greater the differential pressure between inlet and outlet, the lower the volumetric efficiency will be. However, this does not represent a direct power loss, because the expanding fluid does work on the intake stroke by exerting pressure on the piston, which is, in turn, transferred to the drive arrangement.

$$\text{Volumetric efficiency} = \frac{\text{free air}}{\text{piston displacement}}$$

The volumetric efficiency involves all things that cause a compressor to deliver less air than piston displacement. This includes reduction in capacity due to preheating of the fluid, slippage of air past piston seals, and reversal of airflow during valve operation. Actual delivered capacity is determined by measuring the volume of air discharged from the compressor and referring it to suction conditions. Volumetric efficiency is not a measure of compression efficiency. Compression efficiency, when heat is contained, is the ratio of the horsepower theoretically required to the horsepower actually expended in the cylinder (known as indicated horsepower). This indicated horsepower is the brake horsepower at the compressor shaft minus mechanical horsepower lost.

Compressor efficiency (adiabatic) is the ratio of the theoretical horsepower to the shaft horsepower input. It is equal to the product of the compression efficiency times the mechanical efficiency.

TYPES OF COMPRESSORS

One of the simplest designs of the positive-displacement type is the hand pump (see Fig. 12-7). The basic principle of operation is the same in all reciprocating-type compressors, and the understanding of this simple design can be

transferred to other more complex designs of the reciprocating type.

The hand pump consists of a piston secured by a rod to a handle, a cylinder, a cup seal, and a one-way valve or check. With the handle pulled all the way to the top, air enters a small hole in the pump cylinder. As the handle is pushed downward, the air now trapped within the cylinder is compressed. As the piston approaches the bottom, the pressure is sufficient to force open the spring-held check valve and enter the system where it is being used. When the piston reaches the bottom of the cylinder and the pressure on each side of the check valve is equalized, the spring closes the check valve. The upward stroke of the piston creates a partial vacuum inside the cylinder until the inlet port is reached near the top of the stroke.

POPULAR INDUSTRIAL TYPES

A single-stage compressor (see Fig. 12-8) has one or more cylinders and operates on the following principle: On the downward stroke of each piston air is taken in at atmospheric pressure. On the upward stroke the compressed air is discharged into the air receiver. Single-stage compressors usually operate at pressures under 100 psi and for intermittent-type duty.

A two-stage compressor (see Fig. 12-9) has a minimum of two cylinders, generally a low-pressure cylinder (largest) and one high-pressure cylinder (smallest). When the LP (large) cylinder piston goes downward, air is taken in at atmospheric pressure through a one-way valve. On the up stroke, the LP piston discharges the air through an intercooler and into the HP (small) cylinder which is on the down stroke. On the upward stroke of the HP cylinder piston air is discharged into the air receiver. Two-stage compressors are generally recommended for use where operating pressures are above 100 psi. The two-stage compressors are more efficient, since they provide some cooling during compression and deliver air to the receiver at lower discharge temperature for the high pressures. Generally, the normal operating pressure for this design is 175 psi maximum.

Reciprocating compressors of the type shown in Fig.

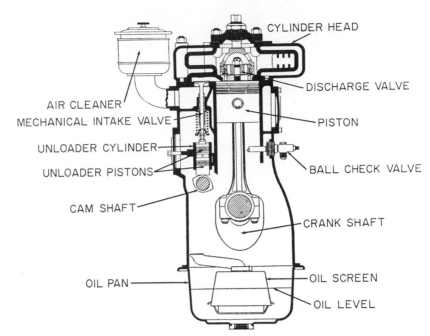

FIGURE 12-8 Single-stage compressor. (Courtesy of Compressed Air & Gas Institute.)

FIGURE 12-9 Two-stage compressor. (Courtesy of Quincy Compressor.)

12-10 operate on the same principle as the hand pump but with refinements toward greater efficiency, durability, and practicability. This shows a horizontal single-stage double-acting compressor which is water-cooled. The rotary motion of a prime mover, such as an electric motor or internal combustion engine, is converted into reciprocating motion by use of a crank and connecting rod. One end of the connecting rod is fastened to the crank pin, the other to a crosshead which reciprocates as the shaft rotates. A piston rod connects the crosshead to the piston, which reciprocates within the cylinder as in the hand pump.

Notice that the compressor in Fig. 12-11 is fitted with one inlet check valve and one discharge check valve on each end of the cylinder. As the piston moves to the left, the inlet valve to the right allows atmospheric pressure to enter that end of the cylinder. In the same direction, air in the opposite end of the cylinder is being discharged out through the discharge valve on the left side and into the system. On the return stroke to the right, air enters through the left side inlet check, and the delivery of air to the system is now through the discharge check valve on the right end of the cylinder.

In other words, a double-acting compressor delivers air to the system on each stroke. In most cases, the discharge is to a receiver or flask from which it can be used as needed in the system. The use of storage devices helps to give smoother operation, and to provide some cooling of the hot discharged air.

Figure 12-12 shows a three-cylinder single-acting air-cooled compound compressor. The compounding is effected by having two cylinders which discharge into a single high pressure cylinder through an air-cooled intercooler. The cylinders are arranged in "W" formation with three connecting rods on a single crank. The locations of the air inlet and discharge valves are in the heads of each cylinder. The lubrication of the cylinder walls results from the splash of oil from the crankcase. Compressors of this type are avail-

FIGURE 12-10 Water-cooled single-stage compressor. (Courtesy of Compressed Air & Gas Institute.)

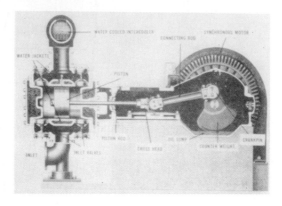

FIGURE 12-11 Double-acting compressor. (Courtesy of Compressed Air & Gas Institute.)

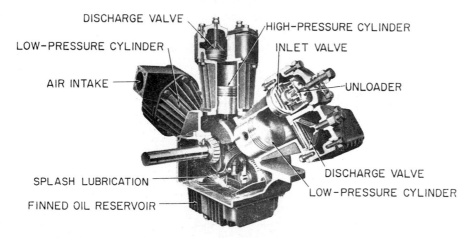

FIGURE 12-12 Three-cylinder, single-acting, air-cooled compound compressor. (Courtesy of Compressed Air & Gas Institute.)

able in single-stage or two-stage designs. The usual pressure range runs as high as 125 psig.

There are also several other cylinder arrangements for "angle-type" reciprocating compressors such as multicylinder vertical two-cylinder Y or V type, four-cylinder radial type, and the two-cylinder angle compressor with one cylinder vertical and one horizontal.

The portable air compressor which is illustrated in Fig. 12-13 has a self-contained compressed air power plant. The unit consists of an air compressor, a prime mover and

an air receiver, and is complete with cooling, lubricating, regulating, and starting systems, all enclosed on a running gear for ready movement.

The prime mover may be an internal combustion engine or an electric motor. Standard discharge pressures for portable compressors essentially match the requirements of portable air tools. Sizes are available as high as 100 hp.

Figure 12-14 illustrates the operating principle of the sliding vane type of rotary compressor. A cylindrical slotted rotor turns within a larger diameter casing. In the rotor slots

FIGURE 12-13 Portable air compressor. (Courtesy of Compressed Air & Gas Institute.)

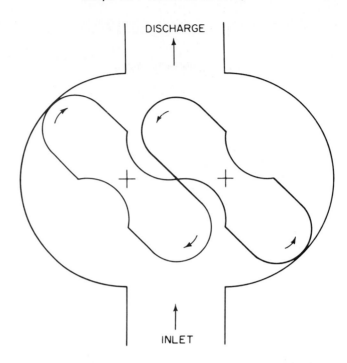

FIGURE 12-15 Two-impeller positive-type compressor.

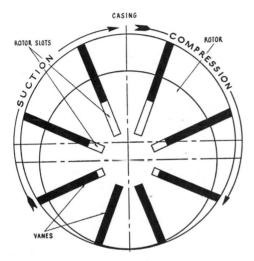

FIGURE 12-14 Sliding vane type of rotary compressor. (Courtesy of Compressed Air & Gas Institute.)

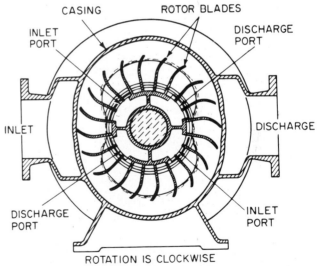

FIGURE 12-16 Liquid-piston-type compressor. (Courtesy of Compressed Air & Gas Institute.)

are rectangular vanes, shown by heavy black lines, which slide in and out of the rotor slots due to centrifugal force. Thus, turning clockwise, air is trapped and compressed between the vanes and discharged through a port on the compression or left-hand side. Rotary sliding vane compressors are used for compressing up to approximately 50 psig in a single-stage, which may be increased to 150 psi in a two-stage, and so on. Sizes cover as high as 700 hp.

Another design of rotary compressor is the two-impeller positive type shown by Fig. 12-15. The two identical impellers are kept in proper rotative position by a pair of external gears. There is no internal contact between the impellers, so no lubrication is required. These compressors are used for lower pressure ranges and for capacities of about 50,000 ft³/min.

Figure 12-16 illustrates a third type of rotary compressor known as the liquid piston type, which employs a liquid, usually water, as the compressing medium. The blades

of the turning rotor form a series of buckets carrying the liquid around the inside of the elliptical casing. Since the liquid following the contour of the inside of the casing surges back into the buckets at the narrow point of the ellipse, the air in the buckets is compressed and discharged through properly located ports. Compressors of this type develop about 75 psi pressure in one stage. The capacity range is up to 5000 ft³/min.

NONPOSITIVE DISPLACEMENT TYPES

Figure 12-17 shows the construction of a four-stage centrifugal compressor. Air enters through the large inlet flange at the left end, and enters the eye of the first stage impeller. The rotative speed of the impeller imparts a high velocity to the air as it moves radially outward into the diffuser, which converts part of the velocity head to static pressure head. The air then moves radially inward to the eye of the next impeller and the process is repeated, increasing the pressure of the air with each stage of compression. Multistage centrifugal air compressors are generally used for handling large volumes of air at lower pressures, but are occasionally designed for pressures of 150 psi or higher. Capacities range as high as 150,000 ft ^{3}/min.

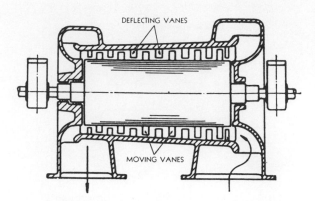

FIGURE 12-18 Multistage axial compressor. (Courtesy of Compressed Air & Gas Institute.)

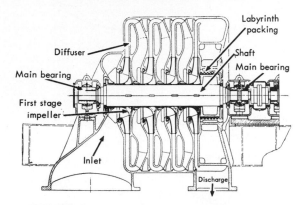

FIGURE 12-17 Multistage centrifugal compressor. (Courtesy of Compressed Air & Gas Institute.)

Figure 12-18 shows a multistage axial-flow, uncooled blower-type compressor with six rows of rotation vanes. Both centrifugal and axial machines inherently operate at high speed. Axial machines are characterized by substantially constant capacity delivery at variable pressures, whereas centrifugal machines deliver practically constant pressure over a considerable range of capacities. These characteristics must be considered when equipment is selected for a given application. Compressors of this type have ratings over 1,000,000 ft ^{3}/min with pressure ranges similar to the centrifugal types. (This material is used through the courtesy of the Committee on Engineering Education of Compressed Air and Gas Institute.)

QUESTIONS

1. What are the two general classes of compressors?
2. List several design types under each class.
3. What is generally considered the maximum pressure and capacity of a single-stage compressor?
4. Draw the compression curve that illustrates the difference between adiabatic and isothermal compression. Explain.
5. Show the modified Boyle's law for adiabatic compression.
6. What effect does an intercooler have in compressing air?
7. What effect does an aftercooler have on compressed air?
8. Why is the actual delivery from a positive displacement compressor less than piston displacement?
9. What is clearance volume?
10. Explain how the volumetric efficiency of a compressor is determined.

13
The Compressed Air System

Compressed air is a major source of industrial power. It is safe, flexible, and easy to use for countless jobs throughout modern industry. A well-planned and developed system can provide excellent trouble-free service, with flexibility to expand for the future. Air-operated equipment is classified as continuous or intermittent duty in operation, much the same as hydraulic equipment.

The type of duty generally indicates the type of compressor controls to use. Constant speed controls are recommended for continuous operations where the compressor operates about 50% of the time. This prevents excessive wear and prevents power losses due to frequent starts and stops.

Dual controls offering automatic stops and starts with a high and low range setting of a pressure switch or continuous speed may be needed in plants working only one shift in 24 hours. If automatic start and stop controls are used, the compressor should have enough capacity to operate only one-third of the time. The number of cubic feet of air per minute depends on the various tools and circuits that require air.

Single-stage compressors are generally used for continuous operation when the maximum pressure is below 80 psi; also for intermittent operation up to 125 psi if less than 50% of the compressor's capacity is being used.

Two-stage compressors are generally used for continuous operation when the maximum pressure is greater than 80 psi; also for intermittent operations above a maximum pressure of 125 psi.

Air compressors are generally rated in terms of cubic feet per minute of free air. *Free air* is defined as the normal air at atmospheric conditions at any given place. Since compressor capacity is given in terms of free air, it is convenient for making calculations as follows:

$$\frac{P_1V_1}{T_1} = \frac{P_2V_2}{T_2}$$

$$\text{cfm} = V_1 = \frac{P_2V_2T_1}{T_2P_1}$$

EXAMPLE

Air is used at the rate of 20 cfm from a receiver at 80°F and 100 psi.

(a) If the barometer reads 14.5 psia and the air inlet temperature is 60°F, how many cfm must the compressor provide?

(b) If the atmospheric pressure dropped to 14.0 psia, how much would the compressor have to deliver?

Solution:

$$\text{(a) } V_1 \text{ (cfm)} = \frac{114.5 \times 20 \times 520}{540 \times 14.5}$$

$$V_1 = ?$$

$$V_2 = 20 \text{ ft}^3 \text{ per min}$$

$$P_1 = 14.5$$

$$P_2 = 100 + 14.5$$

$$T_1 = 60 + 460$$

$$T_2 = 80 + 460$$

$$\text{cfm} = 152 \text{ ft}^3 \text{ of free air/min}$$

$$\text{(b) cfm} = \frac{114.5 \times 20 \times 520}{540 \times 14}$$

$$= 157.5 \text{ ft}^3 \text{ of free air/min}$$

This indicates the difference when normal atmospheric conditions change.

The total capacity required for an air system is determined in a large way by the previous experience with usage of air. The consumption of air by motors or cylinders can be computed as follows (see also Table 13-1):

Cylinders:

$$\text{cfm} = \frac{\text{area} \times \text{length}}{1728} \times \text{no. of strokes per minute}$$

TABLE 13-1
AVERAGE FREE-AIR CONSUMPTION OF TYPICAL TOOLS AND
EQUIPMENT
80–150 psi

Air hammer	approx.	16	cfm
Air hoist (cylinder)	"	1–4	"
Air hoist (motor)	"	2–4	"
Rotary drills $\frac{1}{16}''$–$\frac{5}{8}''$	"	4–7	"
Rotary drills $\frac{1}{4}''$	"	18–20	"
Rotary drills $\frac{3}{8}''$	"	20–40	"
Rotary drills $\frac{1}{2}''$–$\frac{3}{4}''$	"	70	"
Rotary drills $\frac{7}{8}''$–1$''$	"	80	"
Piston drills $\frac{1}{2}''$–1$\frac{1}{4}''$	"	45	"
Piston drills $\frac{7}{8}''$–1$\frac{1}{4}''$	"	75–80	"
Piston drills 1$\frac{1}{2}''$–2$''$	"	89–90	"
Piston drills 2$''$–3$''$	"	100–110	"
Grinders vertical and horizontal	"	10–20	"
1 hp air motor	"	10	"
2 hp air motor	"	15	"
3 hp air motor	"	20	"

EXAMPLE

How much air would a 2-in.-bore single-acting cylinder use that has a stroke length of 12 in. and makes four strokes a minute?

Solution:

$$\text{cfm} = \frac{3.14 \times 12}{1728} \times 4$$

$$= 0.0872$$

$$A = 0.7854D^2$$

$$= 3.14$$

$$(1728 = \text{in.}^3 \text{ in 1 ft}^3)$$

The 0.0872 cfm is at working pressure and would have to be converted to free air for compressor capacity calculations. The following equation is used:

$$\text{cfm} = V_1 = \frac{P_2 V_2 T_1}{T_2 P_1}$$

Table 13-2 may be useful in selecting cylinders for a given application.

When calculating air motors, the displacement per revolution would be needed.

EXAMPLE

If an air motor is operating at 2000 rpm and has a displacement of 0.02 in.³ of air per revolution, how much air is being consumed?

Solution:

$$\text{cfm} = \frac{\text{disp. (in.}^3)/\text{rev}}{1728}$$

$$= \frac{0.02 \times 2000}{1728}$$

$$= 0.023 \text{ ft}^3/\text{min}$$

The general gas law equation would again be used to convert this figure into free air for purposes of compressor capacity (see Table 13-3).

Air lines for the distribution of air power must be properly installed and must be sized to provide an adequate supply of air for each workstation. The two principal factors affecting pressure drop in air systems are leaks and friction (See Tables 13-4 and 13-5).

Air leaks can be prevented through proper maintenance procedures. Because compressed air is not cheap and leaks cut down on production, air distribution systems should be thoroughly inspected at regular intervals. Soap solution and candle flame are two common methods of pinpointing air leaks. Some plants inject essence of peppermint into the air system to find leaks.

The effects of friction have been accurately computed through the years of using compressed air, and charts and tables are available for determining this important loss in air systems. The coefficient of friction depends mainly on the diameter or size of the air line and the velocity of flow. The density or viscosity of the air also affects friction, but changes very little.

The pressure loss due to friction in 100 ft of standard pipe is found by substituting the N value into the following formula:

TABLE 13-2
CYLINDER FORCE VALUES

Cylinder diameter—inches	3/8	1/2	9/16	5/8	3/4	7/8	1	1 1/8	1 1/4	1 3/8	1 1/2	1 3/4	2	2 1/4	2 1/2	3	4	5	6	8	10
Cylinder area—square inches	.11	.20	.25	.31	.44	.60	.78	1.0	1.2	1.5	1.8	2.4	3.1	4.0	4.9	7.1	13	20	28	50	79
Supply air psig										Force delivered by cylinder—pounds											
50	5	10	12	15	22	30	39	50	60	75	90	120	160	200	250	360	650	1000	1400	2500	4000
75	8	15	17	23	33	45	60	75	90	110	120	180	240	300	370	530	1000	1500	2100	3700	6000
100	11	20	25	31	44	60	78	100	120	150	180	240	310	400	490	710	1300	2000	2800	5000	8000
150	16	30	37	45	66	90	120	150	180	220	270	360	460	600	740	1050	2000	3000	4200	7500	12000
							Double-acting cylinder air usage—cubic feet (free basis) per inch of stroke															
50					.002	.003	.004	.005	.006	.008	.009	.013	.016	.021	.026	.037	.068	.10	.15	.26	.41
75					.003	.004	.006	.007	.009	.011	.013	.017	.022	.029	.035	.051	.094	.14	.20	.36	.57
100	.001	.002	.002	.003	.004	.005	.007	.009	.011	.014	.017	.022	.029	.037	.045	.065	.12	.18	.26	.46	.73
150					.006	.008	.010	.013	.016	.020	.024	.032	.041	.053	.065	.094	.17	.26	.37	.66	1.04

Note 1. Divide the above figures by 2 for single-acting cylinders.

Note 2. To allow for loss of force, or reduction in air usage, because of space taken by the rod of double-acting cylinders, subtract the respective values for a cylinder of the same diameter as the rod.

TABLE 13-3
SELECTOR CHART

Selector Chart—Single-stage System (125 psi or less)

System requirements for continuous operation		System requirements for intermittent operations	
Free air needed by system (cfm)	*Hp for compressor*	Free air needed by system (cfm)	*Hp for compressor*
Up to 1.9	1/2	Up to 6.6	1/2
2.0 to 3.0	3/4	6.7 to 10.5	3/4
3.1 to 3.9	1	10.6 to 13.6	1
4.0 to 5.8	1 1/2	13.7 to 20.3	1 1/2
5.9 to 7.6	2	20.4 to 26.6	2
7.7 to 10.2	3	26.7 to 32.5	3
10.3 to 18.0	5	32.6 to 38.0	5

Two-stage System (over 125 psi)

Up to 4.2	1	Up to 14.7	1
4.3 to 6.4	1 1/2	14.8 to 22.4	1 1/2
6.5 to 8.7	2	22.5 to 30.4	2
8.8 to 13.2	3	30.5 to 46.2	3
13.3 to 20.0	5	46.3 to 60.0	5
20.1 to 29.2	7 1/2	60.1 to 73.0	7 1/2
29.3 to 40.0	10	73.1 to 100.0	10
40.1 to 60.0	15	101.0 to 150.0	15
60.1 to 80.0	20	151.0 to 200.0	20

Courtesy of Champion Pneumatic Machinery Co. Inc.

TABLE 13-4
EFFECT OF AIR LEAKS

Size of opening (inches)	Cu ft wasted per month based on 100 psi and nozzle coefficient of 0.65	Cost of air wasted per month based on 10¢ per 1000 cu ft
1/32	45,508	4.56
1/16	182,272	18.21
1/8	740,210	74.01
1/4	2,920,840	292.09
3/8	6,671,890	667.19

TABLE 13-5
FRICTION LOSS IN PIPES:
N VALUES
Nominal Diameter in Inches

Cu ft free air per minute	1/2 in. dia.	3/4 in. dia.	1 in. dia.	1 1/4 in. dia.	1 1/2 in. dia.	2 in. dia.	2 1/2 in. dia.
5	0.127	0.12					
10	0.507	0.78	0.22				
15	1.14	1.76	6.49				
20		3.04	0.87	0.20			
25		5.0	1.36	0.32			
30		7.04	1.96	0.45			
35		9.59	2.66	0.62	0.27		
40		12.53	3.48	0.81	0.36		
45			4.40	1.02	0.45		
50			5.44	1.20	0.56		
60			7.83	1.82	0.80	0.22	
70			10.66	2.47	1.09	0.29	
80			13.92	3.25	1.43	0.38	
90				4.09	1.81	0.48	
100				5.05	2.23	0.60	
110				6.11	2.70	0.72	0.28
120				7.27	3.22	0.86	0.33
130				8.53	3.78	1.01	0.39
140				9.89	4.38	1.17	0.46
150				11.36	5.03	1.34	0.52

(*N* factors derived through experiment.)

P_1 = initial psig

$$\text{pressure drop (psi)} = \frac{N \times 14.7}{P_1 \times 14.7}$$

EXAMPLE

What is the pressure drop for 90 cfm flowing through 100 ft of 1 1/4-in. pipe with the receiver at 40 psig?

Solution:

$$\text{Pressure drop} = \frac{N \times 14.7}{P_1 + 14.7}$$

$$= \frac{4.09 \times 14.7}{40 + 14.7}$$

$$= \frac{60.12}{54.7}$$

$$= 1.09$$

Pressure at point of work:

$$\text{Initial} = \quad 40.00 \text{ psig}$$

$$\text{Drop} = \underline{- \quad 1.09 \text{ psig}}$$

$$38.91 \text{ psig}$$

In long runs of pipe, as pressure drop increases, the friction loss increases. In other words, there is greater friction loss in the last few feet than at the beginning of the line. Since the pressure decreases toward the pipe outlet, the velocity must increase. Pressure drop should be controlled by selecting pipe sizes to prevent a loss greater than 10% between the receiver and point of use. In addition to the proper sizing of pipes, fittings, and valves, a number of techniques are used to ensure proper air supply.

The loop system provides two-way distribution of air to increase efficiency of use (see Fig. 13-1). Long distribution lines may be equipped with receivers or storage tanks near the point of usage to dampen compressor pulsations or to decrease line losses for intermittent operations. The leader system is also effective for an adequate distribution of air to several tools simultaneously (see Fig. 13-2).

When air flows through pipes, the flow is retarded by friction due to the walls of the pipe. Air flow next to the walls of the pipe slows down, while the velocity in the center of the pipe is high. This loss is proportionately greater when the air pressure is at lower values. The loss is much less if the proper pipe diameters are used. Table 13-6 is useful as a guide in determining size of equipment in compressed air work.

For convenience in making calculations, loss through fittings is given in terms of equivalent feet of straight pipe. (This has been developed through actual testing; see Table 13-7.)

EXAMPLE

A circuit requiring 65 ft³/min free air and operating at initial pressure of 100 lb psi could feasibly use a $\frac{1}{2}$-in.-nominal standard pipe size. However, it must be remembered that for every 100 ft of run a pressure drop of 19 psi would occur. This means that the next largest nominal standard pipe size would be selected where long lines are necessary between supply and the pneumatic circuit. In the case mentioned, if the run were only a short distance, say 10

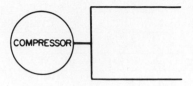

CENTRALIZED SYSTEM – 1 COMPRESSOR

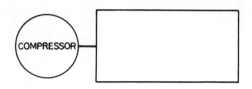

CENTRALIZED LOOP SYSTEM – 1 COMPRESSOR

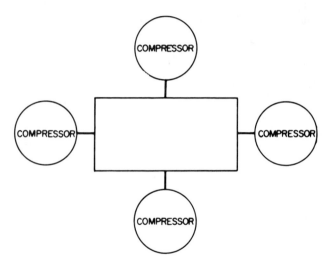

DECENTRALIZED LOOP SYSTEM – 2 OR MORE COMPRESSORS

FIGURE 13-1   Air system piping.

TO STATIONS

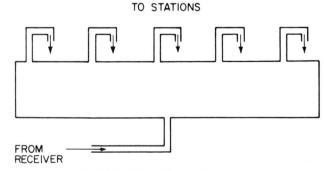

FROM RECEIVER

FIGURE 13-2   Manifold system.

ft, less than 2 psi pressure drop would occur, which could be tolerated. The designer must remember that circumstances sometimes alter a choice to provide adequate supply of air to the circuit under consideration.

TABLE 13-6
MAXIMUM RECOMMENDED FLOW
THROUGH STANDARD WEIGHT PIPE

Nominal Standard Pipe Size							
$\frac{1}{8}''$	$\frac{1}{4}''$	$\frac{3}{8}''$	$\frac{1}{2}''$	$\frac{3}{4}''$	$1''$	$1\frac{1}{4}''$	$1\frac{1}{2}''$
Pressure drop in 100 ft.							
54.0	31.7	25.4	19.0	12.4	9.2	5.94	4.68

Initial psi	Maximum Recommended flow (SCFM—free air)						
10	5.5	9.2	18.6	30.2	51.1	84.	139.	186.
20	6.5	11.	22.	35.7	60.6	100.	165.	220.
30	7.4	12.5	25.	40.7	68.4	112.	187.	250.
40	8.2	13.3	27.6	44.9	76.	125.	207.	276.
50	8.9	15.0	30.0	48.8	82.2	135.	226.	301.
60	9.6	16.1	32.2	52.3	89.0	146	242.	323.
70	10.4	17.7	34.3	56.0	94.	154.	258.	345.
80	10.8	18.2	36.3	59.0	100.	165.	273.	364
90	11.3	19.1	38.2	62.2	105.	172	286.	382
100	12.0	20.0	40.	65.	110	180	300.	400.
110	12.5	20.8	41.8	67.7	114.	188.	312.	417.
120	12.8	21.6	43.4	70.4	119.	195.	325.	434.
130	13.3	22.4	45.	73.0	123.	202.	337.	450.
140	13.8	23.2	46.5	75.5	127.	209.	348.	465.
150	14.2	23.9	48.	77.7	132.	218.	360.	479.

TABLE 13-7
FRICTION LOSS IN PIPE FITTINGS
(EQUIVALENT FEET OF STRAIGHT PIPE)

	$\frac{1}{4}$	$\frac{3}{8}$	$\frac{1}{2}$	$\frac{3}{4}$	1	$1\frac{1}{4}$	$1\frac{1}{2}$	2
Gate Valve (full open)	.30	.30	.35	.44	.56	.74	.86	1.10
Tee (straight through)	.50	.50	.70	1.10	1.50	1.80	2.20	3.00
Tee (Side outlet)	2.50	2.50	3.30	4.20	5.30	7.00	8.10	10.4
90° Ell	1.40	1.40	1.70	2.10	2.60	3.50	4.10	5.20
45° Ell	.50	.50	.78	.97	1.23	1.60	1.90	2.40
Angle Valve (full open)	8.00	8.00	9.30	11.5	14.7	19.3	22.6	29.0
Globe Valve (full open)	14.00	14.00	18.6	23.1	29.4	38.6	45.2	58.0

(The number of fittings should be kept to a minimum to keep down losses.)

Table 13-7 shows the equivalent feet of straight pipe loss for the various pipe fittings that are used in connecting the air circuit. For example, if a right-angle gate valve were used to shut off the branch circuit, the pressure drop through this valve would be equivalent to 9.3 ft of loss in the $\frac{1}{2}$-in. standard pipe size. It is necessary for the circuit designer to calculate all frictional losses in straight-run pipe, fittings, and valves in order to determine how much initial pressure is needed to overcome these resistances and still end up with sufficient pressure to accomplish the work.

The flow of air through pipes is generally expressed in cfm. Flow of air can be measured in lineal feet per minute and expressed as the velocity of flow past a certain point. The relationship is given the following equation:

$$cfm = lineal\ ft/min \times area$$

EXAMPLE

What is the velocity in ft/min of 50 cfm flowing through a hole having an area of $\frac{1}{10}$ ft^2?

Solution:

$$cfm = lineal\ ft/min \times area\ (ft^2)$$

$$Velocity\ (lineal\ ft/min) = \frac{cfm}{area}$$

$$= \frac{50}{1/10}$$

$$= 500\ ft/min$$

C$_v$ CAPACITY COEFFICIENT

1. Originally used as a means of comparing capacity of liquid passing or carrying valves. A valve with a C_v rating of 1 would pass 1 U.S. gallon of water at 60°F, in one minute, with a 1 lb per sq in. loss in pressure across the valve.

    The formula for liquid flow is

    $$C_v = \frac{gpm}{\sqrt{\Delta P/G}}$$

    where gpm = gal/min liquid flow

    ΔP = pressure drop across valve

    G = specific gravity of liquid

2. Research has proven that same C_v rating applicable to liquid or noncompressible fluids is applicable to gases or compressible fluids by revising the formula to take into consideration the various factors applicable to flow of gases. There is some discussion relative to which factors should be used in the formula. For working air, we feel the following is a good formula:

    $$C_v = \frac{cfm}{22.67} \sqrt{\frac{GT}{(P_1 - P_2)P_2}}$$

    where cfm = air flow in ft^3/min *of free air*

    G = specific gravity − 1 for air

    T = *absolute* temperature = °F + 460

$P_1 = $ *inlet* or upstream pressure

$P_2 = $ *outlet* or downstream pressure

$\left.\begin{array}{l} \\ \\ \\ \\ \end{array}\right\}$ *Both* lb/in.^{2} absolute or psig $+$ 14.7

or gauge pressure $+$ 14.7

Remember: cfm *free air* or scfm is displacement or swept volume in cubic feet per minute multiplied by the ratio

$$\frac{\text{gauge pressure} + 14.7}{14.7}$$

3. (Very important) After a given ΔP (or $P_1 - P_2$) is reached, no more flow (cfm) can be pushed through a valve having a given C_v rating. This "terminal capacity" is reached when $P_2 = 0.53 \times P_1$, or when P_2 is 53% of P_1. Therefore, never assume that you can get more speed out of a system just because you would be willing to allow more pressure drop or ΔP.

   For example, with 100 psi (114.7 psia) line pressure, when the pressure in the cylinder drops to 114.7 $\times$ 0.53 or 60.7 psia (46 psi gauge), cfm flow is at its maximum. No higher flow rate can be attained. This does not mean that the pressure will not drop lower or that piston speed will not increase. But if these things happen, they are due to *external* causes, *not* to airflow rate increasing.

Table 13-8 provides a convenient source for capacity constants for various inlet pressures and pressure drops. Table 13-9 groups data relative to flow of air through orifice in cubic feet of air per minute.

The piping installation should be designed to provide compressed air to all parts of the plant in proper condition for the power and control that may be required. Figure 13-3 shows the proper installation of a compressor system with receiver tanks, moisture traps, drains, distribution lines, and branch connections for pneumatic power.

Free air from the atmosphere enters the compressor through a filter, which is essential to keep dust and contamination from entering the system. The air passes from the first compressive step through an intercooler, where it loses some of its heat of compression before entering the second stage. High pressure air discharged from the second stage then passes through the aftercooler to the air drier (a moisture separator is also used). From the drier air passes into the receiver, where it is stored as pressure energy until it is needed by the pneumatic circuit.

When the four-way valve is energized, the air passes from the main line through a filter, regulator, and lubricator combination unit before entering the cylinder to do work.

TABLE 13-8
CAPACITY CONSTANTS FOR VARIOUS
INLET PRESSURES AND PRESSURE DROPS

Inlet pressure P (psig)	Capacity constant (A) for various pressure drops (ΔP)		
	$\Delta P = 2$ psi	$\Delta P = 5$ psi	$\Delta P = 10$ psi
40	0.0986	0.0642	0.0479
50	0.0905	0.0586	0.0433
60	0.0840	0.0543	0.0398
70	0.0788	0.0508	0.0371
80	0.0745	0.0478	0.0348
90	0.0708	0.0454	0.0329
100	0.0676	0.0433	0.0313
110	0.0648	0.0414	0.0299
120	0.0623	0.0398	0.0287
130	0.0600	0.0384	0.0276
140	0.0581	0.0371	0.0267
150	0.0563	0.0359	0.0258

A silencer or muffler is generally used to exhaust the spent air back to atmosphere from the four-way valve as the cylinder reciprocates.

MOISTURE IN AIR

Free air from the atmosphere contains varying amounts of moisture in vapor form. The water vapor content is measured in terms of relative humidity. Relative humidity is the ratio of the amount of water vapor actually present in the atmosphere to that which would be present if the air were saturated.

Vapor in air condenses when the temperature of saturated air is reduced. It is the purpose of an aftercooler to reduce the temperature of the compressed air and separate the moisture out of the air before it enters the system. Moisture in an air system can result in harmful effects, such as washing away lubricants, excessive wear, and corrosion.

When air lines extend outside, pockets of water in the piping may freeze in cold weather. Water also freezes at the exhaust of pneumatic tools, causing sluggish operation.

If the air system provides air for both power systems and control systems, a drier may be needed. Control air must be dry and clean when used in precision instruments. Automatic air control systems are used to regulate power plant equipment, furnaces, pneumatic power systems, hydraulic power systems, data processing machines, computers, and many others.

Air driers use a chemical substance, called a *desiccant,* which is very dry and attracts the moisture. Some driers use a porous filtering material that must be recharged periodically by heating.

Table 13-10 is based on the moisture content of free air at atmospheric conditions and is useful for determining separators, driers, and moisture traps for an installation.

TABLE 13-9
FLOW OF AIR THROUGH ORIFICE IN (FT³ FREE AIR/MIN)

Size of orifice	Receiver gauge pressure																		
	2	5	10	15	20	25	30	35	40	45	50	60	70	80	90	100	125	150	200
$\frac{1}{64}$"	0.038	0.0597	0.0842	0.103	0.119	0.133	0.156	0.173	0.19	0.208	0.225	0.26	0.295	0.33	0.364	0.40	0.486	0.57	0.76
$\frac{1}{32}$"	0.153	0.242	0.342	0.418	0.485	0.54	0.632	0.71	0.77	0.843	0.914	1.05	1.19	1.33	1.47	1.61	1.97	2.33	3.07
$\frac{3}{64}$"	0.342	0.545	0.77	0.94	1.07	1.21	1.4	1.56	1.71	1.91	2.05	2.35	2.68	2.97	3.28	3.66	4.12	5.20	6.9
$\frac{1}{16}$"	0.647	0.965	1.36	1.67	1.93	2.16	2.52	2.80	3.07	3.36	3.64	4.2	4.76	5.32	5.87	6.45	7.85	9.20	12.2
$\frac{3}{32}$"	1.37	2.18	3.06	3.75	4.28	4.84	5.6	6.24	6.84	7.6	8.2	9.4	10.7	11.88	13.1	14.5	17.5	20.8	27.5
$\frac{1}{8}$"	2.435	3.86	5.45	6.65	7.7	8.6	10.0	11.2	12.27	13.4	14.5	16.8	19.0	21.2	23.5	25.8	31.4	36.7	48.7
$\frac{3}{16}$"	5.45	8.72	12.3	15.0	17.1	19.4	22.4	25.0	27.4	30.4	32.8	37.5	42.9	47.5	52.4	58.0	70.0	83.2	110.0
$\frac{1}{4}$"	9.74	15.4	21.8	26.7	30.8	34.5	40.0	44.7	49.0	53.8	58.2	67.0	76.0	85.0	94.0	103.2	125.5	147.0	195.8
$\frac{3}{8}$"	21.95	34.6	49.0	60.0	69.0	77.0	90.0	100.0	110.5	121.0	130.0	151.0	171.0	191.0	211.0	231.0	282.0	330.0	440.0
$\frac{1}{2}$"	39.0	61.6	87.0	107.0	123.0	138.0	161.0	179.0	196.4	215.0	232.0	268.0	304.0	340.0	376.0	412.0	502.0	588.0	782.0
$\frac{5}{8}$"	61.0	96.5	136.0	167.0	193.0	216.0	252.0	280.0	306.8	336.0	364.0	420.0	476.0	532.0	587.0	645.0	785.0	920.0	1220.0
$\frac{3}{4}$"	87.6	133.0	196.0	240.0	277.0	310.0	362.0	400.0	442.0	482.0	521.0	604.0	685.0	765.0	843.0	925.0	1127.0	1322.0	1760.0
$\frac{7}{8}$"	119.5	189.0	267.0	326.0	378.0	422.0	493.0	550.0	601.0	658.0	710.0	722.0	930.0	1004.0	1145.0	1260.0	1531.0	1804.0	
1"	156.0	247.0	350.0	427.0	494.0	550.0	645.0	715.0	786.0	860.0	930.0	1070.0	1215.0	1360.0	1500.0	1648.0	2000.0	2350.0	
$1\frac{1}{8}$"	196.0	310.0	440.0	538.0	620.0	692.0	812.0	900.0	987.0	1082.0	1170.0	1350.0	1530.0	1710.0	1890.0				
$1\frac{1}{4}$"	242.0	384.0	543.0	665.0	770.0	860.0	1000.0	1120.0	1230.0	1345.0	1455.0	1680.0	1900.0	2130.0	2350.0				
$1\frac{3}{8}$"	291.0	464.0	660.0	803.0	929.0	1040.0	1200.0	1350.0	1480.0	1630.0	1760.0								
$1\frac{1}{2}$"	350.0	550.0	780.0	960.0	1102.0	1240.0	1435.0	1600.0	1770.0	1930.0	2085.0								
$1\frac{3}{4}$"	473.0	752.0	1068.0	1304.0	1505.0	1680.0	1960.0												
2"	625.0	985.0	1395.0	1700.0	1967.0	2200.0	2586.0												

Courtesy of Quincy Compressor Co.

FIGURE 13-3 Typical compressed air system.

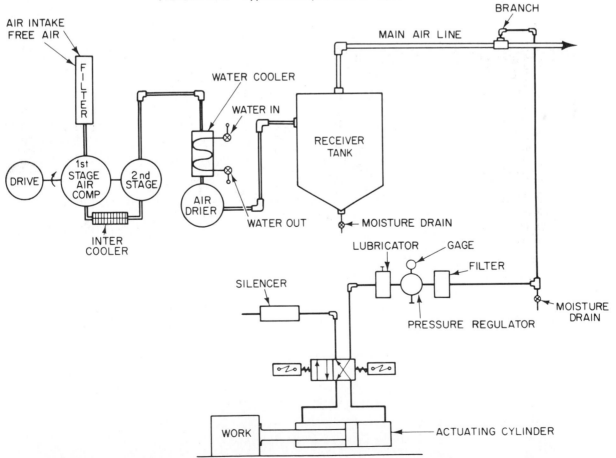

TABLE 13-10
MOISTURE IN AIR
Air Pressure—psig

Temp. °F.	30	40	60	80	100	150
	Pints of water per 1000 cubic feet saturated air (free basis)					
40	0.094	0.075	0.054	0.041	0.036	0.026
60	0.20	0.16	0.12	0.089	0.069	0.050
80	0.36	0.30	0.22	0.18	0.14	0.10
100	0.67	0.55	0.40	0.32	0.27	0.20
120	1.2	0.93	0.70	0.55	0.46	0.32

PIPING THE SYSTEM

Standard schedule 40 pipe, tools, and fittings are adequate for piping up the system. Pipe and fittings should be selected of proper size to maintain less than a 3-psi pressure drop in main distribution and branch lines. Oversized pipe may be a wise investment for growing companies.

Piping should be properly supported and installed to prevent sag and vibration. Lines should be installed at a slight angle slope to facilitate drain for any condensate. At low points and at the ends of the system, "water legs" with drains may be provided to collect and dispose of condensate.

Tables 13-6 and 13-7 are useful in determining the maximum recommended flow through standard pipe and fittings to prevent excessive friction and pressure drop.

Installation should avoid excess fittings and sharp bends that increase pressure drop and loss of power. The compressed air system should be provided with protection against overloads by installing a relief in the system. This is usually located in the receiver. The loop system of distribution is generally recommended where possible to help provide a more even distribution of pneumatic power to all circuits.

SUMMARY OF SYSTEM DESIGN FACTORS

A careful study of immediate and future needs should be made before the compressed air system components are selected. Volume and pressure requirements are essential, as well as a detailed analysis of each pneumatic device to be operated.

If 100 cfm at 100 psig are required and a 50-cfm compressor is installed, the pressure will soon drop lower than the effective amount needed, since the volume produced is only half the amount required for effective tool operation. In such a case, the compressor will run continuously, trying to keep up, and will not reach the unload or cutoff point until the air demand is below the normal work load.

A safety factor of approximately 25% is desirable for efficient operation to reduce maintenance costs. This applies to constant speed control units particularly. With automatic start and stop service, the percentage of intermittent use should be approximately one-third running time and two-thirds down time. This provides for fewer starts per hour for the electric motors, and prevents overheating with subsequent kickout of the starter thermal overloads. It is far more practical to oversize than to undersize a compressed air system, and higher benefits will result over a period of time.

If altitude location is not considered, the unit may fall far short of actual requirements. Losses run approximately 3% for each 1000 ft of elevation above sea level because of rarefied air at lower barometric pressure. This condition also affects cooling of electric motors and has serious implications on the sizing of engine-driven units.

An orifice $\frac{1}{4}$ in. in diameter passing air at 80 psig requires approximately 20 hp. This indicates the importance of accurate calculations in designing air systems. A listing of all application requirements including the suggested safety factory of 25% can eliminate costly errors.

The following questions may be helpful in determining specific needs of the compressed air system:

1. What elevation is to be considered?

2. What is the ambient operating temperature?

3. Is the compressor to run one, two, or three shifts per day?

4. What are the actual uses of compressed air?

5. Should the compressor be constant speed, automatic start and stop, or dual control?

6. If motor driven, what is the electrical service available?

7. Are special electrical controls required?

8. In the case of dual units, is an electrical automatic alternator advisable?

9. What is the maximum pressure requirement?

10. Will the unit operate in a hazardous, dusty, or gaseous atmosphere where totally enclosed fan cooling and electrical equipment are necessary?

11. Will oil and moisture in the air present a problem?

12. What accessories are needed, such as coolers, silencers, filters, or driers?

QUESTIONS

1. Explain the use of constant-speed controls.

2. Explain the use of dual controls.

3. What are some of the important considerations in selecting a single-stage compressor? When should a multistage compressor be used over a single-stage compressor?

4. Why are compressors rated in terms of cubic feet of free air per minute?

5. Air is used at the rate of 24 cfm from a receiver at 80°F and 100 psi. If the barometer reads 14.6 psia and the inlet temperature is 56°F, how many cfm must the compressor provide?

6. How many cfm of air would a 4-in.-bore single-acting cylinder use that has a stroke length of 12 in. and makes 800 strokes a minute?

7. How much air would the compressor take in if the cylinder in Question 6 operates at 110 psi?

8. If air at 100 psi escapes from an opening of $\frac{1}{16}$ in., what would the approximate cost be if it continued to leak for a period of 6 months?

9. How is pressure drop controlled in a pneumatic installation?

10. Explain the advantages of the loop system of air distribution over the header system.

11. What size of nominal standard pipe would be selected for a pressure of 100 psi at a maximum flow of 300 ft³/min?

12. An air motor uses 10 cfm at 3000 rpm. What is its displacement?

13. How are leaks detected in an air system?

14. How are friction losses determined in an air system?

15. Explain why friction loss increases as the pressure drop increases.

16. How are friction losses controlled in an air system?

17. Diagram a loop system of air distribution and show how a connection is made for a branch line.

18. Diagram or explain the header system of air distribution.

19. What is the velocity in feet per minute of 60 cfm flowing through a hole $\frac{1}{12}$ ft² in area?

20. What effect does moisture have on an air distribution system?

21. What is the principle of a chemical air drier?

22. List 10 factors that are important in the selection of a compressed air system.

14

Controlling Pneumatic Power

Pneumatic circuits are easily applied for both linear and rotary motion.

HIGHLIGHTING PNEUMATICS VERSUS HYDRAULICS

The main difference between applying pneumatic and hydraulic fluid power is the compressibility factor of the two media.

The weight of the hydraulic fluid is of concern when sudden acceleration or deceleration is experienced. Since force is equal to mass times acceleration when a hydraulic motor reverses under load, the resultant pressure created will be many times that created if the fluid were air.

Hydraulic fluid averages about 58 lb/ft^3. Dry air weighs 0.0746 lb/ft^3 at atmospheric pressure and 72°F. It increases to about 5 lb/ft^3 at 1000 psi and about 15 lb/ft^3 at 3000 psi. It is apparent that pneumatic circuits are less concerned with shock loads in the system.

The kinetic energy of moving fluid is shown by

$$EK = \tfrac{1}{2} mV^2$$

$$\text{where } EK = \text{kinetic energy (ft-lb)}$$

$$m = \text{mass of fluid (slugs)}$$

$$V = \text{velocity of fluid (ft/sec)}$$

Pressure drop and temperature effects are also quite different when the two media are compared.

Hydraulic fluids exchange pressure energy for velocity and heat due to friction when fluid bypasses over a relief valve or flow control. The increase in temperature changes the viscosity of the fluid, which may have an effect on the operation of a machine during warm-ups and so on.

In pneumatic circuits when pressure energy is exchanged for velocity energy, generally a decrease in temperature results (frost or snow at the exhaust side of an air motor). Pneumatic power is immediately available for quick response at low temperatures. Air compressed to 45 psia, if suddenly released to the atmosphere, travels approximately 17,000 ft/min and about 35,000 ft/min if compressed to 50 psi. In an actual circuit velocity is a function of conductor size in area.

Pneumatics will usually require the use of lubricating devices for the prevention of wear on integral moving parts.

Operating pressures for industrial pneumatic applications are generally between 28 in. Hg (mercury) and 150 psi, which limits the force capabilities unless very large cylinders are used. Aircraft pneumatic systems have found that high-pressure pneumatic systems have certain weight advantages, as mentioned previously. Aircraft systems use pressures up to 5000 psi and higher for some military uses. Missile launching devices may reach 30,000 psi.

Industrial uses of high-pressure pneumatic systems have been limited because of the cost of producing air pressure above 150 psi in large quantities.

PNEUMATIC CONTROLS

Branch lines leading to each pneumatic circuit are generally equipped with a filter regulator and lubricator combination unit (see Fig. 14-1).

FIGURE 14-1 Combo unit. (Courtesy of Bastian-Blessing Company.)

Filters

The purpose of the filter is to remove contamination from the air before it reaches the directional valve and actuating cylinder or air motor. Air line filters are generally fitted with a filter element that removes contaminants in the range 25 to 50 microns (see Fig. 14-2).

Air flow entering the filter is directed downward with a swirling motion that forces moisture and heavier particles to the wall of the bowl. When removed from the main airstream, these contaminants are pulled out of suspension and fall to the bottom of the bowl and become trapped. When the drain cock is opened, they are forced out of the filter. Pressure drop through the filter should be less than 2 psi in most designs.

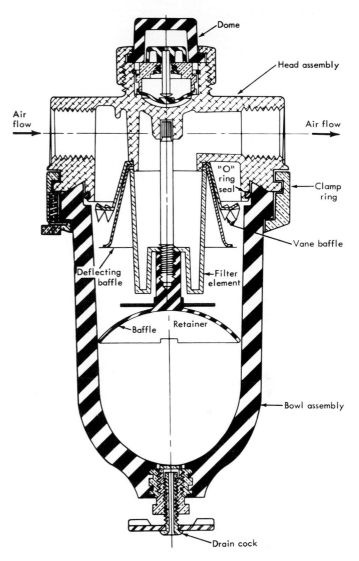

FIGURE 14-2 Air line filter. (Courtesy of Bastian-Blessing Company.)

There are a number of different designs from which to select a filter to keep circuits operating without trouble. The filter design in Fig. 14-3 drains contaminants during circuit operation and automatically closes when the airflow stops. This helps to prevent sediment in the bowl from getting back into the system.

Regulators

The maximum system pressure is controlled by the compressor control system, which generally maintains a pressure range. In other words, the compressor may start automatically when the system pressure drops to 120 psi, and automatically stop again when the pressure in the receiver reaches 140 psi. The compressor is generally controlled with pressure switches and protected by a relief valve.

In order to maintain a continuous pressure at a work

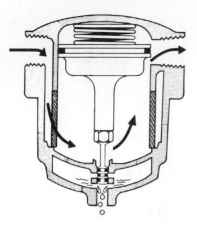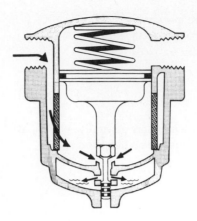

FIGURE 14-3 Filter design.
(Courtesy of WABCO.)

station and eliminate compressor fluctuations, or to utilize pressures lower than compressor output, a pressure control valve is used.

The pressure regulator is fitted with an adjustable spring that allows the valve to hold a given pressure on the downstream side (see Fig. 14-4). The force of the spring on top of the valve spool is set for the required pressure downstream. This force holds the valve open until the pressure downstream starts to exceed the spring force. The pressure downstream acting against the area on the bottom of the spool causes it to move up, throttling the supply air to control the proper pressure downstream.

A number of excellent designs are available for accurate pressure control of pneumatic circuits. It should be understood that a regulator at each work station helps to keep the entire system stable and provides the correct air supply for each job without wasting power.

Figure 14-5 shows a regulator with a diaphragm operating against the adjustable spring. When the diaphragm

overcomes the adjusted spring force, it allows the push rod to move up, and the spring-loaded valve at the bottom throttles the supply air to the controlled pressure side.

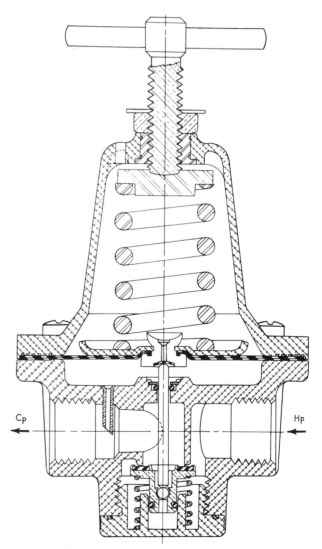

FIGURE 14-5 Air line regulator. (Courtesy of Bastian-Blessing Company.)

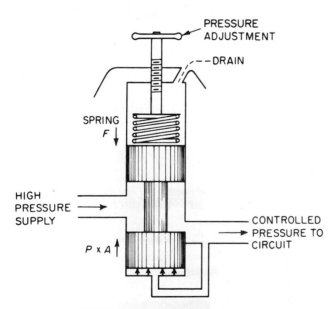

FIGURE 14-4 Pressure regulator.

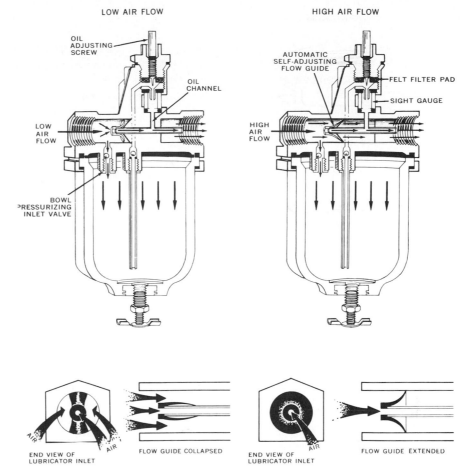

FIGURE 14-6 Lubricators. (Courtesy of WABCO.)

Lubricators

Lubricators of various designs are available to assure proper lubrication of integral moving parts of pneumatic components. Figure 14-6 explains the principle of operation. Figure 14-7 shows a complete installation.

Mufflers

Mufflers (or silencers) are used to control the noise caused by rapidly exhausting (expanding) air. Dust filters may be used at air intakes to minimize the entry of contaminants into the system.

CONDITIONING OF THE COMPRESSED AIR

From the brief overview of the devices associated with preparation of air for use as a power transmission medium we can now look more closely at each requirement and the devices used to accomplish each task.*

Pneumatic systems and components discussed up to

*Figures 14-8 through 14-23 have been provided courtesy of Parker Hannifin Corporation, with associated technical matter from Parker, *Industrial Pneumatic Technology*. Copyright 1980.

FIGURE 14-7 Application of filter, regulator, and lubricator. (Courtesy of Bastian-Blessing Company.)

this point require compressed air free of contamination. No matter how well a system is designed or how expensive or sophisticated a particular component may be, contaminated air will interfere with components and systems operation. In general, the more contaminated the air, the less dependable a pneumatic system will be.

Even the air we breathe may not be appropriate for a specific pneumatic system. Air must be conditioned; it must be decontaminated before it is used in the pneumatic system. Some contaminants are literally built into a pneumatic system. Thus it is important that we look at the sources.

We will find the contaminents in a pneumatic system come from three basic sources: built-in, generated, and ingested. Built-in dirt occurs in newly fabricated systems where components or piping are dirty or where installation practices are below standard. As a system is assembled, pipes, valves, and storage tanks become a collector for rust, paint chips, dust, sealant tape, cigarette butts, and grit. Many harmful dirt particles are invisible to the unaided eye and cannot be removed by wiping with a rag or blowing off with an air hose.

A second source of dirt is that generated within the working system itself. As a system operates, moving parts in contact with other surfaces naturally begin to wear, generating wear particles. The use of incorrect types of fluid conductors may cause rust flakes or particles to form. These are eventually carried down through lines to a tool station or into a control system.

A third source of contaminants are those added to the system. Should a valve break down, the maintenance person may replace the component or repair it on the spot. In either case, the mechanic will more than likely be working in a dirty environment, which may allow contamination of the system as soon as a line is disassembled.

Dirt can also be added to a system by means of a cylinder. After a time, a cylinder rod wiper wears its outer sealing edge, so that it can no longer wipe off fine particles of dirt. This condition allows dirt to be drawn into the cylinder each time it is stroked. Quick disconnects are another potential source for externally generated contaminants.

Generally, the contaminants found in a system may be divided into three groups: abrasive dirt, soft dirt, and entrained liquids. Hard dirt may come from inside or outside the plant. Hard dust, grinding compounds, and foundry sand are just a few examples. This dirt is abrasive, affecting the proper operation of components. This type of dirt can wedge into clearances between moving parts, causing faulty operation or drastic component failure.

Dried paint, soft dust, or some types of threaded pipe joint compound can be considered soft dirt. This type of contaminant can cause orifices to plug or ports to cake, and is usually larger than clearances, which could cause faulty operation of components.

Entrained liquids usually enter the pneumatic system through entrainment in the air. In large quantities, moisture can wash away lubricants and, in any quantity, will cause rust to form in contact with the metals generally used in pneumatic systems. Oil carried over from top end compressor lubrication can cause resilient seals to deteriorate. This is especially true if synthetic lubricants are used with standard seals. Without effective seals, a pneumatic system wastes its stored energy. Also, operational problems occur because seals become swollen, making certain types of valve shifting erratic.

The size of contaminants is important. To measure contaminants we use the micrometer scale. As we found in our study of oil contamination, 1 micrometer (micron) is equal to one millionth of a meter or thirty-nine millionths of an inch. A single micrometer is invisible to the human eye without magnification and is so small that it is difficult to imagine. To bring the size more down to earth, some everyday objects will be measured using the micrometer scale. For example, an ordinary grain of table salt measures 100 micrometers (μm), and the average human hair measures 70 μm in diameter. For comparative purposes, 25 μm is approximately one thousandth of an inch. The lower limit of visibility for an unaided human eye is 40 μm. In other words, the average person can see individual particles measuring 40 μm and larger; many of the harmful dirt particles in a pneumatic system are below 40 μm.

Let's look next a little more closely at air line filters. The first line of defense for industrial compressed air is the compressor intake filter and the aftercooler. Proceeding from the aftercooler the air enters a receiver tank, where further cooling takes place and we can expect a reduction in moisture content. Additional moisture may be removed as the air cools in the distribution system, settling in the water legs. The last stage of air preparation takes place at the individual pieces of equipment that are served by the plant's compressed air system. An air line filter is a device that is placed in the air line at the workstation to be protected. By doing this, it removes particulate matter from all the air that passes through the filter element.

You will note that an air line filter consists basically of a housing with inlet and outlet ports, deflector, shroud, filter element, baffle plate, filter bowl, and drain. Filtration through an air line filter takes place in two stages. In the first stage, air enters the inlet port and flows through the openings in the deflector plate (A) in Fig. 14-8, which causes a swirling action. Entrained liquids and large (heavy) particles are forced out to the bowl's interior wall (B) by the centrifugal action of the swirling air. They then run down the sides of the bowl. Shroud C assures that swirling action occurs at low flow rates and uniformly distributes the airflow across the entire length of the elements (D). It also prevents debris from concentrating at any point on the element.

The baffle (E) separates the lower portion of the bowl into a *quiet zone,* where the removed liquid and particles

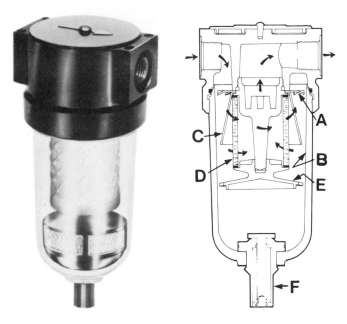

FIGURE 14-8 Filter. (Copyright 1980, Parker Hannifin Corporation, Cleveland, Ohio.)

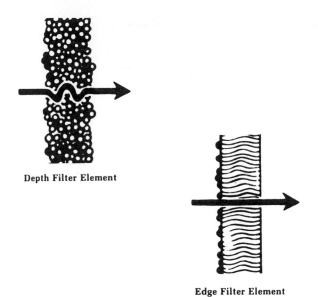

FIGURE 14-9 Filter element types. (Copyright 1980, Parker Hannifin Corporation, Cleveland, Ohio.)

collect unaffected by the swirling air and are therefore not reentrained into the flowing air. In the second stage of filtration, air flows through element D, where smaller particles that are still airborne are filtered out and retained. The filtered air then passes downstream.

Because of the need for clean air at the workstation the function of the mechanical filter element in a pneumatic system is to remove particles of dirt from the airstream. This is done by forcing airflow to pass through porous materials. Typical filter materials include cloth, felt, porous sintered metal, wire mesh, ceramics, polyethylene, plastics, and resin-impregnated paper. Some are throwaway types, while others may be cleaned and reused. Filter elements are generally divided into two types, depth and edge. With a depth-type element, air is forced to pass through an appreciable material thickness. Dirt is trapped because of the tortuous path the air must take. Depth elements in pneumatic systems are frequently made of porous bronze or plastic. Edge-type elements offer an airstream a relatively straight flow path. Dirt is caught on the surface or edge of the element that faces the airflow. Edge-type elements in pneumatic systems are often made of resin-impregnated paper ribbon (see Fig. 14-9).

Because of its construction, an air line filter element may have many pores of various sizes. Many of the pores are small. A few pores are relatively large. If it has no consistent hole or pore size, an air line filter element is given a nominal rating. The nominal rating is an element rating given by the filter manufacturer, indicating the expected average hole size in the element. For example, a depth element with a nominal rating of 40 μm indicates that the majority of the pores in the element are 40 μm in size

and that a large percentage of the 40-μm particles will be trapped.

Most air line filter elements used in industrial pneumatic systems have nominal ratings ranging from 50 to 5 μm. Since dirt in a pneumatic system comes in all sizes, shapes, and materials, no guarantee is made as to what size particles will be removed from a compressed airstream. If we have removed the contaminants, we must also remove the liquids. This can be done with an oil removal filter.

Oil removal filters (Fig. 14-10) are designed to meet the need for delivering oil-free air for a number of industries.

FIGURE 14-10 Oil removal filter. (Copyright 1980, Parker Hannifin Corporation, Cleveland, Ohio.)

A partial list of the industries requiring this type of filtration includes:

1. Dental and medical
2. Chemical and pharmaceutical
3. Bottling, packaging, and food processing
4. Many types of industrial paint spraying

Special applications include pneumatic control instrumentation, pneumatic measuring and gaging, protection of air bearings, paint spraying, blow molding machinery, moving-part logic devices, printing processes, paper separation, production and packaging of fine chemicals, photographic processes, absorbtion air dryers, or any place where oil-type compressors are used but oil-free air is required. The principal feature of such a device is that it removes liquid aerosols and submicronic particles down to approximately 0.3 μm. The efficiency of removing such liquid aerosols is as high as 99.9%. Usually, a prefilter is added to increase life and efficiency.

An oil removal filter works as follows. The contaminated air enters the element interior and is forced through a membrane thickness of borosilicate glass fibers, sometimes coated with epoxy. The flow then passes through an outer structural support, and at this stage most of the submicrometer oil particles carried by the air have been removed. The droplets coalesce and are blotted from the filter surface by layers of nonwoven glass felt and rayon cloth. The drops can gravitate to the filter sump, where they may be drained periodically.

The liquids and particles collected in the quiet zone must be drained before their level reaches a height where they would be reentrained into the air. This can be accomplished manually or with an automatic drain.

The manual drain is standard on most filters. This type requires that maintenance personnel periodically check each unit and open the drain if the contaminant level is high. These drains may consist of a petcock type, similar to a car radiator drain cock, or other manually operated types which are of a spring-loaded design that must be pressed to open (Fig. 14-11).

The automatic drain is a device placed in the quiet zone of the filter's bowl (Fig. 14-12). It typically has a float that when raised, directly or through a pilot action, causes a drain port to open, expelling liquid and entrained contaminants.

The method of operation is such that when system pressure is applied, it enters under the diaphragm and lifts it and the spool upward (Fig. 14-13), sealing the spool nose to the bleed seal. The bleed seal and pin continue to rise, lifting the float and float gasket from the top body orifice. Pressure begins to enter above the diaphragm, causing it to reverse and move down. At some point in the downward

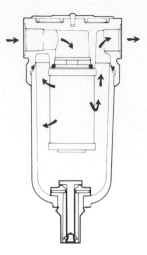

FIGURE 14-11 Filter drain. (Copyright 1980, Parker Hannifin Corporation, Cleveland, Ohio.)

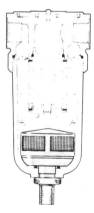

FIGURE 14-12 Automatic drain. (Copyright 1980, Parker Hannifin Corporation, Cleveland, Ohio.)

motion, the bleed seal and nose separate after the float/float gasket have sealed the top body orifice. Thus air trapped in the chamber above the diaphragm can vent through the spool and become slightly decreased in pressure.

When the forces above and below the diaphragm come into balance, the diaphragm, spool, and pin have moved to a position allowing the float gasket to form a seal at the top body orifice, and the spool nose has closed the bleed seal. Pressure above the diaphragm remains slightly less than that below. As the liquid level in the bowl rises, the float moves

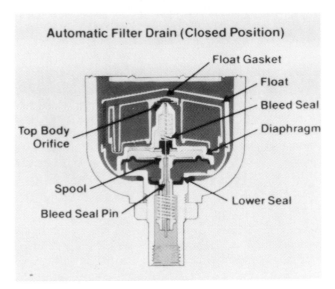

FIGURE 14-13 Automatic filter drain (closed position). (Copyright 1980, Parker Hannifin Corporation, Cleveland, Ohio.)

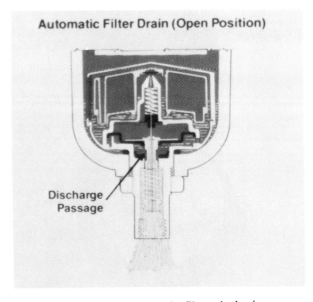

FIGURE 14-14 Automatic filter drain (open position). (Copyright 1980, Parker Hannifin Corporation, Cleveland, Ohio.)

upward, lifting its gasket off the top of the body orifice, allowing system pressure to enter above the diaphragm. This forces the diaphragm/spool downward, opening the discharge passage past the lower seal and discharging the liquid. After discharge, the float/float gasket drops and seals the top body orifice. As pressure above the diaphragm bleeds out through the clearance between the bleed seal pin and the spool nose, the spool/diaphragm continues to rise until the spool nose seals against the bleed seal. The cycle is ready to repeat. Both types of drains are found at the bottom

of the filter bowl. These bowls can be made of many different materials (see Fig. 14-14).

Many of the filters manufactured make use of polycarbonate for bowl materials. This is generally a good material for filters and lubricators because it is transparent and tough. Properly designed, fabricated, and maintained, it is suitable for use in normal industrial workplace environments, but should not be located in areas where it could be subjected to impact blows or temperatures outside the rated range.

As with most plastics, exposure to certain chemicals can cause damage. For example, polycarbonate bowls should not be exposed to chlorinated hydrocarbons, ketones, phosphate esters, and certain alcohols. They should not be used in air systems where compressors are lubricated with fire-resistant fluids such as phosphate ester or diester types. Metal bowls resist the action of most solvents but should not be used where strong acids or bases are present or in salt-laden atmospheres (see Fig. 14-15).

For most applications suitable for polycarbonate bowls, a bowl guard is recommended. This perforated metal shield fits around the bowl exterior to protect the bowl from mechanical damage as well as to contain the bowl parts in case of bowl rupture.

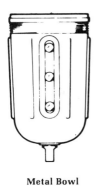

Metal Bowl

Bowl Guard

FIGURE 14-15 Metal bowl and bowl guard. (Copyright 1980, Parker Hannifin Corporation, Cleveland, Ohio.)

SELECTING A FILTER

When selecting a filter, these three steps should be followed:

1. Determine the maximum flow requirement where the filter will be used. This will be in terms of cfm.

2. Determine the allowable pressure drop at the needed flow.

3. Refer to the flow curves on the filter to find one that offers the particular flow rate at the predetermined pressure drop.

Remember, if an element is sized too small, the result will typically be a high initial pressure drop at the rated flow. Also, the element will *load up* quickly at the start of operations, with a resulting increase in pressure drop. This can lead to possible air starvation downstream. When the pressure drop reaches 10 psi, the filter element should be changed to prevent its rupture. If the filter is too large, the low swirl velocity generated may result in poor contaminant and condensate separation.

EXAMPLE

The ABC Machinery Company is using a radial-type piston motor which develops a maximum output of 10 hp (7.5 kW) at 800 rpm. Its air consumption, as read directly from catalog data, is 200 cfm. Select a filter that will work with a maximum pressure drop of 2 psi. The data given for the motor are at 85 psi.

Solution:

From motor catalog data it is evident that we need a filter that will pass 200 cfm at 85 psi, with a 2-psi pressure drop. As we go through a filter catalog we come upon two graphs (Fig. 14-16) showing the flow characteristics of a basic 1-in. port body with 1-in. pipe and 1½-in. pipe ports. Since the primary pressure of 85 psi is not given, it must be estimated as shown in Fig. 14-17. Working with the 08F53A graph, we find a 2-psi drop for 200 cfm if 1-in. plumbing is used. A 1.5-psi drop is evident for the 08F73A if 1½-in. plumbing is used. Either may be selected, but the former has better water removal efficiency.

EXAMPLE

A 6-in.-bore, 3-in. rod, 10-in. stroke cylinder extends in 4 sec and retracts in 3 sec. The working pressure required at the cylinder is 50 psi. The filters primary pressure is 100 psi. What filter is needed?

Solution:

The cfm must be calculated first.

$$\text{cfm} = \frac{V \text{ (in.}^3) \times \text{compression ratio}}{\text{time to fill cylinder} \times 28.8}$$

For extension:

$$\text{cfm} = \frac{(28.27 \times 10) \times \dfrac{50 + 14.7}{14.7}}{4 \times 28.8}$$

$$= 10.8$$

For retraction:

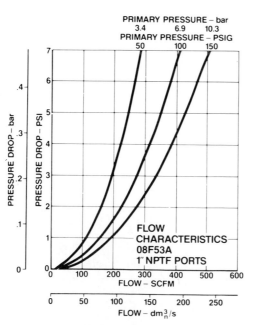
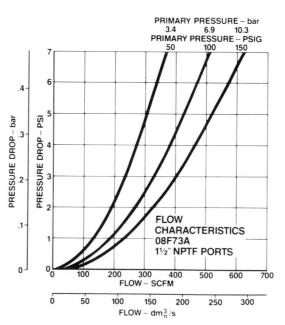

FIGURE 14-16 Filter performance characteristics. (Copyright 1980, Parker Hannifin Corporation, Cleveland, Ohio.)

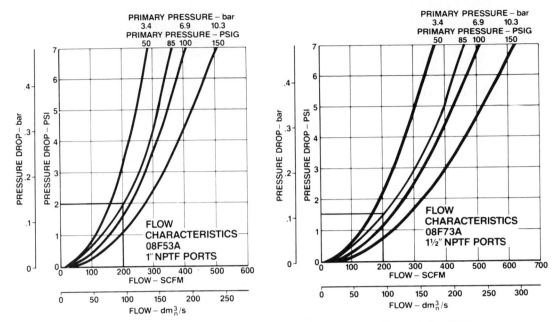

FIGURE 14-17   Filter performance characteristics. (Copyright 1980, Parker Hannifin Corporation, Cleveland, Ohio.)

$$cfm = \frac{(\pi/4)\ (6^2 - 3^2) \times 10 \times \dfrac{50 + 14.7}{14.7}}{3 \times 28.8}$$

$$= 10.8$$

The filter must be able to pass 10.8 cfm for either direction of cylinder motion, with a primary pressure of 100 psi. Assume an allowable pressure drop of 1.5 psi. Looking through the catalog,

we find the graph illustrated in Fig. 14-18. The filter selected will be the 04F11B with $\frac{1}{4}$-in. ports which will deliver 10.8 scfm with a 1.2-psi pressure drop.

It is common practice to select a larger filter to facilitate ease of piping. A 6-in.-bore cylinder typically would have 1-in. ports. If this larger filter is selected, the component's contaminant removal efficiency must be examined.

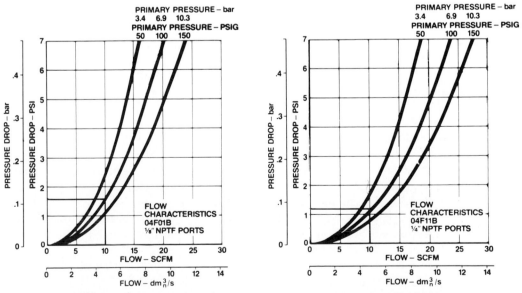

FIGURE 14-18   Filter performance characteristics. (Copyright 1980, Parker Hannifin Corporation, Cleveland, Ohio.)

LUBRICATION

When the air is filtered it may also be desirable to add a lubricant to the system. This can be done with an in-line lubricator. Some forms of lubrication found in a pneumatic system are accidental. They are caused by condensed moisture and oil picked up during compression being carried downstream to the working components. This type of lubrication is very crude and usually causes more problems than it cures. Deliberate lubrication can be provided by periodic injection of oil through an extra port or grease through a fitting, by drip-type oilers, or by automated pressure injection of the lubricant.

Most common today is the introduction of oil droplets into the air supply by devices called air line lubricators. When injected into the flowing airstream, the oil in the lubricator can take several forms: drops, spray, or mist.

Drops injected into an air line are often done by a pulse-type lubricator. This method injects a given minute amount of lubrication directly into the pneumatic component. Oil injection is directly proportional to cycle rate. This type may require an additional line to be run to each pneumatic component requiring lubrication.

Another method is to spray the oil into the flowing airstream. This spray of oil may be obtained in an air line by applying a standard air line mist-type lubricator (Fig. 14-19). This type delivers oil particles ranging in size from 0.01 to 500 μm in diameter.

Air flowing through the unit goes through two paths. At low airflow rates, the majority of airflows through the venturi section (A). The rest of the air slightly deflects and flows past the restrictor (J). The velocity of air flowing through the venturi section creates a lower pressure at the throat section (B). This lower pressure allows oil to be forced from the reservoir through the pickup tube (C), past the check ball (D), to the metering block assembly (E). This is where the oil flow rate is controlled by the metering screw (F). Rotation of the metering screw in a counterclockwise direction increases the oil flow rate; conversely, rotation in a clockwise direction decreases it. Oil then flows through the clearance between the inner and outer sight domes (G), where drops are formed and drip into the throat section (B). Here it is broken into fine particles and mixed with the swirling air to be carried to the outlet, where it joins air bypassing the restrictor disk (J). As the airflow increases, the restrictor disk (J) deflects, allowing the additional air to bypass the venturi section. This also generates an increased pressure drop across the venturi and will increase the oil delivery rate proportional to the increased airflow rate. The check ball (D) assures that when there is no airflow, oil in the passageways is held in place, shortening the time required to resume oil delivery when flow is reestablished.

For the standard air line mist lubricator (Fig. 14-20), the bowl can be filled while the lubricator is at pressure in

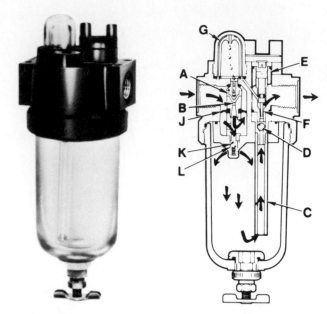

FIGURE 14-19   Air line mist-type lubricator. (Copyright 1980, Parker Hannifin Corporation, Cleveland, Ohio.)

FIGURE 14-20   Air line lubricator. (Copyright 1980, Parker Hannifin Corporation, Cleveland, Ohio.)

a working system. This is made possible by the action of the check ball (K). When the fill cap is partially opened, air in the bowl escapes and pressure forces the check ball to nearly seal at (L). *Only* then may the cap be removed and oil poured into the lubricator. When the fill cap is replaced, a small amount of air bleeds past the check ball (K) causes pressure to build up in the bowl, allowing the system to repressurize.

There is another type of lubricator (Fig. 14-21), the recirculating or micromist, flow in-line lubricator. In this type the oil is misted very finely for use downstream. The oil particles coming from the outlet range in size from 0.01 to 2 μm in diameter. These smaller droplets (compared to

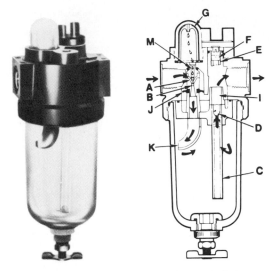

FIGURE 14-21 Recirculating lubricator. (Copyright 1980, Parker Hannifin Corporation, Cleveland, Ohio.)

the standard lubricator) will stay in suspension over a much greater distance, allowing longer lines. Also, elbows and vertical rungs will not wet (coalesce) out the droplets as quickly as in the standard unit. This is because the larger droplets are returned to the bowl by recirculating the airflow into the bowl.

The operation of the recirculating lubricator is shown in Fig. 14-21. Air flowing through the unit goes through two paths. At low airflow rates, the majority of the air flows through the venturi section (A). The rest of the air slightly deflects and flows past the restrictor disk (J). The velocity of the air flowing through the venturi section (A) creates a lower pressure, which allows oil to be forced from the reservoir through the pickup tube (C) past the check ball (D), to the metering block assembly (E). It is at this point that the oil delivery rate is controlled by a metering screw (F).

Rotation of the metering screw in the counterclockwise direction increases the oil flow rate, whereas clockwise rotation decreases it. Oil flows through the clearance between the inner and outer sight domes (G), where drops are formed and drip into the nozzle tube (M). Here it is broken into fine particles as it expands into the low-pressure venturi. From there, the atomized oil flows through the curved baffle plate (K) and is deflected against the interior wall of the reservoir. This action causes larger oil particles to coalesce and fall back into the reservoir, where it can recirculate through the system. The remaining mist of fine particles (typically less than 2 μm) is carried through the opening (I), where it joins and mixes with air that bypassed the restrictor disk. As airflow increases, the restrictor disk deflects, allowing more of the inlet air to bypass the venturi section. This also generates an increased pressure drop across

the venturi and will increase the oil delivery rate proportionate to the increased airflow rate.

The check ball (D) prevents reverse oil flow down the pickup tube when airflow stops. Thus oil delivery can resume virtually immediately when airflow restarts.

It should be noted that this type of lubricator, as described, can *only* be filled with the supply pressure *shut off*. If this is unacceptable, options for refilling with auxiliary pressure supply equipment are usually obtainable (refill check valve and auto-fill accessory).

SIZING A LUBRICATOR

The lubricator, like the filter, is sized on airflow rate. Once again, the maximum flow requirements of the system must be determined and the pressure drop (usually in the range 1 to 5 psi) for various body sizes and plumbing must be determined. Then catalog data are checked and a lubricator is selected.

EXAMPLE

A die grinder employs a vane motor. At maximum output the motor needs 16.3 cfm at 50 psi. What lubricator should be used?

Solution

The lubricator selected must be able to pass 16.3 cfm at a reasonable pressure drop. The pressure drop selected will be 5 psig. The graphs in Fig. 14-22 are those typically found in a catalog for micromist lubricators.

The pressure drop for a $\frac{1}{2}$-in.-body-size lubricator is 2.3 psi, 1.5 psi, and 1.3 psi for $\frac{1}{4}$ in., 3/8 in., and $\frac{1}{2}$ in. ported, respectively. All lubricators have a reasonable pressure drop; they all fall in the range 1 to 5 psi. Since the plant engineer prefers to buy a lubricator with the same-size ports as the motor, the $\frac{1}{4}$-in. model is selected.

THE LUBRICATOR IN A SYSTEM

Lubricators are necessary to ensure that oil is carried through the circuit with the air. Proper lubrication is needed to coat resilient seals to reduce friction and significantly extend their life. Most pneumatic cylinders and motors require lubrication to decrease friction and prevent scoring, thus decreasing operating temperature and preventing scoring. This will increase the life of the motor or cylinder.

Because a little oil is good does not mean that a lot of oil is better. Lubricators are often adjusted so that too much oil is injected into the stream. Puddles of oil develop in the piping and cause problems. Also, air exhausting with large amounts of entrained air may cause enough oil to become airborne in the plant that the air in working areas will not meet federal pollution standards. Consider that the oil put into the air system must be taken out before the air is returned to the free atmosphere. The OSHA regulation

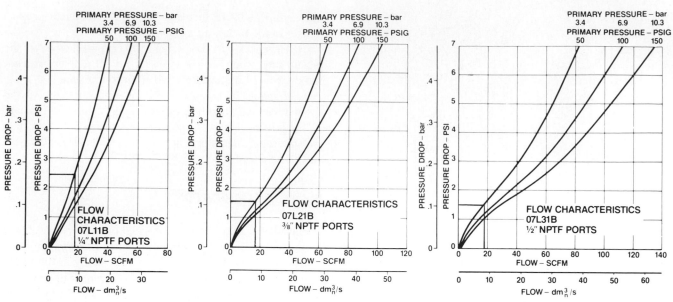

FIGURE 14-22 Lubricator performance characteristics. (Copyright 1980, Parker Hannifin Corporation, Cleveland, Ohio.)

states that air must not contain more than 5 mg of oil mist particles per cubic meter of air. This is equal to approximately an ounce of oil in 200,000 ft^3, which is equivalent to a building 10 by 100 by 200 ft. If a plant uses dozens of lubricators, it may take a very long time for the oil to settle out. The limit of 5 mg/m^3 may quickly be exceeded when the plant is operating. This federal requirement in the United States makes it necessary that you use some kind of device to reclassify the oil before the air carrying it is exhausted into the atmosphere. Typical in-line air filters do a

relatively good job of reclassifying oil. If almost total reclassification is necessary, a coalescing type of filter can be employed. It is common to place all of these units together to form a filter-regulator-lubricator (FRL). An FRL unit combines the three components into a pre-piped package for easy installation. Branches of a pneumatic system are generally equipped with FRLs such as those shown in Fig. 14-23, so that individual actuators and workstations receive filtered, regulated, and lubricated air meeting their specific requirements.

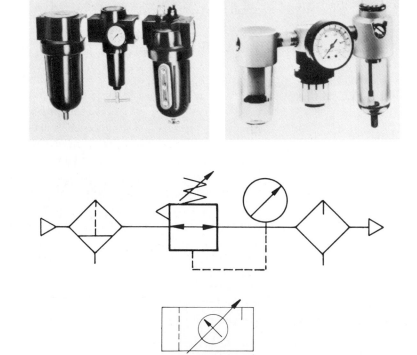

FIGURE 14-23 Filter-lubricator-regulators, complete symbol, and simplified symbol. (Copyright 1980, Parker Hannifin Corporation, Cleveland, Ohio.)

A pneumatic system may be equipped with the best filters and lubricators available and they may be positioned in the system where they do the most good, but if filters are not drained and changed at appropriate intervals for such services, or lubricator reservoirs refilled, the money spent for their use may have been wasted and the desired protection lost.

FRL DESIGN CONSIDERATIONS

When designing systems protected with FRLs, the following rules should be considered:

1. Size units for the maximum flow rate in the portion of the circuit they serve.

2. These devices should be placed as close as practical to the component being serviced.

3. Their placement should also be in an area of easy accessibility. This is especially true for units requiring maintenance, such as filling, adjusting, or cleaning.

4. Install a shutoff valve (possibly a lockout-type shutoff exhaust) ahead of the unit to facilitate maintenance of the unit.

5. Whenever possible, locate lubricators at an elevation higher than or equal to the points being lubricated.

FUNCTIONAL TYPES OF PNEUMATIC DIRECTIONAL CONTROL VALVES

Pneumatic directional control valves are functionally described by the number of their flow ports, the number of valve positions, and the internal connections of their flow ports in each position. The most common designations are listed below.

Two-Way Valve

This is an on-off type of device. This valve is usually provided with two external flow ports: a supply port and an exhaust port. A normally open two-way valve permits flow in its normal or "at rest" position and blocks flow when shifted. The normally closed two-way valve blocks flow in its normal position and permits flow when actuated (see Figs. 14-24 and 14-25).

Three-Way Valve

This device is a flow path selector. One flow port is internally connected to either of two other flow ports. It may be

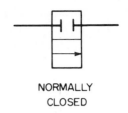

NORMALLY
CLOSED

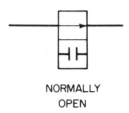

NORMALLY
OPEN

FIGURE 14-24 Symbols: two-position and two-way valves.

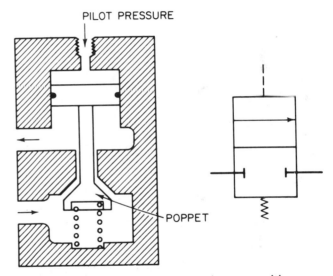

FIGURE 14-25 Pilot-operated, two-position, two-way air valve; normally closed.

used alternately to pressurize or exhaust one port, or to direct pressure alternately from one source to either of two other ports. Three-way valves may be constructed either normally open or normally closed. These valves can be used as pilot relays to operate other valves. Three-way valves may be used singly to control single-acting cylinders, or in pairs to control double-acting cylinders (see Figs. 14-26 and 14-27).

Four-Way Valve

This is a flow-reversing device. Usually, four external flow ports are provided—a supply port, two working ports, and an exhaust port. In one position, the valve allows air to flow from the supply source to one of the working ports. Simultaneously, air is allowed to flow from the other working

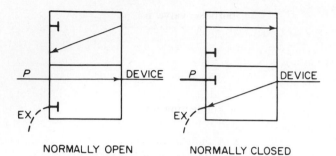

NORMALLY OPEN   NORMALLY CLOSED

FIGURE 14-26 Two-position, three-way air valves.

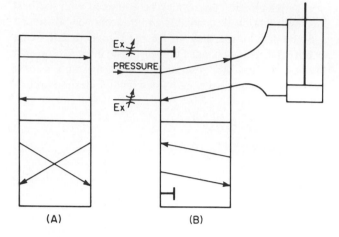

(A)   (B)

FIGURE 14-28 (a) Two-position, four-way air valve; (b) two-position, five-way air valve with adjustable exhaust ports.

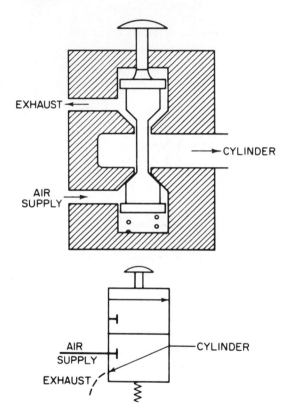

FIGURE 14-27 Two-position, three-way, manually operated air valve.

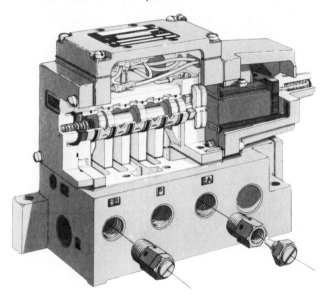

FIGURE 14-29 Lapped-spool-type precision air valve with exhaust throttle valves. (Courtesy of Numatics, Inc.)

FIGURE 14-30 Speed control with two exhaust ports.

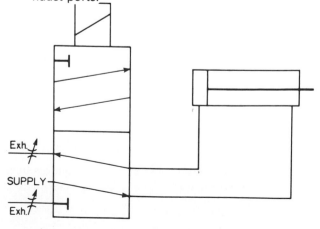

port to exhaust. When the valve is shifted, the flow paths are reversed. That is, supply is directed to the second working port, while the first working port is open to exhaust. The most common application of four-way valves is to control the motion of double-acting cylinders (see Fig. 14-28).

Four-way valves are also available with five external ports. This permits use of either dual supply or dual exhaust. These five-ported versions provide the same basic control of flow paths as four-ported versions do. In addition, dual supply ports permit the use of different pressures for cylinder retraction. Dual exhaust ports conveniently permit individual speed control of the exhaust from each of the working ports (see Figs. 14-29 and 14-30).

The foregoing discussions of pneumatic directional control valves have been confined to two position devices: a normal and one actuated position. Four-way valves are also available in three-position varieties (either four- or five-ported). These three-position types have a normal and two actuated positions. The *normal position* may permit the following flow arrangements:

1. Both working ports connected to supply pressure.

2. All ports blocked (closed center)—this is a "hold" position, and permits "inching" of a cylinder.

3. Both working ports open to exhaust—this provides a "float" condition, and enables positioning of the cylinder by external forces.

Other Valve Types

In addition to the basic directional control valves described, other devices provide optimum flexibility of pneumatic circuitry. Speed (flow) control valves are self-contained devices which provide "free flow" in one direction, and an adjustable restriction to flow in the reverse direction. Free flow is accomplished by means of a built-in check valve. Restricted, or metered, flow in the reverse direction is accomplished by means of a variable orifice that bypasses the check valve. Speed control of pneumatic cylinders and equipment may require special design considerations, because of the compressibility factor of air. A pneumatic cylinder performing work, for instance, could "jump" forward if the work resistance suddenly diminished. This rapid partial stroking of the cylinder is caused by the stored energy of the compressed air on the driving side of the cylinder. A similar occurrence of "jump" could be encountered if a directional valve is shifted before the cylinder completes its stroke, or if the valve is shifted too soon after the cylinder stroke is completed. Speed controls are most often installed to allow free flow of working air into the cylinder and to "meter out" or restrict the air flowing from the cylinder. Some applications may require use of the speed controls to "meter in," or perhaps a combination of both "meter out" and "meter in" control.

A special-purpose valve peculiar to pneumatic circuitry is the quick-exhaust valve. The quick-exhaust valve is used close-connected to a cylinder. Air from a directional control valve (which may be remotely positioned) passes through the connecting lines and the quick-exhaust valve into a cylinder. When the directional control valve is shifted, only the air in the connecting lines between the directional valve and the quick-exhaust valve is exhausted through the directional valve. A built-in valving mechanism allows the air from the cylinder to exhaust directly through the quick-exhaust valve to atmosphere (Fig. 14-31). This eliminates the need for exhaust air from the cylinder to flow through long or restricted lines to the main control valve. The quick-exhaust valve permits increased cylinder velocities and/or smaller directional control valves.

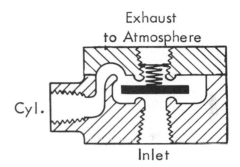

FIGURE 14-31   Quick exhaust valve.

Pneumatic time-delay valves may be used to adjust the timing of various phases of a machine cycle. This is accomplished with a pilot-operated directional control valve that receives pilot pressure from an external source and either meters airflow into or out of the pilot chamber to create delay (Fig. 14-32).

Other special-purpose devices include mufflers or dust filters (Fig. 14-33), shuttle valves (Fig. 14-34), sequence valves, and deceleration valves. Note that the shuttle valve of Fig. 14-34 can be structured in an insertable cartridge or designed with a housing with threaded port connections.

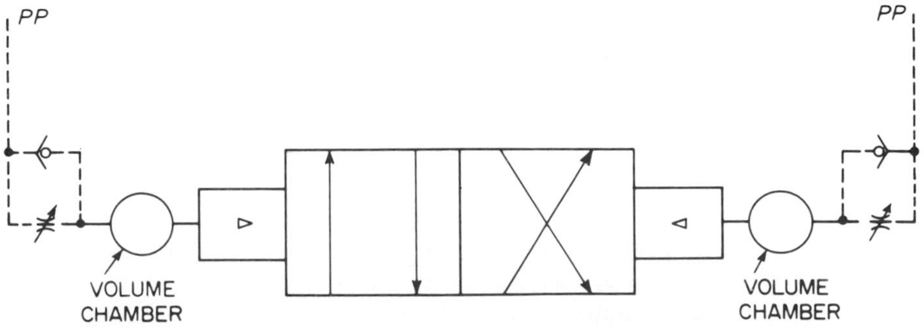

FIGURE 14-32   Time-delay valves: volume chambers.

VOLUME CHAMBER

VOLUME CHAMBER

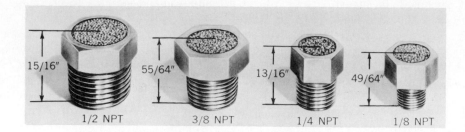

FIGURE 14-33 Mufflers or dust filters. (Courtesy of Air-Mite, Inc.)

FIGURE 14-34 Shuttle valves. (Courtesy of Kepner Products Company, Villa Park, Ill.)

BASIC DESIGNS OF PNEUMATIC DIRECTIONAL CONTROL VALVES

Carefully consider the basic types of valve design when selecting directional control valves. Each design provides advantages for certain applications. The basic designs are shear action (D-slide or rotary disk), metal-to-metal spool, packed spool, and poppet type.

Shear Action (D-Slide or Rotary Disk)

An element slides across the flow path. Frequently, pressure unbalance is used to force the element against a mating surface to effect a metal-to-metal seal. This valve readily provides two-, three-, or four-way action, but does not lend itself to the large number of flow path variations which are often required in modern pneumatic circuitry. Some important advantages of the shear action type of design are as follows:

1. Valve mechanisms "wear in" with use, enabling these valves to provide millions of trouble-free cycles, even under adverse conditions.

2. Valve materials may be selected which are not affected by contaminants in compressed air lines.

3. The shear action of this valve tends to exclude foreign material from lodging between the mating surfaces.

4. The valve is not seriously affected by temperature extremes.

5. The valve can be used as a throttling device to control flow rates.

Consider the following factors of shear-action-type valves to avoid circuit problems:

1. Large forces may be required to shift the valve slide, especially at high pressures when designs are used which are not pressure-balanced.

2. Units generally require long stroke lengths to obtain full flow capabilities.

3. Valve must be continuously lubricated for maximum life.

4. The sealing member (D-slide or rotary disk) can be forced from its mating surface if the pressure under the plate (working pressure) exceeds supply pressure. Valves may not be suitable for cylinder applications where large external forces could create such "reverse" pressures.

5. This design usually does not provide "bubble-tight" sealing ability.

Metal-to-Metal Spool

A variation of the shear action design is the metal-to-metal spool-type valve. This valve has cylindrical flow and sealing paths rather than flat (single-plane) paths. This design depends on close fits between a metal spool and its matching bore. Useful features of metal-to-metal spool-type valves include the following:

1. The design is capable of providing almost any flow path configuration desired, in a variety of porting and actuating configurations.

2. External forces required to shift a balanced spool are low.

3. Spool lands can be made wide enough to overlap bore openings and thus prevent undesirable interconnection of flow ports while the valve is being shifted from one flow path arrangement to another. This ability to eliminate "crossover" flow during shift reduces the chance of shift failure when a pilot-operated valve at low supply pressures is being used.

To avoid circuit problems with metal-to-metal spool valves, consider these factors:

1. Foreign matter may get between mating parts, causing wear and leakage, and even possibly sticking of the spool.

2. This type of unit requires good filtration and lubrication.

3. Oxidized airborne lubricant from a compressor or other foreign air line material may accumulate in the small valve clearances, and cause the spool to "varnish" in place. This occurs particularly when valves are idle, as over weekends.

Packed Spool

Packed spool valve designs have seals to provide leakproof sealing ability. Usually, the seals are an integral part of the moving spool and cross over the flow passages. Selection of the seal materials is critical. The elastomer must be resilient to provide tight sealing, yet tough enough to resist abrasion for millions of trouble-free cycles. It must be dimensionally stable—and remain resilient—under operating temperature extremes. It must also maintain its dimensional stability in the presence of various contaminant fluid particles in air lines. Unwanted shrinking of the seal could cause loss of sealing interference, and excessive swelling could cause sticking of the spool. Advantages of the packed spool type of valve design are:

1. A design is available for most porting and actuating configurations.

2. Forces on the spool are balanced. Shifting forces are usually only slightly higher than for metal-to-metal designs.

3. Resilient seals provide "leaktight" sealing ability.

4. Resilient seals make the valve less vulnerable to abrasion by foreign material.

5. Valve spools have limited contact areas and less tendency to "varnish" in place than metal-to-metal designs.

6. Lubrication requirements are not as important as for metal-to-metal types.

7. A spool may be replaced without changing mating parts.

Consider the following factors to avoid circuit problems with packed spool valve designs:

1. Improper selection of seal material may mean that the elastomers can adhere to the surfaces against which they seal.

2. Environmental temperatures and air line fluids must not be allowed to affect dimensional stability of the seals excessively.

Poppet Type

Poppet-type valves are desirable for use in circuits where high flow capacity and rapid response are required. These valves usually have resilient seals to provide tight sealing and help absorb the kinetic energy of the moving member.

The poppet type is easily adopted for three-way valve designs. Four-way valve designs are achieved with a dual poppet arrangement—that is, really two three-way valves in a common body. Some important advantages of the poppet-type valve design are:

1. Poppet valves offer rapid cycling capabilities. Lightweight moving members require only a short stroke to provide maximum flow opening.

2. Short stroke of the valve gives minimum wear and maximum life capabilities.

3. Self-cleaning resilient seats provide "bubble-tight" sealing capabilities.

4. Lubrication requirements are not so important as with metal-to-metal valves.

5. Dimensional stability to the sealing member is not as critical as with packed spool valves.

To avoid circuit problems with poppet-type valve designs, consider the following:

1. Poppet designs with direct operated actuators are bulky in large port sizes.

2. The poppet valve design allows flow from the pressure port to escape to exhaust as the valve is shifting. This "crossover" flow may drop the supply pressure below the minimum operating pressure, causing the valve to fail to complete its shift.

3. Some flow path configurations are not available in poppet valves.

FIGURE 14-35 High-speed poppet valve. (Courtesy of Schrader Bellows, Akron, Ohio.)

SOLENOID VALVES

Figure 14-35 shows the actual valve, which is a poppet-type, two-position, three-way valve that operates at very high speeds.

Figure 14-36 shows the internal construction of the valve. Solenoid differential pilot valves can provide about 3000 cycles/min.

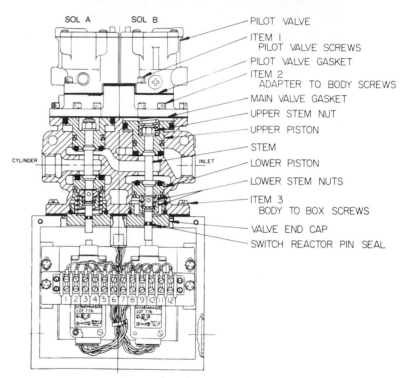

FIGURE 14-36 Cutaway view of poppet structure. (Courtesy of Schrader Bellows, Akron, Ohio.)

QUESTIONS

1. Discuss the advantages and disadvantages between using air or oil fluid power.

2. Does an air system have a relief valve? If yes, where is it located?

3. Describe the design and operation of a filter-regulator-lubricator combination unit.

4. What are the main differences between hydraulic controls and air controls?

5. What is the pressure drop through a four-way valve that is passing 30 cfm per hour (60°F = 14.7 psia); specific gravity of air = 1; temperature = 72°F, and the pressure downstream is 80 psi. (The C_v factor for the valve is 5.)

6. What is the main purpose of a pressure regulator?

7. Why are lubricators used with pneumatic circuits?

8. How is noise controlled with air circuitry?

9. Explain the function of a two-position four-way valve.

10. Why do some four-way valves have five external ports?

11. What is the normal position of any control valve?

12. What kind of valve is used to control the speed of cylinders or air motors?

13. What is the purpose of a quick-exhaust valve?

14. Name the construction types of directional control valves available. Name several considerations when the valve types available are being selected.

15. What is a shuttle valve used for?

15
Using Pneumatic Power

The application of pneumatics is expanding in two main categories: pneumatic power and pneumatic control. Pneumatic power uses include punching, stamping, shearing, trimming, pressing, lifting, hoisting, clamping, and many others. The application of pneumatics for jobs of this nature involves the same calculations as those covered throughout Section One and in earlier chapters of Section Two.

Both actuating cylinders and air motors can be applied to provide linear and rotary motion. The same general design of cylinders described in Section One is used for pneumatic circuits.

EXAMPLES OF APPLICATIONS

Air presses are popular for light press assembly and similar work. The press shown in Fig. 15-1 can develop about $7\frac{1}{2}$ tons of force. Air hoists (Fig. 15-2) are convenient to apply for lifting parts and machine members in manufacturing processes. The linear motor shown in Fig. 15-3 is air-powered and hydraulically controlled for feed speed. The large cylinder on the left provides the power, and the small cylinder directly behind it contains hydraulic oil.

When the forward solenoid is energized and the cylinder extends, fluid from the rod side of the small hydraulic cylinder is metered back to the large area side through a

FIGURE 15-1 Air press. (Courtesy of Miller Fluid Power Corporation, Bensenville, Ill.)

1. Overhead plant rail
2. Air hoist eye
3. 3-way air control valve
4. Case-hardened (50–54 Rockwell C), hard chrome plated piston rod
5. Hook swivels 360° on ball joint that also permits lateral movement in any direction
6. Check valve
7. Flexible air hose is all you need to connect hoist to your shop air supply

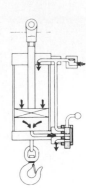

1. Lifting the load. Shifting 3-way valve control lever to lift position opens cap end port to exhaust. Air pressure at head end lifts load.

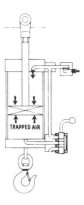

2. Stopping the load. Shifting 3-way valve control lever to neutral position traps pressurized air in cap end of hoist. This results in a balance between the forces at cap end and head end, thus stopping and holding the load.

FIGURE 15-2   Air hoist. (Courtesy of Miller Fluid Power Corporation, Bensenville, Ill.)

FIGURE 15-3 Linear motor. (Courtesy of Schrader Bellows, Akron, Ohio.)

needle valve. When the air cylinder is reversed, oil in the large area side of the hydraulic cylinder is returned to the rod side. A small spring-loaded accumulator (shown with the short stem) provides for the difference in volume of the two sides of the hydraulic cylinder because of the space taken by the rod. Linear motors without the hydraulic control can be applied as air cylinders would be, offering the simplicity of mounting obtained by having the control valve integrally mounted.

Precise feed speeds can be obtained with linear motors, which makes them valuable for uses such as feeder gates, fan controls, drilling and tapping machines, and the positioning of valves to control water systems, chemicals, etc.

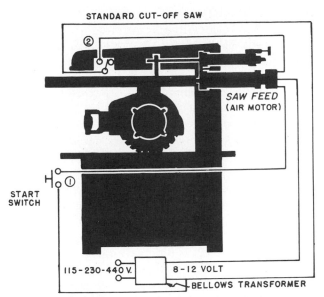

FIGURE 15-4 Standard cut-off saw.

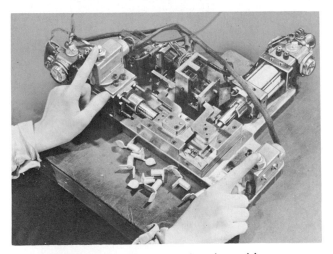

FIGURE 15-5 Safety two-hand machine operation.

Precise feed control is given to the cutoff saw in Fig. 15-4 with a linear motor. In Fig. 15-5, both hands must be used to punch out the port, which protects the operator from injury and possible loss of fingers.

AIR POWER EMERGENCY POWER UNIT

The emergency power unit shown in Fig. 15-6 uses air over oil boosters to provide a crushing force to crimp the casing and drilling rod closed in case of an oil gusher. The operator trips an emergency blowout preventer, as he retires from the scene, to actuate the boosters and crush the casing around the drill rod. The system develops 5000 psi. Air is supplied by the tanks to operate the boosters. The four valves are also air-operated with cylinders.

FIGURE 15-6 Emergency power unit.

ROTARY AIR MOTORS

Rotary air motors provide a smooth source of power and are not subject to damage due to overloads. Air motors can stall out for long periods of time, and there are no heat problems. Air motors stop and start with quick response. Pressure regulating and metering of flow provide infinitely variable torque and speeds.

The two most popular types are the vane and piston design. The radial piston designs are low speed, high torque motors and are generally fitted with four, five, and six reciprocating pistons (see Fig. 15-7).

$$\text{hp} = \frac{T \times \text{rpm}}{5252}, \text{ where torque, } T, \text{ is measured in lb. ft}$$

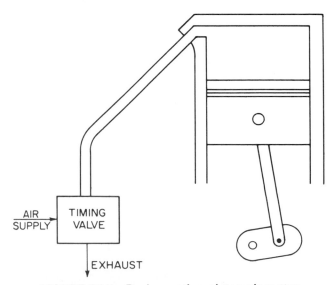

FIGURE 15-7 Reciprocating piston air motor.

Power developed is dependent on inlet pressure, number of pistons, the area of the pistons, stroke, and speed. Horsepower ranges are available up to about 25 hp. The timing valve is generally linked to the crank shaft and automatically supplies air to the cylinder or exhausts the spent air from the cylinder. The large reciprocating piston air motor is limited in speed by the inertia of the mechanical parts.

Axial piston air motors are smaller and develop greater speeds but are only available in smaller units up to about 3 hp. They are equipped with built-in lubrication.

Vane air motors are high-speed units and deliver more horsepower per pound than piston motors. Some vane motors operate up to 25,000 rpm, depending on the diameter of the rotor. Generally, vane motors are equipped with a governor to prevent overspeed when loss of load occurs. Air motor speeds are easy to adjust with an ordinary globe or needle valve.

VANE-TYPE AIR MOTORS

Figure 15-8 shows a rotary vane air motor and performance curve for its operating range.

ROBOTS AND AUTOMATIC MACHINERY

Both pneumatic and hydraulic power transmission systems are widely used for robotic devices and automatic machinery. A robotic device is normally considered a mechanism that will provide a predetermined function repetitively with accuracy commensurate with the needs of the task at hand. Robots may be programmed to pick and place (Fig. 15-9) with fingers, grippers, or other hand-like mechanisms.

Eight typical robotic or automatic machine functions are shown in Figs. 15-10 through Fig. 15-17. Most of these motions are based on a cylinder/rotary actuator unit as shown in Fig. 15-18. Figure 15-19 illustrates an assembly of the multimotion devices that can provide needed mechanical motions in a predetermined, easily controlled pattern. The harmonic motion concept is illustrated in Fig. 15-20.

All of the control concepts presented in Sections One and Two are basic to the design and maintenance of robots and other automatic machinery. Each mechanical movement is a response to the control of a linear cylinder, rotary actuator, or fluid motor. The limitation of robots and automatic machinery is primarily that of the skills of the designer and personnel who must understand and maintain the machinery.

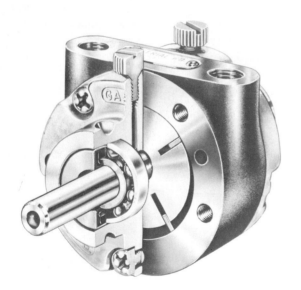

(a)

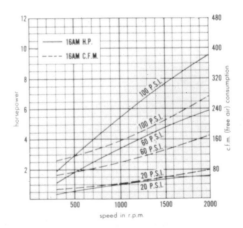

(b)

FIGURE 15-8 (a) Rotary vane air motor; (b) Performance curves. (Courtesy of Gast Manufacturing Co., Benton Harbor, Mich.)

FIGURE 15-9 Pick-and-place robot. (Courtesy of Schrader-Bellows, Akron, Ohio.)

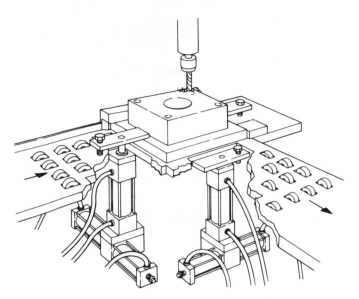

FIGURE 15-11 Clamping. (Courtesy of PHD, Inc., Fort Wayne, Ind.)

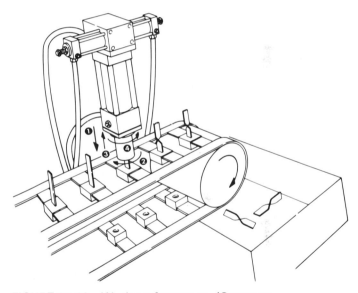

FIGURE 15-12 Work performance. (Courtesy of PHD, Inc., Fort Wayne, Ind.)

FIGURE 15-13 Spraying. (Courtesy of PHD, Inc., Fort Wayne, Ind.)

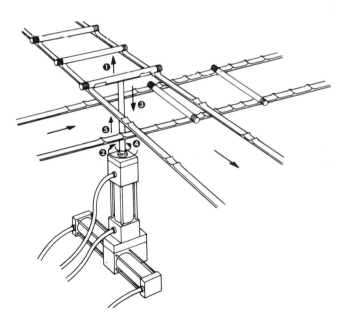

FIGURE 15-10 Transfer function. (Courtesy of PHD, Inc., Fort Wayne, Ind.)

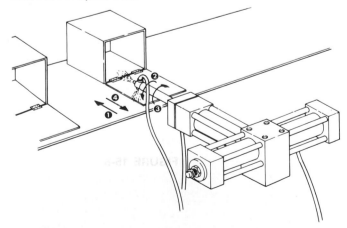

PNEUMATIC CONTROL SYSTEMS FOR ROBOTS AND AUTOMATIC MACHINERY

In this age of spaceships and computers, pneumatic control is taking on real significance for future developments. There are many divisions of control, which range from direct mechanical connection between the operator and the parts op-

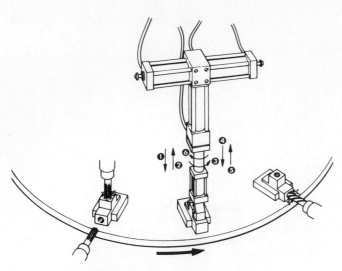

FIGURE 15-14 Parts turnaround. (Courtesy of PHD, Inc., Fort Wayne, Ind.)

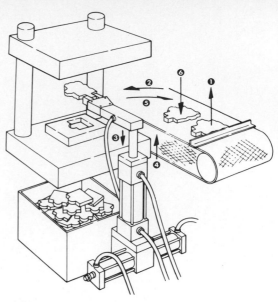

FIGURE 15-16 Parts handling. (Courtesy of PHD, Inc., Fort Wayne, Ind.)

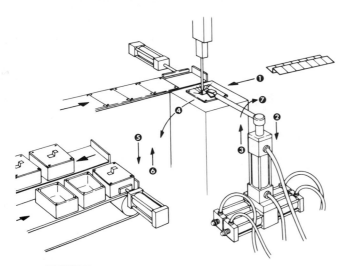

FIGURE 15-15 Multiposition assembly. (Courtesy of PHD, Inc., Fort Wayne, Ind.)

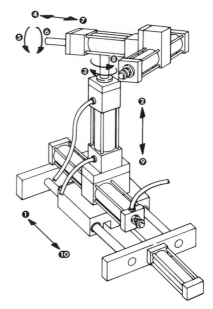

FIGURE 15-17 Multimotions. (Courtesy of PHD, Inc., Fort Wayne, Ind.)

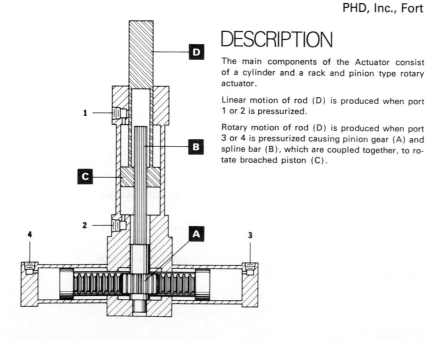

DESCRIPTION

The main components of the Actuator consist of a cylinder and a rack and pinion type rotary actuator.

Linear motion of rod (D) is produced when port 1 or 2 is pressurized.

Rotary motion of rod (D) is produced when port 3 or 4 is pressurized causing pinion gear (A) and spline bar (B), which are coupled together, to rotate broached piston (C).

FIGURE 15-18 Rack and pinion for rotary motion. (Courtesy of PHD, Inc., Fort Wayne, Ind.)

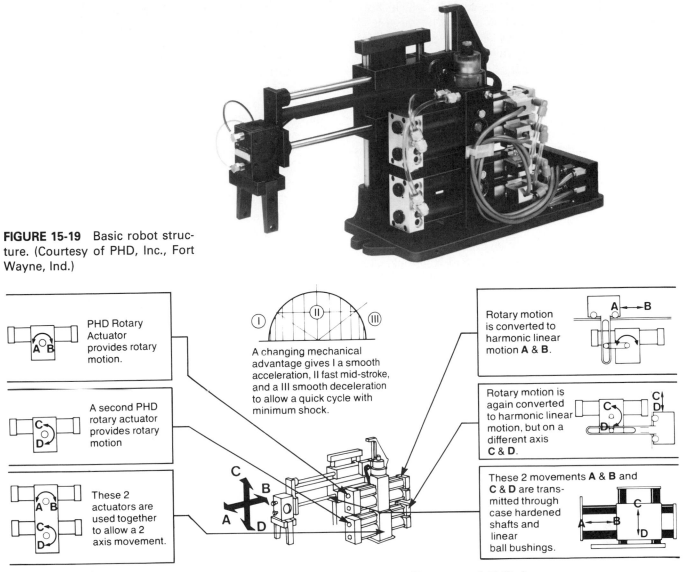

FIGURE 15-19 Basic robot structure. (Courtesy of PHD, Inc., Fort Wayne, Ind.)

PHD Rotary Actuator provides rotary motion.

A second PHD rotary actuator provides rotary motion

These 2 actuators are used together to allow a 2 axis movement.

A changing mechanical advantage gives I a smooth acceleration, II fast mid-stroke, and a III smooth deceleration to allow a quick cycle with minimum shock.

Rotary motion is converted to harmonic linear motion **A** & **B**.

Rotary motion is again converted to harmonic linear motion, but on a different axis **C** & **D**.

These 2 movements **A** & **B** and **C** & **D** are transmitted through case hardened shafts and linear ball bushings.

FIGURE 15-20 Harmonic motion concept. (Courtesy of PHD, Inc., Fort Wayne, Ind.)

erated to "logic control," which is a relatively new field holding much promise as it is further developed and applied.

Logic control involves pneumatic devices such as actuators and pilot valves that provide means for obtaining the control functions of AND, OR, NOT, MEMORY, and TIME.

For practical purposes, logic might be compared to an actuating cylinder that advances but is blocked because the previous phase of the cycle failed to eject the machined part. If the cylinder stops and the machine jams, no logic is involved. However, if the control system senses the obstruction and automatically retracts, causing the previous failure to recycle, then logic is involved.

Logic, then, is the control system's ability to deter-

mine an alternative course of action when the course directed is proved undesirable. Logic control can become extremely complex when the number of functions and components used increases.

The art of employing electricity for control has gained much experience throughout the years. People working in electronics have developed symbols and diagrams that can easily be read to trace down any small malfunctions within a highly complex control system.

Pneumatic control has many advantages and some disadvantages compared to electrical control. One of the major factors in favor of air is that it does not have the self-destructive characteristics of other systems that are subject to resistance and heat.

Specifically, air-piloted valves can operate indefinitely without malfunction compared to solenoids, which have a certain life expectancy and can burn out if overheated.

Air can provide a positive holding power for long periods of time. It will hold while the control system performs other functions or is static. Air performs the function of work as well as control and does not require additional equipment to convert its energy to motive power.

It is reasonable to believe that if some of the diagraming techniques used for electrical control were employed for logic circuits, an improved method of control is possible that can be relatively uncomplicated.

Through the use of thermoplastic tubing and recently designed small components, air controls can be easily installed in compact metal boxes just as electrical switches, relays, and so on, can.

The following example of logic control has been made available through the courtesy of Edward Holbrook, Clippard Instrument Laboratory, Inc. He is one of the outstanding authorities in pneumatic logic control systems.

PNEUMATIC LOGIC CONTROL

The purpose of this section is not to produce all the answers or to offer pat solutions, but to indicate methods whereby control answers can be attained. Too, much is yet being learned about pneumatic systems that, in time, will lead to easier methods. Also, it is not intended to be a deep technical treatise, since it would serve no useful purpose and much of the mathematics is still more approximate than accurate. It will be obvious that many variations and/or combinations can be made to suit the conditions that confront the engineer.

The diagrams illustrate means for obtaining the control functions of AND, OR, NOT, MEMORY, and TIME so that circuits can be assembled with a minimum of time. For the most part, these functions are produced with standard types of devices, such as two-way, three-way, four-way, shuttle, regulator valves, etc. However, the author would caution that varying spring values, friction differences, lubrication or lack of it, etc. can create problems that can be irritating, giving reactions that make a system inoperable or unduly sensitive. Most, if not all, of these problems can be solved, but without experience and prior knowledge the answer is not always readily apparent. A device can sometimes be improperly located relative to upstream and downstream volumes, for example, thus causing false signals that cause damage if sequencing is necessary and is used as the means for protection.

Much has been made of the comparison between pneumatic and electrical circuitry to a decided advantage, but the engineer must not lose sight of the fact that air as a medium operates differently, so that both its good and bad qualities must be taken into account.

Figure 15-21 illustrates the type of symbols used in the circuitry diagrams which are based on the block and hinge principle, with the control block diagram including the hinge to indicate flow paths.

A valuable point is to remember that all the control functions we get by pressuring lines can be obtained by venting lines. This can often be a means of simplifying an otherwise complicated system. It can be noted that the AND functions illustrated by Figure 15-22, when the lines are pressurized, become OR functions when these lines are vented. That is, in Figure 15-22a, all four incoming lines have to be filled prior to getting a pressure signal out at line 5, yet venting 1, 2, 3, or 4 will shut off the pressure at line 5.

The OR circuit, Figure 15-23b, becomes an AND circuit when the incoming signals are exhausted. That is, lines 1, 2, 3, or 4 will create pressure signal at 5, but *all* four incoming signals, if pressurized, must be vented before line 5 can be vented.

Figure 15-23b illustrates one of the many uses for NOT circuitry. If supply were introduced through valve A, air would fill lines 4, 5, and 6. Air in 6 would trip B to block line 1 and connect line 2 to the atmosphere. If, then, valve A is shifted rapidly, it would connect 4 to the atmosphere and fill line 1. Air would *not* flow through valve B until the pressure dropped below the spring value which, if very low (3 to 5 psi), would insure the full release of line 5 before 2 could be pressurized.

The shuttle valve OR function of Figure 15-23b is obvious. Figure 15-23d illustrates the use of two four-way valves to produce the effect of a three-way electric switch combination. As shown, supply flows through B into 1 and through A into 2. If A is moved, 2 will vent into 5 and 4 through B to the atmosphere, while the supply in 4 is directed to 3. If, from the position shown, B is shifted, line 1, therefore 2, is vented while the supply in 1 is directed to 3. No matter what position A or B is in, the other valve will alter the air condition in lines 2 and 3.

In Figure 15-23e, supply through lines 5 and 6 is maintained on both sides of valve G. If any one of bleeder valves A, B, or C is tripped, pressure in line 4 will drop (replacement retarded by choke H), causing a pressure differential across valve G which will shift to the left. Valves D, E, or F will cause G to shift to the right.

To indicate the "false signal" problem, refer to Fig. 15-24a and assume this to be a one-shot cycle. The signal (50 psi or more) coming into line 1 is from a cam valve that, when depressed, will be held for a period of time. The outgoing signal at 2 is to shift a double-piloted four-way valve, and we want that valve to be free to return *before* the signal is vented from line 1—hence the limited memory function. We will also assume that the three-way valve (B) available to us has a 40 psi return spring and that the pressure to shift the double-piloted four-way valve, due to detent or friction, is 10 psi.

DIAGRAM SYMBOLS

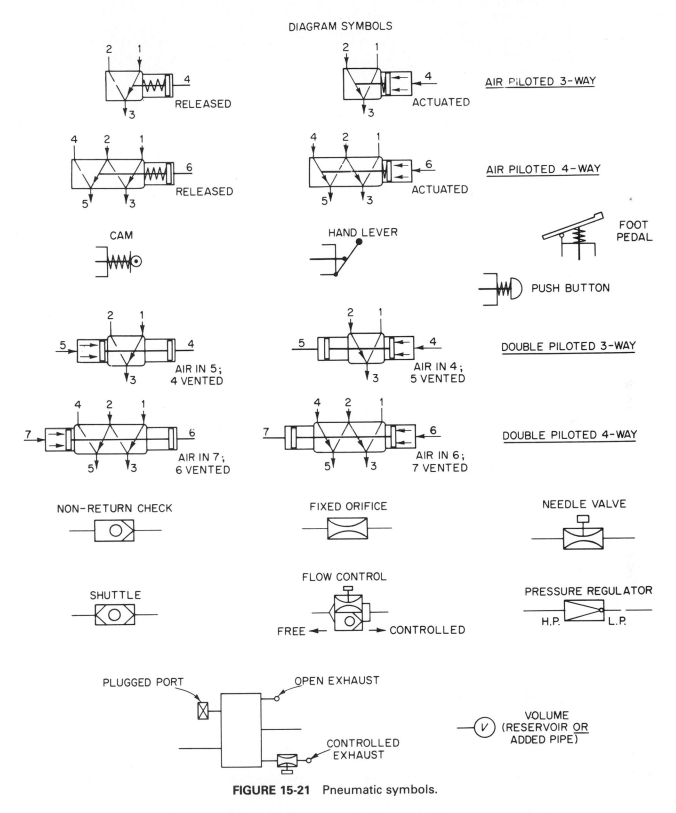

FIGURE 15-21 Pneumatic symbols.

Air enters line 1, 1-b and through B to line 2, to shift the four-way valve. Pressure, past flow control B, builds up above 40 psi to shift B and vent 2. This gives us a satisfactory circuit function. Later, the cam valve supplying line 1 is relieved and 1 begins to vent. Just below 40 psi, valve B shifts, and any air in 1 or 1-b expands into line 2. If we have placed valve B such that line 2 is 2 ft long and line 1 is that or greater, the pressure resulting from this

"AND" FUNCTION

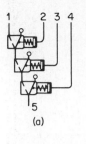

1,2,3 AND 4 MUST BE
PRESSURIZED TO GET A
SIGNAL AT 5

(a)

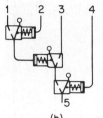

SAME AS (a) EXCEPT TO ILLUSTRATE
THAT ANY ONE OF THE FOUR
INCOMING SIGNALS CAN EMERGE AS
THE OUTPUT

(b)

BOTH VALVES MUST BE ACTUATED
TO RELEASE 4 AND FILL 5

(c)

ALL 4 VALVES MUST BE ACTUATED TO PRODUCE
A WORKING SIGNAL AT 1

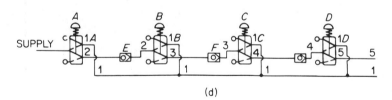

(d)

FIGURE 15-22   AND function.

expansion may exceed 10 psi so that the four-way valve, attached to line 2, will shift and, if the sequenced circuit is critical, damage can result. If the values are known, the easiest means to control this false signal is to be assured that line 2 is about four times the length of line 1 so that the balanced pressure, after expansion, is less than 10 psi, or below that which it takes to shift the four-way valve. The ideal signal valve for B would be one that resets at a much lower psi than it takes to trip, so that timing is easier and the critical positioning requirement is removed. Control functions like this can often occur in pneumatic circuitry and, although troublesome, can usually be removed by properly associating its position with that of other valves with which it must work—or by giving full consideration to the proper design of signal valves.

If the signal is momentary (line 3 of Fig. 15-24b), the small volume will be filled then as indicated in line 23a, valve B will trip, causing flow between 1 and 2. Flow control A allows free flow in, but choked flow out, so this con-

nection between 1 and 2 can be timed to meet the requirements.

Figure 15-24c illustrates a use for the limited memory in creating two complete cycles of a cylinder from the actuation of one valve. A is depressed to put air in line 1, through C and past shuttle (OR) H to extend the cylinder. B times the action at the end of which C is tripped to vent 5 and fill 2, through E and past shuttle I to return the cylinder. D times this eventually to trip E to vent 7 and fill 3, through G past H to extend the cylinder again. When F allows G to trip, 10 vents and 4 fills, past shuttle I to return the cylinder. In the illustrated circuit, air also enters 9, delayed by choke J, finally to trip and reset A to vent the systems ready for the next cycle.

Figure 15-24d is obvious, as a double-piloted valve is shifted one way or the other by a momentary signal. Figure 15-24e illustrates one use where, when A is actuated to put a momentary signal in line 2, C shifts to put the supply to X. B will put the supply to Y.

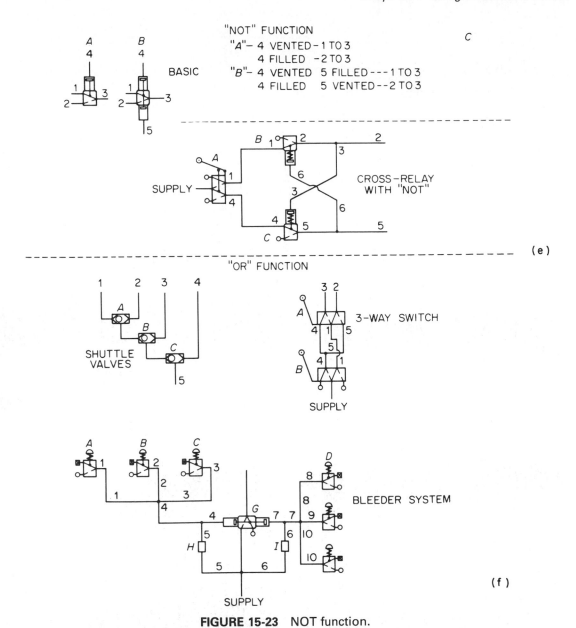

FIGURE 15-23 NOT function.

Figure 15-24f shows the lockup relay arrangement where a momentary signal from A into line 1 past shuttle B into line 2 shifts valve C to put supply into 3, 3a, and 3b so that when 1 is vented, air in 2 is replaced and C is locked up by its own supply. E will trip D to vent 3b and unlock the relay.

The AND and OR functions, as obtained by exhausting air, have already been explained. Figure 15-25a illustrates a limited memory or time function with venting. With air in line 1, A would be tripped and so would D, so that air in 3 is in line 5. If line 1 is vented, D would trip to block line 3 and vent 5. A is also released so that air in 3a flows, timed, through flow control B and eventually, past shuttle C, to trip D and restore the connection between 3 and 5.

Figure 15-25b represents a full memory function actuated by exhausting air. With A as shown and air in 1 and 2, B is tripped and 4 is connected to 5. If line 1 is momentarily dropped, A will shift to vent 3, releasing B to block 4 and vent 5.

Figure 15-25c represents the use of full memory in producing alternate cycling from one signal valve. When A is actuated, air in 1 flows through C to 2 to shift E, putting air in 5 to extend the cylinder—line 4 is vented. Air in 2 also trips B, so that air in line 5 is blocked. When A is released, venting line 1 and 2, B is released, so that air in 5 shifts C to connect 1 to 3 and 2 to exhaust. On the next trip of A, air is directed into line 3 to shift E, vent 5, and fill 4 to return the cylinder. D blocks 4 until A is released, then shifting C back. A, B, C, and D can all be small valves,

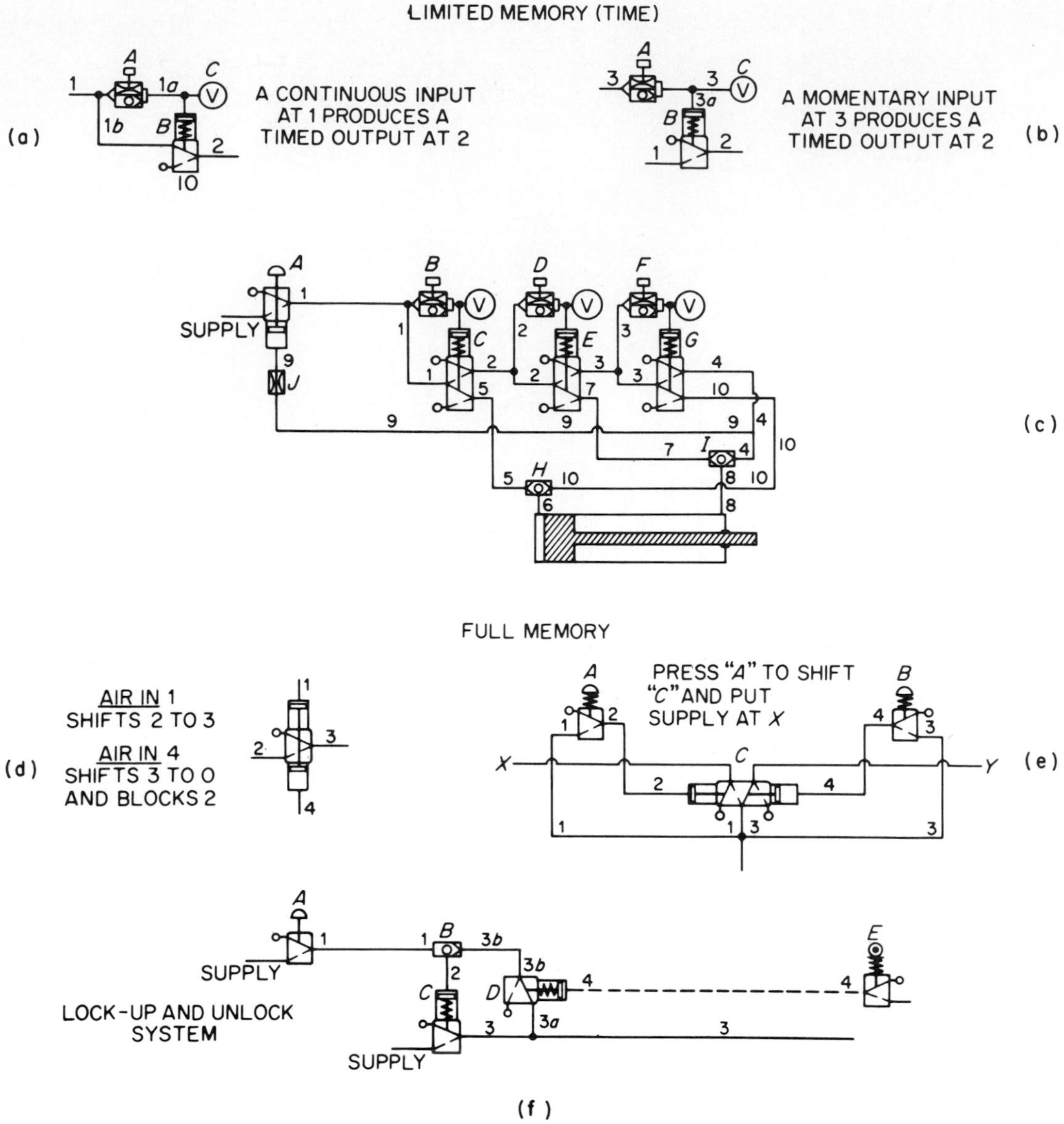

LIMITED MEMORY (TIME)

(a) A CONTINUOUS INPUT AT 1 PRODUCES A TIMED OUTPUT AT 2

(b) A MOMENTARY INPUT AT 3 PRODUCES A TIMED OUTPUT AT 2

(c)

FULL MEMORY

(d) AIR IN 1 SHIFTS 2 TO 3
AIR IN 4 SHIFTS 3 TO 0 AND BLOCKS 2

(e) PRESS "A" TO SHIFT "C" AND PUT SUPPLY AT X

(f) LOCK-UP AND UNLOCK SYSTEM

FIGURE 15-24 Limited memory (time).

whereas E is the power valve of a size sufficient to handle the cylinder involved. Figure 15-26 illustrates a typical machine using a logic circuit for the processing of soap.

Figure 15-27 illustrates the use of "full memory" and "OR" to produce a three-station control for handling waste cotton material. When chutes call for material, the air motors are started in the order in which the demands are made, even when all three call for material near the same time. Assume that station I is being filled; station III calls for material, and then station II before station I is filled. With station I operating, both LV-1 and LV-2 have been tripped, so that RV2 is shifted to put air to air motor I. This also puts air past SV5 and SV8 to hold RV3 and RV5 to the left. Air is also directed past SV6 to RV4 and SV9 to RV6.

As the chute fills, LV-2 is released, but air on the upper side of SV-1 holds RV2 in place.

Material drops in station III to trip LV-5 to put air to RV6 and, prior to acting on LV-6, material drops in chute II to trip LV-3 and put air to RV4. If the material drops faster in station III, LV-6 will trip to put air to SV-7, but air on the other side of RV6 holds it in place. Air is also directed to SV-3 and RV2 is held (locked up). Air also goes to SV-6 to hold RV4 left.

When station I is filled, LV-1 is released, unlocking RV2 and releasing RV3 and RV5. The latter will shift to start air motor III, so when LV-4 is finally tripped it will have to wait its turn to start air motor II.

Figure 15-28 is an automatic bag control system. The

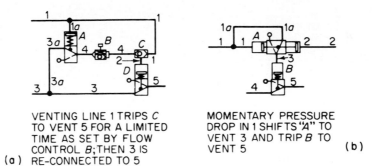

(a) VENTING LINE 1 TRIPS *C* TO VENT 5 FOR A LIMITED TIME AS SET BY FLOW CONTROL *B*; THEN 3 IS RE-CONNECTED TO 5

MOMENTARY PRESSURE DROP IN 1 SHIFTS "*A*" TO VENT 3 AND TRIP *B* TO VENT 5   (b)

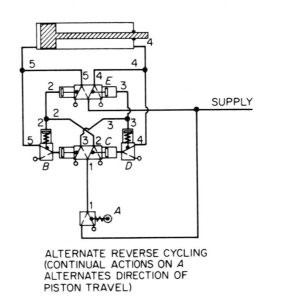

ALTERNATE REVERSE CYCLING (CONTINUAL ACTIONS ON *A* ALTERNATES DIRECTION OF PISTON TRAVEL)   (c)

FIGURE 15-25   Cycling.

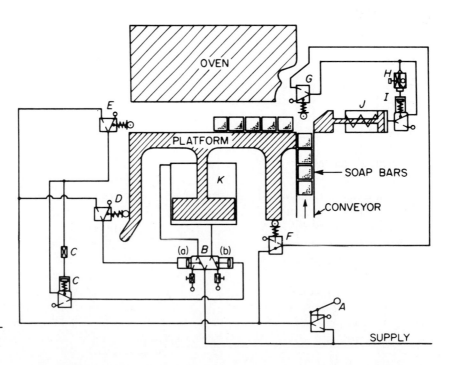

FIGURE 15-26   Logic circuit controls soap machine.

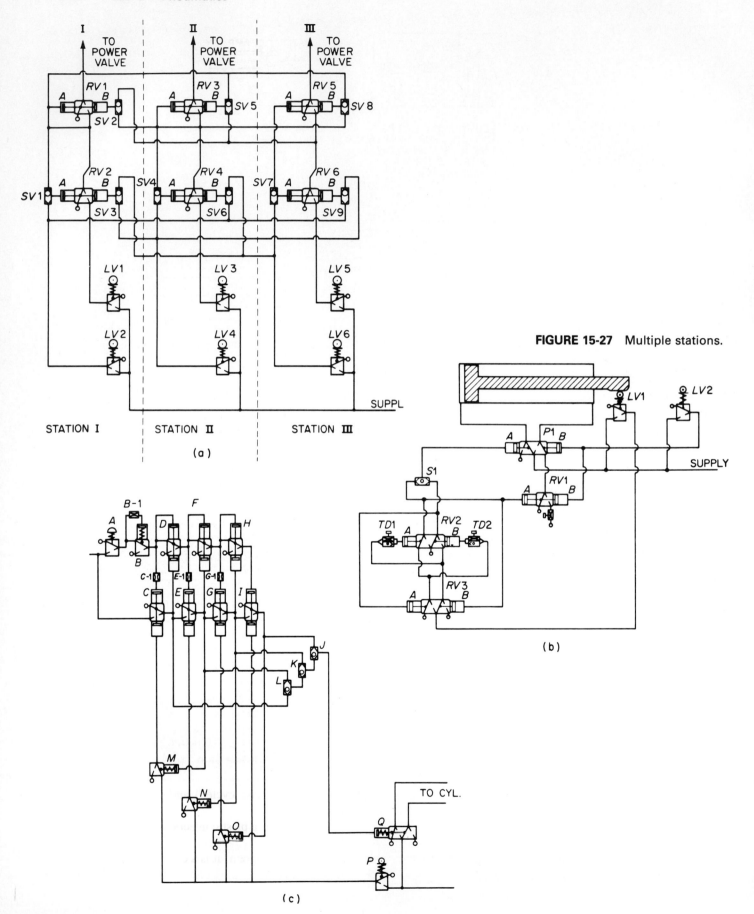

FIGURE 15-27 Multiple stations.

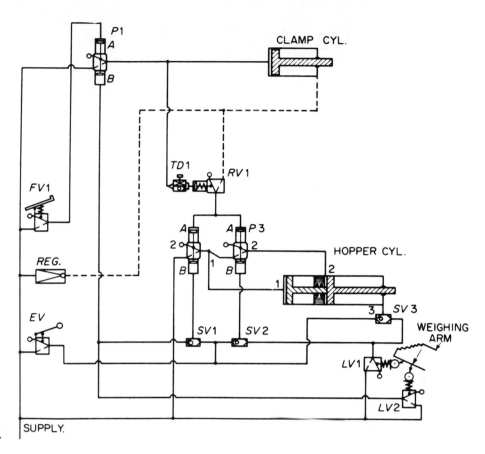

FIGURE 15-28 *Circuit continued.*

initial signal clamps the bag in place, opens the hopper door for pouring, closes off at the designated weight, unclamps, and drops the bag. The operator holds up the empty bag and tramps FV-1 to put air to P1-A to shift P-1 and clamp the bag in place. Air, timed by TD1, finally trips RV1 to put reduced pressure from the regulator to P2-A and P3-A—tripping both valves to fill lines 1 and 2 to open the hopper doors wide. As the weighing arm approaches near filling, LV-1 is tripped to put air past SV-2 to P3-B, which overcomes the lower pressure in P3A so P3 shifts to vent line 2. Air also passes SV3 into line 3 and the right-hand cylinder section, shifting it back to a point where the left cylinder-section rod stops it at two-thirds closed position, reducing the flow. When filled, LV-2 is tripped to put air past SV-1 to P2-B, overcoming lower pressure in P2-A to shift P2 and vent line 1, allowing the three-position cylinder to close off all flow fully. Air also goes to P1-B to vent the clamp cylinder and release the filled bag. RV1 is also tripped to vent P2-A and P3-A, so that when the weighing arm rises and releases LV-1 and LV-2 the hopper remains closed. The clamp cylinder is returned by the reduced pressure from regulator's being on the rod side. This system uses all the five major control functions.

Figure 15-29 is a molding machine control system. To start the cycle, PB-1 puts a momentary signal to P1-A to shift P-1 and extend cycle 1. This releases LV-1. Cycle 1 actuates LV-2 to put air through RV1 to P2-A and P5-A. P5 shifts to connect P-2 exhaust to P6-A. P2 shifts to extend cycle 2. Air also flows through TD-1 to RV2-A. LV-3 is released. Eventually, pressure in RV2-A overcomes the set pressure in RV2-B, and air goes to P3-A and P4A to shift P3 and P4. This turns on the vibrators and extends cycle 3. Air also travels to P2B to shift P2, exhausting it through P5, P6, and the controlled exhaust of P6.

The pressure in P6A is much higher than P6B, allowing it to shift to exhaust. When the cylinder and P6A exhaust to a pressure slightly less than that in P6B, P6 shifts to block pressure in cycle 2 (about 5 psi). Incidentally, TD2 delayed the action on P3 and P4.

Although LV-4 was tripped as cycle-2 piston extended, its supply, coming from P3, was not available, so no action occurred. As the piston retracts, LV-4 is tripped, shifting P4 to cut off vibrators, and shifts P5 rapidly to vent the blocked 5 psi in cycle 2 for a rapid release. On the return, LV-3 is tripped and shifts P-1 to return cycle 1. When cycle 1 returns, it trips LV-1 finally to, timed by TD3, shift P-3 to release cycle 3 and remove the supply from LV-3 and LV-4.

Figure 15-30a is a plastic injection molding machine. Figure 15-30b and 15-30c illustrate the JIC symbols and the electric-ladder-type diagram. It is a continuous cycling system. Starting when TD4 completes its time and shifts RV-

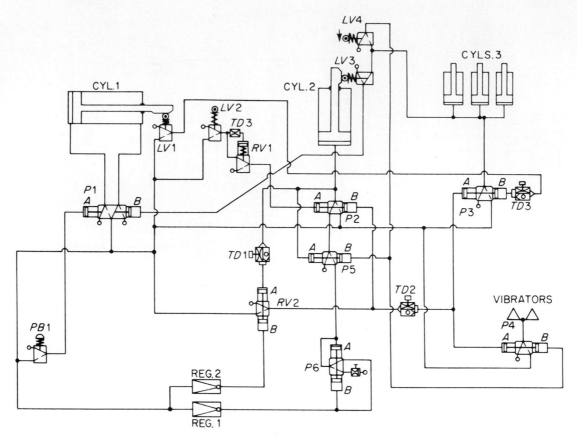

FIGURE 15-29   *Circuit continued.*

1 to vent line 15, P-3 shifts to cut off the air blast; RV3-B is vented; RV4-B is vented, and RV1-B is vented. Supply flows into line 2 through RV-4 into 3 and, after timing by TD3, P-1 is shifted to extend the clamp cylinder; when the pressure in line 4 exceeds the set pressure (regulator 2) in line 6, RV2 is shifted to put the pressure in 3 into line 9 and 9a through RV3 into line 10b, past SV-1 into 10 to shift P-2 and extend the injection cylinder. Air also enters

TD1 finally to trip RV3 to vent 10b and 10 to retract the injection cylinder and also put air in line 12. TD2 times out and finally shifts RV4 to vent line 3 and re-trip P-1 to return the clamp cylinder. Air also enters line 13, delayed slightly by C-1, and trips RV1 to vent 2 and fill 15. This trips RV 4 to vent 13, trips RV3 to vent 12, and trips P-3 to turn on the air blast. TD4 times out to stop the blast and start the cycle again.

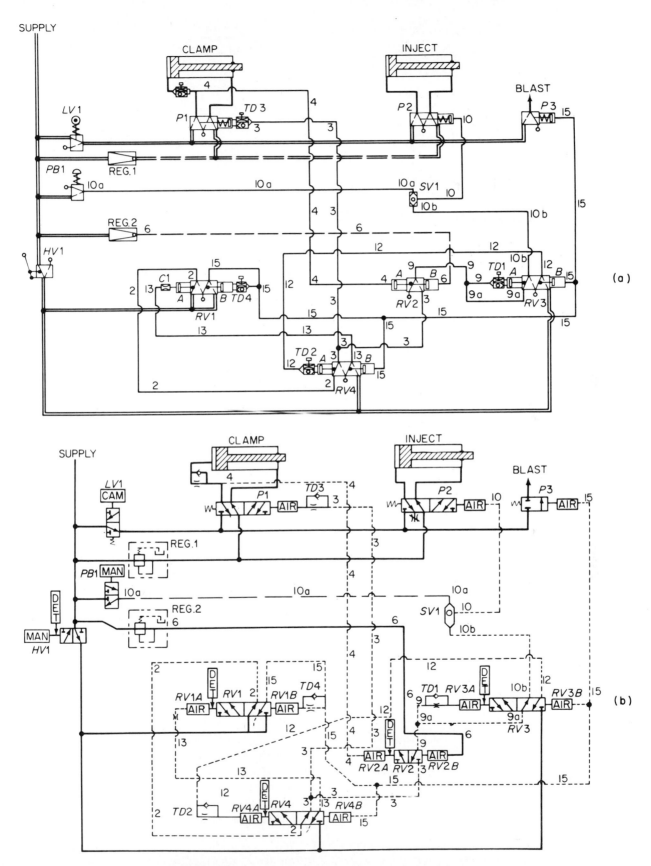

FIGURE 15-30 Plastic injection molding machine circuit.

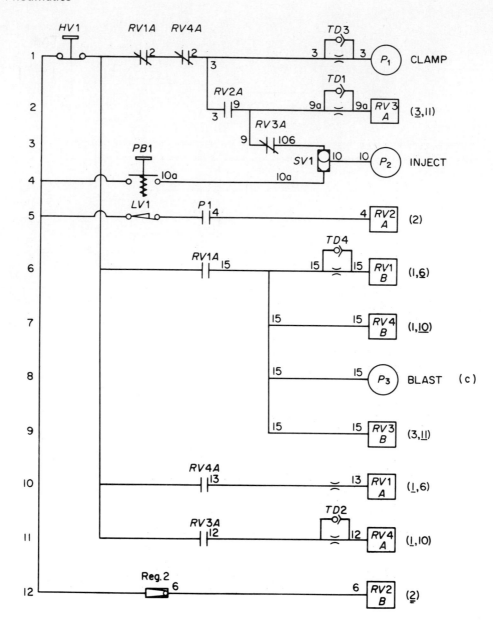

QUESTIONS

1. What are some of the limitations in using pneumatics for applications requiring large forces?

2. What types of applications operate most successfully with pneumatic power?

3. Draw the graphical diagram for a pneumatic circuit using linear motion, and list all of the necessary components. Draw

the graphical diagram for a pneumatic circuit using rotary motion and list the components necessary.

4. What are some of the advantages of using logic systems for control?

5. What is *fluidics,* and what are your predictions concerning the future of this field?

The following material on pp. 317–321 is a presentation by an expert in the industry. It is presented for your information as a typical industrial problem together with a solution.

BY_____ DATE_____ SUBJECT _____ SHEET NO. ___OF ___

CHKD. BY _____ DATE_____ _____ JOB NO. _____

_____ _____ _____

THE "HOW OF PNEUMATIC
CONTROL DESIGN"

E. L. HOLBROOK

THE PROBLEM

3 CYLINDERS TO OPERATE CONTINUOUSLY WHEN
TURNED ON WITH A SEQUENCE AS FOLLOWS

A - CYL. 1 → OUT
B - CYL. 2 → OUT
C - CYL. 3 → OUT
D - CYL. 2 ← IN
E - CYL. 1 ← IN
F - CYL. 2 → OUT
G - CYL. 3 ← IN
H - CYL. 2 ← IN
I - CYL. 1 → OUT ETC.

CONDITIONS
CYLINDERS ARE HYDRAULIC EACH
WITH A DOUBLE AIR PILOTED 4-WAY VALVE

THERE IS SPACE FOR CAM VALVES

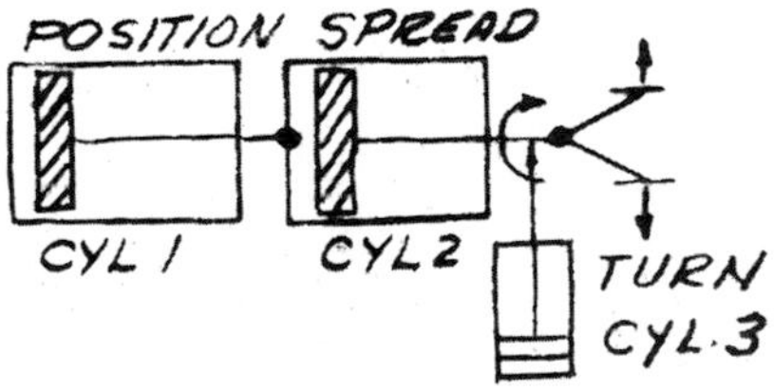

POSITION SPREAD

CYL 1   CYL 2   TURN CYL. 3

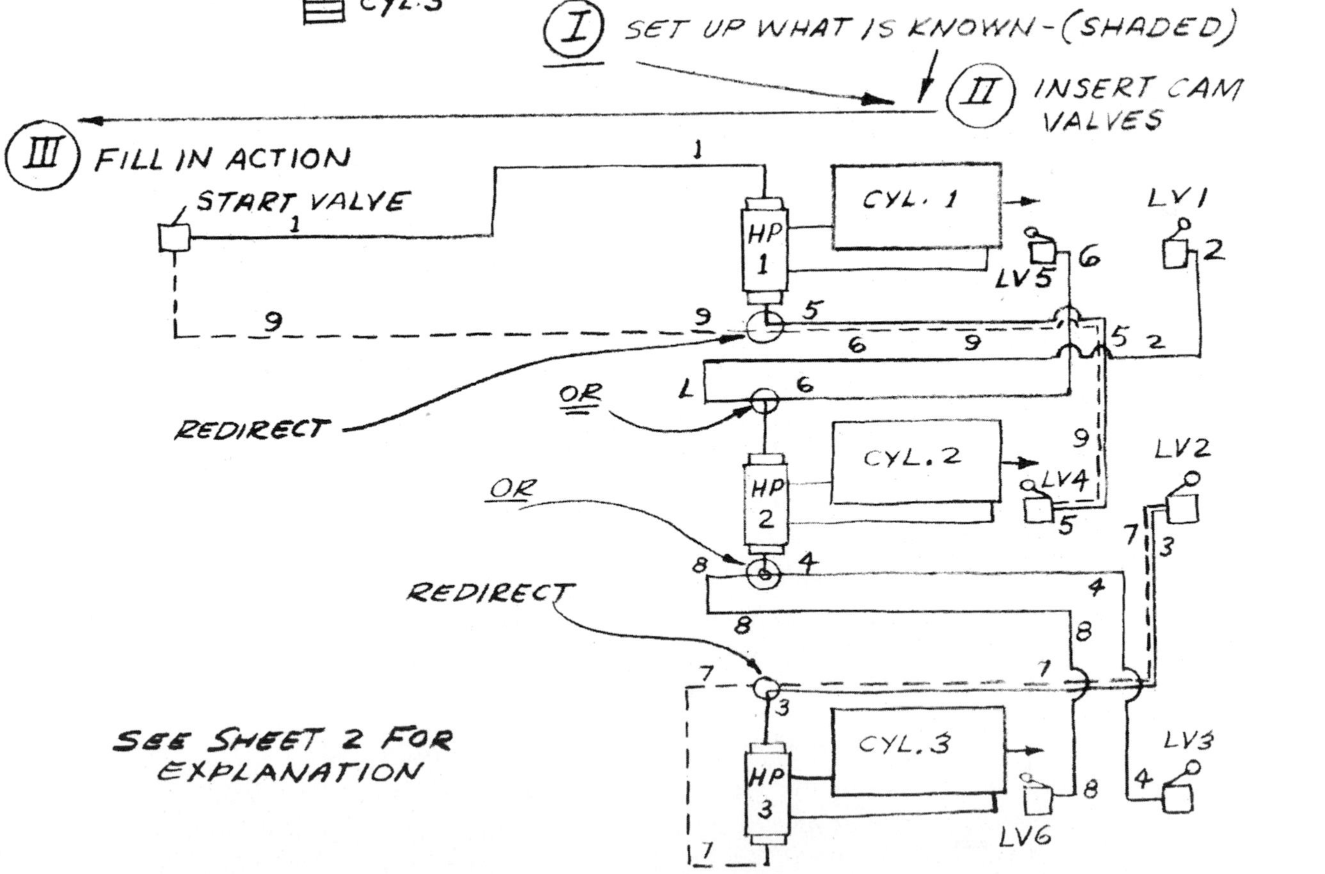

Ⓘ SET UP WHAT IS KNOWN - (SHADED)

Ⓘ Ⓘ INSERT CAM VALVES

Ⓘ Ⓘ Ⓘ FILL IN ACTION

START VALVE

CYL. 1   LV1

HP 1   LV5   6   2

REDIRECT   OR   L

CYL. 2   LV2

HP 2   LV4

OR

REDIRECT

CYL. 3   LV3

HP 3   LV6

SEE SHEET 2 FOR EXPLANATION

BY_____   DATE _____   SUBJECT_____   SHEET NO. ___OF ___

CHKD. BY _____   DATE _____   _____   JOB NO. _____

_____   _____   _____   _____   _____

HOW OF PNEUMATIC CONTROL DESIGN

E. L. HOLBROOK

(I) SETTING UP WHAT IS KNOWN · MEANS THE
CYLINDERS & VALVES -- HP1, HP2, HP3 & CYLS 1, 2 & 3.

(II) 6 CAM VALVES CAN BE INSERTED. WE MIGHT HAVE USED
8 (2 AT LV4 POS. & 2 AT LV2 POS.) - THEN MADE
THEM ALTERNATELY "LIVE" OR "DEAD" - HOWEVER, THERE
WAS A SPACE LIMITATION AND, IN ANY EVENT, IT
WOULD NOT EASE THE PROBLEM

(III)   FILL IN THE ACTION

A   THE "START VALVE" CAN FILL LINE 1 TO TRIP HP1 - CYL 1 OUT →
B   LV 1 FILLS LINE 2 TO TRIP HP 2   ----   CYL. 2 OUT →
C   LV 2 FILLS LINE 3 TO TRIP HP 3   ----   CYL 3 OUT →
D   LV3 FILLS LINE 4 TO TRIP HP 2   ----   CYL. 2 IN ←
E   LV5 FILLS LINE 5 TO TRIP HP 1   ----   CYL 1 IN ←
F   LV6 FILLS LINE 6 TO TRIP HP 2   ----   CYL. 2 OUT →
G   LV2 FILLS LINE 7 TO TRIP HP 3   ----   CYL. 3 IN ←
H   LV6 FILLS LINE 8 TO TRIP HP 2   ----   CYL. 2 IN ←
I   LV5 FILLS LINE 9 TO TRIP HP 1   ----   CYL. 1 OUT →^(AGAIN.)

THE ABOVE, OBVIOUSLY, CANNOT WORK DUE TO INTERFERING
SIGNALS.

LV2 & LV5 MUST, ALTERNATELY, SHIFT HP2 - ^(OUT) NEED OR - A SHUTTLE
LV6 & LV3 " , " , " HP2 ^(IN) " " - " "
LV2 MUST OR HP3 OUT & IN - NEED RE-DIRECTED SIGNAL,
LV5 " " HP1 " " " - " " " ")

CERTAIN LINES MUST BE CHANGED FROM CONTINUOUS TO
MOMENTARY SIGNALS TO FREE POWER VALVES TO SHIFT.

LINE 9 AND LINE 5 TO FREE HP1 (OUT & IN)
LINE 2 AND LINE 6 TO FREE HP2 (OUT)
LINE 4 AND LINE 8 TO FREE HP2 (IN)

NEXT STEP - INSERT ALL VALVES THEN LOCATE.
POINTS TO SIGNAL RE-DIRECT VALVES TO POSITIONS
REQUIRED AT THE PROPER TIME FOR ACTION

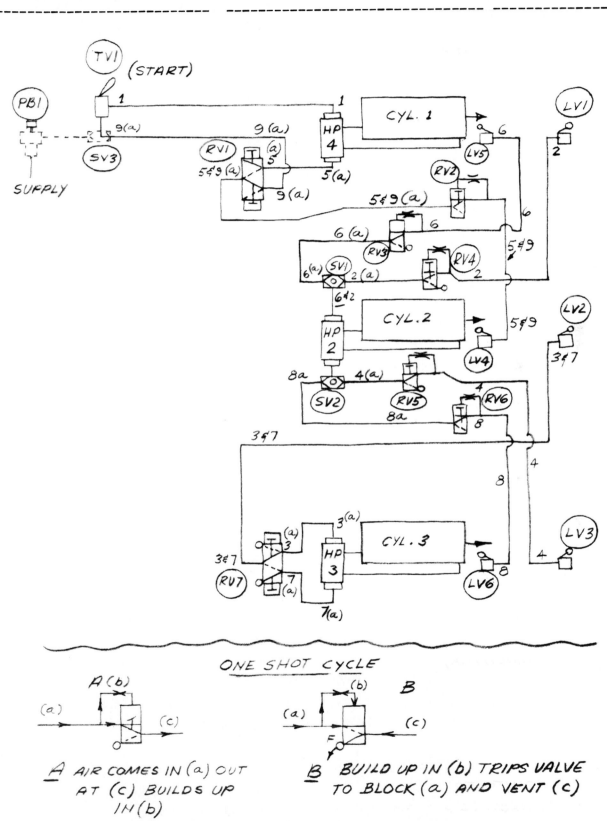

ONE SHOT CYCLE

A AIR COMES IN (a) OUT
AT (c) BUILDS UP
IN (b)

B BUILD UP IN (b) TRIPS VALVE
TO BLOCK (a) AND VENT (c)

BY_____ DATE _____ SUBJECT_____ SHEET NO. ___OF ___

CHKD. BY _____ DATE _____ _____ JOB NO. _____

_____ _____ _____

SINCE ALTHOUGH LV4 IS TRIPPED (AIR IN LINE 9) RV2 (ONE-SHOT) WOULD
VENT LINE 9(a) - NO AIR FOR INITIAL START - INSERT PB1 & SHUTTLE

A MOMENTARY TRIP OF PB1, PASSED SV3, THRU TV1 TO HP1 - CYL 1 OUT →

B CYL 1 TRIPS LV1, AIR IN 2, THRU RV4, 2(a), PASSED SV1 TO
        HP2 (RV4 TRIPS TO VENT 2(a) - - - - - CYL 2 OUT →

C LV2 TRIPPED BY CYL. 2, AIR IN LINE 3 (&7) THRU
        RV7 TO LINE 3(a), HP3 -   - - - - - CYL 3 OUT →

D LV3 TRIPPED, AIR IN LINE 4, THRU RV5, LINE 4(a)
        PASSED SV2 TO HP2 (RV5 VENTS 4(a) CYL. 2 IN ←

E LV4 TRIPPED, AIR IN 5 (& 9) THRU RV2, THRU RV1 (SHOULD HAVE
        BEEN SHIFTED) TO LINE 5, HP1 - - - - CYL 1 IN ←

F LV5 TRIPPED, AIR IN 6, THRU RV3 TO 6(a), PASSED SV1
        TO HP2 (RV3 TRIPS TO VENT 6(a)) - - CYL. 2 OUT →

G LV2 TRIPPED, AIR IN LINE 7 (&3), THRU RV7 (SHOULD HAVE BEEN
        SHIFTED) TO LINE 7(a) TO HP3   - - - - CYL. 3 IN ←

H LV6 TRIPPED, AIR IN 8, THRU RV6 TO 8(a) PASSED SV2
        TO HP2 (RV6 SHIFTS TO VENT 8(a)) - - - - CYL. 2 IN ←

I LV4 TRIPPED, AIR IN 9 (&5) THRU RV2 TO 9(a) (&5(a)), THRU
        RV1 (SHOULD HAVE BEEN SHIFTED) TO 9(a)
    PASSED SV3, THRU OPEN TV1 (CLOSED IF STOP REQUIRED) TO HP1 - CYL. 1 OUT →

NOTE - SIGNALS NEEDED TO SHIFT, CORRECTLY, RV1 & RV2
AND A CHECK TO REMOVE UNNECESSARY VALVING.

① COULD USE LINE 1 TO SHIFT RV1 TO CONN. 5 & 9(a) TO 5(a)

② COULD USE LINE 4(a) TO SHIFT RV7 TO CONN. 3 & 7 TO 7

③ CAN USE ONE VALVE IN LINE 6 & 2 (BEYOND SV1) IN PLACE
        OF RV3 & RV4

④ CAN USE LINE 6 (OR 6(a)) TO RESHIFT RV1

⑤ CAN USE LINE 8 TO RESHIFT RV7

⑥ SINCE BOTH SIGNALS (FROM LV6 & LV3) MUST FREE RV7,
    MUST MAINTAIN SEPARATE ONE-SHOT FUNCTIONS (RV5 & RV6)

MAKE ABOVE CHANGES & CHECK

BY _____ DATE _____ SUBJECT _____ SHEET NO. ___OF___

CHKD. BY _____ DATE _____ _____ JOB NO. _____

_____ _____ _____

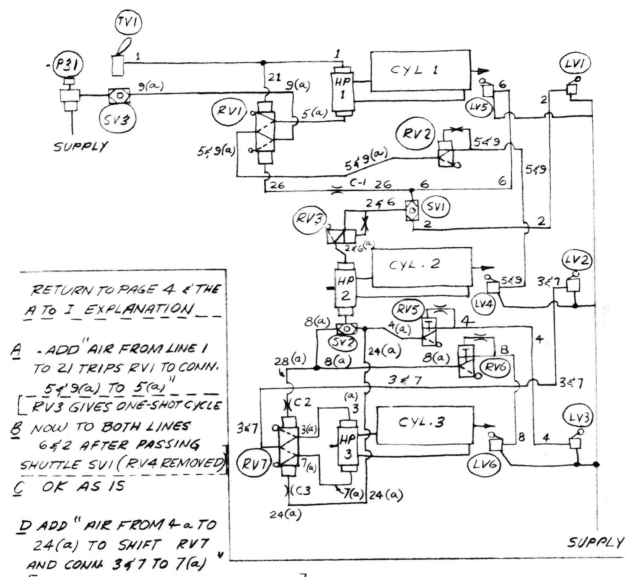

RETURN TO PAGE 4 & THE
A TO I EXPLANATION ___

A - ADD "AIR FROM LINE 1
TO 21 TRIPS RV1 TO CONN.
5 & 9(a) TO 5(a)"
[RV3 GIVES ONE-SHOT CYCLE

B NOW TO BOTH LINES
6 & 2 AFTER PASSING
SHUTTLE SV1 (RV4 REMOVED)

C OK AS IS

D ADD "AIR FROM 4-a TO
24(a) TO SHIFT RV7
AND CONN. 3 & 7 TO 7(a) "
[CHOKE C-3 GIVES SLIGHT DELAY]

E OK AS IS (RV1 WAS SHIFTED AT A)

F ADD" AIR FROM LINE 6 TO 26 (DELAYED BY CHOKE C-1) RESHIFTS
RV1 TO CONN. 5 & 9 (a) TO 9(a) "

G OK AS IS (RV7 WAS SHIFTED IN D)

H ADD " AIR FROM 8(a) TO 28(a) DELAYED BY CHOKE C-2 RESHIFTS
RV7 TO CONN. 3 & 7 TO 3(a) "

Q.E.D.

E.L. HOLBROOK

National Fluid Power Association
GLOSSARY OF TERMS FOR FLUID POWER

SECTION 1

PRIMARY TERMS

FLUID POWER: Energy transmitted and controlled through use of a pressurized fluid.

FLUID POWER SYSTEM: A system that transmits and controls power through use of a pressurized fluid within an enclosed circuit.

HYDRAULICS: Engineering science pertaining to liquid pressure and flow.

HYDRODYNAMICS: Engineering science pertaining to the energy of liquid flow and pressure.

HYDROKINETICS: Engineering science pertaining to the energy of liquids in motion.

HYDROPNEUMATICS: Pertaining to the combination of hydraulic and pneumatic fluid power.

HYDROSTATICS: Engineering science pertaining to the energy of liquids at rest.

PNEUMATICS: Engineering science pertaining to gaseous pressure and flow.

SECTION 3

FLUID POWER LAWS AND RELATED TERMS

(Letter symbols at end of Section)

BERNOULLI'S LAW: If no work is done on or by a flowing frictionless liquid its energy due to pressure and velocity remains constant at all points along the streamline.

BOYLE'S LAW: The absolute pressure of a fixed mass of gas varies inversely as the volume, provided the temperature remains constant.

CHARLES' LAW: The volume of a fixed mass of gas varies directly with absolute temperature, provided the pressure remains constant.

CONTINUITY EQUATION: The mass rate of fluid flow into any fixed space is equal to the mass flow rate out. Hence, the mass flow rate of fluid past all cross sections of a conduit is equal.

DARCY'S FORMULA: A formula used to determine the pressure drop due to flow friction through a conduit.

$$h_f = \frac{fLv^2}{2Dg}$$

HAGEN POISEUILLE LAW: The friction factor of Darcy's Formula is a ratio of 64 to the Reynolds Numbers when flow is laminar.

$$f = \frac{64}{R_c}$$

PASCAL'S LAW: A pressure applied to a confined fluid at rest is transmitted with equal intensity throughout the fluid.

REYNOLDS NUMBERS: A numerical ratio of the dynamic forces of mass flow to the shear stress due to viscosity. Flow usually changes from laminar to turbulent between Reynolds Numbers 2,000 and 4,000.

$$N_r = \frac{\rho v D}{\mu} \qquad N_r = \frac{vD}{v}$$

TORICELLI'S THEOREM: The liquid velocity at an outlet discharging into the free atmosphere is proportional to the square root of the head.

$$v = \sqrt{2gh}$$

LETTER SYMBOLS

D = Internal diameter of conduit — feet

f = Friction factor

g = Acceleration due to gravity — nominal 32.2 feet per second squared

h = Head — feet

h_f = Head loss — feet

L = Length of conduit — feet

N_r = Reynolds Number

v = Mean velocity of flow — feet per second

(Greek Letters)

μ (Mu) Absolute viscosity — pounds mass per foot second

ν (Nu) Kinematic viscosity — square feet per second

ρ (Rho) Mass density of fluid — slugs per cubic foot

SECTION 5

FLOW TERMS

CAVITATION: A localized gaseous condition within a liquid stream which occurs where the pressure is reduced to the vapor pressure.

FLOW, Laminar: A flow situation in which fluid moves in parallel lamina or layers.

Flow, Metered: Flow at a controlled rate.

Flow, Steady State: A flow situation wherein conditions such as pressure, temperature, and velocity at any point in the fluid do not change.

Flow, Streamline: (See FLOW, Laminar)

Flow, Turbulent: A flow situation in which the fluid particles move in a random manner.

Flow, Unsteady: A flow situation wherein conditions such as pressure, temperature and velocity at points in the liquid change.

SHOCK WAVE: A pressure wave front which moves at a supersonic velocity.

SURGE: A momentary rise of pressure in a circuit.

SECTION 6

MENSURATION TERMS

AIR, STANDARD: Air at a temperature of 68°F, a pressure of 14.70 pounds per square inch absolute, and a relative humidity of 36% (0.0750 pounds per cubic foot). In gas industries the temperature of "standard air" is usually given as 60°F.

ANILINE POINT: The lowest temperature at which a liquid is completely miscible with an equal volume of freshly distilled aniline (ASTM Designation D611-55T).

BULK MODULUS: The measure of resistance to compressibility of a fluid. It is the reciprocal of the compressibility.

COMPRESSIBILITY: The change in volume of a unit volume of a fluid when subjected to a unit change of pressure.

FLASH POINT: The temperature to which a liquid must be heated under specified conditions of the test method to give off sufficient vapor to form a mixture with air that can be ignited momentarily by a specified flame.

FLOW RATE: The volume, mass, or weight of a fluid passing through any conductor per unit of time.

FLUID FRICTION: Friction due to the viscosity of fluids.

HEAD: The height of a column or body of fluid above a given point expressed in linear units. Head is often used to indicate gage pressure. Pressure is equal to the height times the density of the fluid.

Head, Friction: The head required to overcome the friction at the interior surface of a conductor and between fluid particles in motion. It varies with flow, size, type and condition of conductors and fittings, and the fluid characteristics.

Head, Static: The height of a column or body of fluid above a given point.

Head, Static Discharge: The static head from the centerline of the pump to the free discharge surface.

Head, Static Suction: The head from the surface of the supply source to the centerline of the pump.

Head, Total Static: The static head from the surface of the supply source to the free discharge surface.

Head, Velocity: The equivalent head through which the liquid would have to fall to attain a given velocity. Mathematically it is equal to the square of the velocity (in feet) divided by 64.4 feet per second square.

$$h = \frac{v^2}{2g}$$

HYDRAULIC HORSEPOWER: Horsepower computed from flow rate and pressure differential.

> Hydraulic Horsepower = .000583Qp
>
> Q = Flow rate — gallons per minute
>
> p = Pressure — pounds per square inch

LIFT: The height of a column or body of fluid below a given point expressed in linear units. Lift is often used to indicate vacuum or pressure below atmospheric.

Lift, Static Suction: The lift from the centerline of the pump to the surface of the supply source. (See Head, Static Sunction)

MICRON: A millionth of a meter or about 0.00004 inch.

NEUTRALIZATION NUMBER: A measure of the total acidity or bascity of an oil; this includes organic or inorganic acids or bases or a combination thereof (ASTM Designation D974-58T).

NEWT: The standard unit of kinematic viscosity in the English system. It is expressed in square inches per second.

POISE: The standard unit of absolute viscosity in the c.g.s. (centimeter-gram-second) system. It is the ratio of the shearing stress to the shear rate of a fluid and is expressed in dyne seconds per square centimeter; 1 centipoise equals .01 poise.

POUR POINT: The lowest temperature at which a liquid will flow under specified conditions (ASTM Designation D97-57).

PRECIPITATION NUMBER: The number of milliliters of precipitate formed when 10 ml. of lubricating oil are mixed with 90 ml. of ASTM precipitation naptha and centrifuged under prescribed conditions (ASTM Method D91-61).

PRESSURE: Force per unit area, usually expressed in pounds per square inch.

Pressure, Absolute: The sum of atmospheric and gage pressures.

Pressure, Atmospheric: Pressure exerted by the atmosphere at any specific location. (Sea level pressure is approximately 14.7 pounds per square inch absolute.)

Pressure, Back: The pressure encountered on the return side of a system.

Pressure, Breakloose: The minimum pressure which initiates movement.

Pressure, Breakout: (See Pressure, Breakloose)

Pressure, Burst: The pressure which causes rupture.

Pressure, Charge: The pressure at which replenishing fluid is forced into a fluid power system.

Pressure, Cracking: The pressure at which a pressure operated valve begins to pass fluid.

Pressure, Differential: The difference in pressure between any two points of a system or a component.

Pressure, Gage: Pressure differential above or below atmospheric pressure.

Pressure Head: The pressure due to the height of a column or body of fluid. It is usually expressed in feet.

Pressure, Operating: The pressure at which a system is operated.

Pressure Override: The difference between the cracking pressure of a valve and the pressure reached when the valve is passing full flow.

Pressure, Pilot: The pressure in the pilot circuit.

Pressure, Precharge: The pressure of compressed gas in an accumulator prior to the admission of a liquid.

Pressure, Proof: The non-destructive test pressure in excess of the maximum rated operating pressure.

Pressure, Rated: The qualified operating pressure which is recommended for a component or a system by the manufacturer.

Pressure, Shock: The pressure existing in a wave moving at supersonic velocity.

Pressure, Static: The pressure in a fluid at rest.

Pressure, Suction: The absolute pressure of the fluid at the inlet of a pump.

Pressure, Surge: The pressure existing from surge conditions.

Pressure, System: The pressure which overcomes the total resistances in a system. It includes all losses as well as useful work.

Pressure, Vapor: The pressure, at a given fluid temperature, in which the liquid and gaseous phases are in equilibrium.

Pressure, Working: The pressure which overcomes the resistance of the working device.

PRESSURE DROP: (See Pressure, Differential)

REYN: The standard unit of absolute viscosity in the English system. It is expressed in pound-seconds per square inch.

SPECIFIC GRAVITY (LIQUID): The ratio of the weight of a given volume of liquid to the weight of an equal volume of water.

STOKE: The standard unit of kinematic viscosity in the c.g.s. (centimeter-gram-second) system. It is expressed in square centimeters per second; 1 centistoke equals .01 stoke.

SURFACE TENSION: The contractile surface force of a liquid in contact with a fluid by which it tends to assume a spherical form and to present the least possible surface. It is expressed in pounds per foot or dynes per centimeter.

VACUUM: Pressure less than atmospheric pressure. It can be expressed in absolute or gage pressure.

VISCOSITY: A measure of the internal friction or the resistance of a fluid to flow.

Viscosity, Absolute: The ratio of the shearing stress to the shear rate of a fluid. It is usually expressed in centipoise.

Viscosity, Kinematic: The absolute viscosity divided by the density of the fluid. It is usually expressed in centistokes.

Viscosity, SAE Number: The Society of Automotive Engineers arbitrary numbers for classifying fluids according to their viscosities. The numbers in no way indicate the viscosity index of fluids.

Viscosity, SUS: Saybolt Universal Seconds (SUS), which is the time in seconds for 60 milliliters of oil to flow through a standard orifice at a given temperature (ASTM Designation D88-56).

VISCOSITY INDEX: A measure of the viscosity-temperature characteristics of a fluid as referred to that of two arbitrary reference fluids (ASTM Designation D567-53).

SECTION 7

CIRCUITRY AND RELATED TERMS

CIRCUIT: An arrangement of interconnected component parts.

Circuit, Pilot: A circuit used to control a main circuit or component.

Circuit, Pressure Control: Any circuit whose main purpose is to adjust or regulate fluid pressure in the system or any branch of the system.

Circuit, Regenerative: A circuit in which pressurized fluid discharged from a component is returned to the system to reduce power input requirements. On single rod end cylinders the discharge from the rod end is often directed to the opposite end to increase rod extension speed.

Circuit, Safety: A circuit which prevents accidental operation, protects against overloads, or otherwise assures safe operation.

Circuit, Sequence: A circuit which establishes the order in which two or more phases of a circuit occur.

Circuit, Servo: A circuit which is controlled by automatic feed back; i.e., the output of the system is sensed or measured and is compared with the input signal. The difference (error) between the actual output and the input controls the circuit. The controls attempt to minimize the error. The system output may be position, velocity, force, pressure, level, flow rate, or temperature.

Circuit, Speed Control: Any circuit where components are arranged to regulate speed of operation.

Circuit, Synchronizing: A circuit in which multiple operations are controlled to occur at the same time.

Circuit, Unloading: A circuit in which pump volume is returned to reservoir at near zero gage pressure whenever delivery to the system is not required.

CYCLE: A single complete operation consisting of progressive phases starting and ending at the neutral position.

Cycle, Automatic: A cycle of operation

which once started is repeated indefinitely until stopped.

Cycle, Manual: A cycle which is manually started and controlled through all phases.

Cycle, Semi-automatic: A cycle which is started upon a given signal, proceeds through a predetermined sequence, and stops with all elements in their initial position.

DIAGRAM, Combination: A drawing utilizing a combination of graphical, cutaway and pictorial symbols showing interconnected lines.

Diagram, Cutaway: A drawing showing principle internal parts of all components, controls and actuating mechanisms, all interconnecting lines and functions of individual components.

Diagram, Graphical: A drawing or drawings showing each piece of apparatus including all interconnecting lines by means of approved (ASA, JIC) standard symbols.

Diagram, Pictorial: A drawing showing each component in its actual shape according to the manufacturer's installation.

Diagram, Pressure-Time: A graphical presentation of pressure plotted against time for a complete cycle of the equipment.

Diagram. Schematic: (See Diagram, Graphical)

PHASE: A distinct functional operation during a cycle. Some typical sequential phases are: neutral, rapid advance, feed or pressure stroke, dwell and rapid return.

Phase, Dwell: The phase of a cycle where a specified motion is stopped for a pre-determined length of time.

Phase, Feed: The phase of a cycle where work is performed on the workpiece.

Phase, Neutral: The phase of a cycle from which the work sequence begins.

Phase, Rapid Advance: The phase of a cycle where tools or workpiece approach at high speed to the feed position.

Phase, Rapid Return: The phase of a cycle where tools or workpiece returns at high speed to the cycle starting position.

SECTION 9

GENERAL TERMS

BACK CONNECTED: Where connections are made to normally unexposed surfaces of components.

FRONT CONNECTED: Where connections are made to normally exposed surfaces of components.

POSITION REVERSAL: A reversal of direction of movement initiated by a signal given at some predetermined point of the movement.

POSITIVE POSITION STOP: A structural member which accurately stops motion.

POSITIVE SAFETY STOP: A structural member which confines maximum travel to safe limits.

PRESSURE REVERSAL: A reversal of direction of movement initiated by a signal responsive to rise in pressure.

SECTION 11

FLUIDS AND RELATED TERMS

ADDITIVE: A chemical compound or compounds added to a fluid to change its properties.

AIR, Compressed: Air at any pressure greater than atmospheric pressure.

Air. Free: Air under the pressure due to atmospheric conditions at any specific location.

Air, Pressure: (See Air, Compressed)

Air, Standard: (See MENSURATION SECTION)

FLUID: A liquid or a gas.

Fluid, Emulsion, Water-Oil: A stabilized mixture of two immiscible components, water and oil. It may contain additives. There are two types: OIL IN WATER and WATER IN OIL.

Fluid, Emulsion, Oil in Water: A dispersion of oil in a continuous phase of water.

Fluid, Emulsion, Water in Oil: A dispersion of water in a continuous phase of oil.

Fluid, Fatty Oil: A fluid composed of fats derived from animal, marine or vegetable origin. It may contain additives.

Fluid, Fire Resistant: A fluid difficult to ignite which shows little tendency to propagate flame.

Fluid, Halogenated: A fluid composed of halogenated organic materials. It may contain additives.

Fluid, Hydraulic: A fluid suitable for use in a hydraulic system.

Fluid, Non-Flammable: (Deprecated)

Fluid, Organic Ester: A fluid composed of esters which are compounds of carbon, hydrogen, and oxygen only. It may contain additives.

Fluid, Petroleum: A fluid composed of petroleum oil. It may contain additives.

Fluid, Phosphate Ester: A fluid composed of phosphate esters. It may contain additives.

Fluid, Phosphate Ester Base: A fluid which contains a phosphate ester as one of the major components.

Fluid, Polyglycol: A non-aqueous fluid composed of polyglycols or polyglycol derivatives. It may contain additives.

Fluid. Silicate Ester: A fluid composed of organic silicates. It may contain additives.

Fluid, Silicone: A fluid composed of silicones. It may contain additives.

Fluid, Water-Glycol: A fluid whose major constituents are water and one or more glycols or polyglycols.

FLUID STABILITY: Resistance of a fluid to permanent change in properties.

Fluid Stability, Chemical: Resistance of a fluid to chemical change.

Fluid Stability, Hydrolytic: Resistance of a fluid to permanent changes in properties caused by chemical reaction with water.

Fluid Stability, Oxidation: Resistance of a fluid to permanent changes caused by chemical reaction with oxygen.

Fluid Stability, Thermal: Resistance of a fluid to permanent changes caused solely by heat.

INHIBITOR: Any substance which slows or prevents such chemical reactions as corrosion or oxidation.

SECTION 21

FLUID CONDITIONERS AND RELATED TERMS

AIR BREATHER: A device permitting air movement between atmosphere and the component in which it is installed.

BLEEDER: A device for removal of pressurized fluid.

Bleeder, Air: A bleeder for the removal of air.

CONTAMINANT: Detrimental matter in a fluid.

COOLER: A heat exchanger which removes heat from a fluid.

Cooler, Aftercooler: A device which cools a gas after it has been compressed.

Cooler, Intercooler: A device which cools a gas between the compressive steps of a multiple stage compressor.

Cooler, Precooler: A device which cools a gas before it is compressed.

FILTER: A device whose primary function is the retention by a porous media of insoluable contaminants from a fluid.

FILTER ELEMENT: The porous device which performs the actual process of filtration.

FILTER MEDIA: The porous materials which perform the actual process of filtration.

Filter Media, Depth: Porous materials which primarily retain contaminant within a tortuous path.

Filter Media, Effective Area: The total funtional area of the porous media.

Filter Media, Surface: Porous materials which primarily retain contaminants on the influent face.

FLUID CONDITIONER: A device which controls the physical characteristics of a fluid.

HEAT EXCHANGER: A device which transfers heat through a conducting wall from one fluid to another.

LUBRICATOR: A device which adds controlled or metered amounts of lubricant into a fluid power system.

MUFFLER: A device for reducing gas flow noise. Noise is decreased by back pressure control of gas expansion.

RESERVOIR: A container for storage of liquid in a fluid power system.

Reservoir, Atmospheric: A reservoir for storage of fluid media at atmospheric pressure.

Reservoir, Sealed: A reservoir for storage of fluids isolated from atmospheric conditions.

Reservoir, Sealed, Pressure: A sealed reservoir for storage of fluids under pressure.

SEPARATOR: A device whose primary function is to isolate undesirable fluids and.or contaminants by physical properties other than size.

SILENCER: A device for reducing gas flow noise. Noise is decreased by tuned resonant control of gas expansion.

STRAINER: A coarse filter.

TANK: A container for the storage of fluid in a fluid power system.

Tank, Vacuum: A container for gas at less than atmospheric pressure.

SECTION 31

COMPRESSORS — INTENSIFIERS — PUMPS

COMPRESSOR: A device which converts mechanical force and motion into pneumatic fluid power.

Compressor, Multiple Stage: A compressor having two or more compressive steps in which the discharge from each supplies the next in series.

Compressor, Single Stage: A compressor having only one compressive step between inlet and outlet.

INTENSIFIER: A device which converts low pressure fluid power into higher pressure fluid power.

POWER UNIT: A combination of pump, pump drive, reservoir, controls and conditioning components which may be required for its application.

PUMP: A device which converts mechanical force and motion into hydraulic fluid power.

Pump, Axial Piston: A pump having multiple pistons disposed with their axes parallel.

Pump, Centrifugal: A pump which produces fluid velocity and converts it to pressure head.

Pump, Centrifugal, Concentric: (See Diffuser Centrifugal Pump).

Pump, Centrifugal, Diffuser: A centrifugal pump in which fluid enters at the center of the impeller, is accelerated radially, and leaves through vanes arranged to provide a gradually enlarging flow passage.

Pump, Centrifugal, Peripheral: A centrifugal pump in which fluid enters, follows, and leaves the periphery of the impeller.

Pump, Centrifugal, Spiral: (See Pump, Centrifugal, Volute)

Pump, Centrifugal, Volute: A centrifugal pump in which fluid enters at the center of the impeller, is accelerated radially, and leaves through a gradually enlarging flow passage.

Pump, Fixed Displacement: A pump in which the displacement per cycle cannot be varied.

Pump, Gear: A pump having two or more intermeshed rotating members enclosed in a housing.

Pump, Hand: A hand operated pump.

Pump, Multiple Stage: Two or more pumps in series.

Pump, Radial Piston: A pump having multiple pistons disposed radially actuated by an eccentric element.

Pump, Reciprocating Duplex: A pump having two reciprocating pistons.

Pump, Reciprocating Single Piston: A pump having a single reciprocating piston.

Pump, Screw: A pump having one or more screws rotating in a housing.

Pump, Vane: A pump having multiple radial vanes within a supporting rotor.

Pump, Variable Displacement: A pump in which the displacement per cycle can be varied.

SECTION 36

ACCUMULATORS
AND AIR RECEIVERS

ACCUMULATOR: A container in which fluid is stored under pressure as a source of fluid power.

Accumulator, Hydropneumatic: An accumulator in which compressed gas applies force to the stored liquid.

Accumulator, Hydropneumatic, Bladder: A hydropneumatic accumulator in which the liquid and gas are separated by an elastic bag or bladder.

Accumulator, Hydropneumatic, Diaphragm: A hydropneumatic accumulator in which the liquid and gas are separated by a flexible diaphragm.

Accumulator, Hydropneumatic, Non-Separator: A hydropneumatic accumulator in which the compressed gas operates dirctly on liquid within the pressure chamber.

Accumulator, Hydropneumatic, Piston: A hydropneumatic accumulator in which the liquid and gas are separated by a floating piston.

Accumulator, Mechanical: An accumulator incorporating a mechanical device which applies force to the stored fluid.

Accumulator, Mechanical, Spring: A mechanical accumulator in which springs apply force to the stored fluid.

Accumulator, Mechanical, Weighted: A mechanical accumulator in which the gravitational force acting upon weights applies force to the stored fluid.

AIR RECEIVER: A container in which gas is stored under pressure as a source of pneumatic fluid power.

PRESSURE VESSEL: A container which holds fluid under pressure.

SECTION 41

CONDUCTORS

CHANNEL: A fluid passage, the length of which is large with respect to its cross-sectional area.

CONDUCTOR: A component whose primary function is to contain and direct fluid.

CONDUIT: Any confining element employed to transfer a fluid.

HOSE: A flexible line.

Hose, Wire Braided: Hose consisting of a flexible material reinforced with woven wire braid.

LINE: A tube, pipe, or hose for conducting fluid.

Line, Drain: A line returning leakage fluid independently to the reservoir or vented manifold.

Line, Exhaust: A line returning power or control fluid back to the reservoir or atmosphere.

Line, Pilot: A line which conducts control fluid.

Line, Suction: A supply line at sub-atmospheric pressure to a pump, compressor, or other component.

Line, Working: A line which conducts fluid power.

LINES: Two or more fluid power lines.

Lines, Joining: Lines which connect in a circuit.

Lines, Passing: Lines which cross but do not connect in a circuit.

MANIFOLD: A conductor which provides multiple connection ports.

Manifold, Vented: A manifold which is open to the atmosphere and returns fluid to the reservoir.

NIPPLE: A short length of pipe or tube.

PASSAGE: A machined or cored fluid-conducting path which lies within or passes through a component.

PIPE: A line whose outside diameter is standardized for threading. Pipe is available in Standard, Extra Strong, Double Extra Strong or Schedule wall thicknesses.

SUBPLATE (BACK PLATE): An auxiliary ported plate for mounting components.

TUBE: A line whose size is its outside diameter. Tube is available in varied wall thicknesses.

SECTION 42

CONNECTORS AND CLOSURES

BUSHING: A short externally threaded connector with a smaller size internal thread.

CAP: A cover for fluid passages.

CLOSURE: A cap or a plug.

CONNECTOR: A device for joining a conductor to a component port or to one or more other conductors.

COUPLING: A straight connector for fluid lines.

Coupling, Quick Disconnect: A coupling which can quickly join or separate lines.

CROSS: A connector with four ports arranged in pairs, each pair on one axis, and the axes at right angles.

ELBOW: A connector that makes an angle between mating lines. The angle is always 90 degrees unless another angle is specified.

FITTING: A connector or closure for fluid power lines and passages.

Fitting, Compression: A fitting which seals and grips by manual adjustable deformation.

Fitting, Flange: A fitting which utilizes a radially extending collar for sealing and connection.

Fitting, Flared: A fitting which seals and grips by a preformed flare at the end of the tube.

Fitting, Flared, AN: A United States Air Force—Navy 37° flared tube fitting Design Standard.

Fitting, Flareless: A fitting which seals and grips by means other than a flare.

Fitting, Reusable Hose: A hose fitting that can be removed from a hose and reused.

Fitting, Welded: A fitting attached by welding.

JOINT: A line positioning connector.

Joint, Rotary: A joint connecting lines which have relative operational rotation.

Joint, Swivel: A joint which permits variable operational positioning of lines.

PLUG: A closure which fits into a fluid passage.

Plug, Dryseal Pipe: A plug made with a thread which conforms to Dryseal Pipe Thread Standards.

Plug, Short Pipe Thread: A plug which conforms in all respects to standard pipe threads except that the full thread has been shortened one full thread from the small end.

Plug, Standard Pipe Thread: A plug with American (National) tapered pipe threads.

Plug, Straight Thread: A plug with straight thread conforming to unified thread standards.

REDUCER: A connector having a smaller line size at one end than the other.

TEE: A connector with three ports, a pair on one axis with one side outlet at right angles to this axis.

UNION: A connector which permits lines to be joined or separated without requiring the lines to be rotated.

WYE (Y): A connector with three ports, a pair on one axis with one side outlet at any angle other than right angles to this axis. The side outlet angle is usually 45°, unless another angle is specified.

SECTION 43

PORTS AND THREADS

PORT: An internal or external terminus of a passage in a component.

Port, AND10050: A United States Air Force—Navy Aeronautical Design Standard in which a straight thread port is used to attach tube fittings to various components. It employs an "O" ring seal compressed in a special cavity.

Port, Bleed: A port which provides a passage for the purging of gas from a system or component.

Port, Control: A port which provides a passage for control fluid.

Port, Cylinder: A port which provides a passage to or from an actuator.

Port, Discharge: A port which provides a passage for fluid power to the system.

Port, Drain: A port which provides a free passage for the removal of slippage fluid from a hydraulic system or moisture from a pneumatic system.

Port, Exhaust: A port which provides a passage to the atmosphere.

Port, Fill: A port which provides a passage for filling purposes.

Port, Inlet: A port which provides a passage from the source of fluid.

Port, Pipe: A port which conforms to pipe thread standards.

Port, Plain "O" Ring: A port which uses an "O" ring in a groove located on the port face.

Port, Pressure: A port which provides a passage from the source of fluid.

Port, Reservoir: (See Port, Tank)

Port, Return: (See Port, Tank)

Port, SAE: A straight thread port used to attach tube and hose fittings. It employs an "O" ring compressed in a wedge-shaped cavity. A standard of the Society of Automotive Engineers.

Port, Suction: A port which provides a passage for atmospheric charging of a pump or compressor.

Port, Tank: A port which provides a passage to the fluid source.

Port, Vent: A port which provides a passage to the atmosphere.

PIPE THREADS: Screw threads for joining pipe.

Pipe Threads, Dryseal: Pipe threads in which sealing is a function of root and crest interference.

SECTION 46

PANELS — COMPARTMENTS — ENCLOSURES

COMPARTMENT: A space within the base, frame or column of the equipment.

ENCLOSURE: A housing for components.

PANEL: A plate or a surface for mounting components.

Panel, Control: A grouping of components mounted on a panel or integrally built into an assembled unit having a single mounting surface.

Panel, Mounting: A panel on which a number of components may be mounted.

SECTION 51

ACTUATORS — CYLINDERS — MOTORS

ACTUATOR: A device which converts fluid power into mechanical force and motion.

CYLINDER: A device which converts fluid power into linear mechanical force and motion. It usually consists of a movable element such as a piston and piston rod, plunger or ram, operating within a cylindrical bore.

Cylinder, Adjustable Stroke: A cylinder equipped with adjustable stops at one or both ends to limit piston travel.

Cylinder, Cushioned: A cylinder with a piston-assembly deceleration device at one or both ends of the stroke.

Cylinder, Double Acting: A cylinder in which fluid force can be applied to the movable element in either direction.

Cylinder, Double Rod: A cylinder with a single piston and a piston rod extending from each end.

Cylinder, Dual Stroke: A cylinder combination which provides two working strokes.

Cylinder, Piston: A cylinder in which the movable element has a greater cross-sectional area than the piston rod.

Cylinder, Plunger: A cylinder in which the movable element has the same cross-sectional area as the piston rod.

Cylinder, Ram: (See Cylinder, Plunger)

Cylinder, Retractable Stroke: An adjustable-stroke cylinder in which the stop can be temporarily changed to permit full retraction of the piston assembly.

Cylinder, Single Acting: A cylinder in which the fluid force can be applied to the movable element in only one direction.

Cylinder, Single Rod: A cylinder with a piston rod extending from one end.

Cylinder, Spring Return: A cylinder in which a spring returns the piston assembly.

Cylinder, Tandem: Two or more cylinders with interconnected piston assemblies.

Cylinder, Telescoping: A cylinder with nested multiple tubular rod segments which provide a long working stroke in a short retracted envelope.

MOTOR: A device which converts fluid power into mechanical force and motion. It usually provides rotary mechanical motion.

Motor, Fixed Displacement: A motor in which the displacement per unit of output motion cannot be varied.

Motor, Linear: (See CYLINDER)

Motor, Rotary: A motor capable of continuous rotary motion.

Motor, Rotary, Limited: A rotary motor having limited motion.

Motor, Variable Displacement: A motor in which the displacement per unit of output motion can be varied.

SECTION 56

CYLINDER PARTS

(Wherever the term "piston rod" appears it is intended to also mean "plunger" or "ram").

ANGLE: A mounting device which is angular in cross section. It is usually made from a 90° structural angle.

BACK END: (See CAP)

BASE: (See SIDE)

BLIND END: (See CAP)

BLIND HEAD: (See CAP)

CAP: A cylinder end closure which completely covers the bore area.

CLEVIS: A "U" shaped mounting device which contains a common pin hole at right angle or normal to the axis of symmetry through each extension. A clevis usually connects with an eye.

END: Either of two envelope surfaces at right angle or normal to the piston rod centerline.

EYE: A mounting device consisting of a single extension which contains a mounting pin hole at right angle or normal to the axis of symmetry. An eye usually connects with a clevis.

FLANGE: A mounting device consisting of a plate or collar extending past the basic cylinder profile to provide clearance area for mounting bolts. A flange is usually at right angle or normal to the piston rod centerline.

FOOT: (See LUG)

FRONT END: (See HEAD)

FRONT FACE: (See HEAD)

FRONT HEAD: (See HEAD)

HEAD: The cylinder end closure which covers the differential area between the bore area and the piston rod area.

HINGE: (See CLEVIS and EYE)

LUG: A mounting device consisting of a block extending past the basic cylinder profile. The block usually has a tapped or through mounting hole at right angles to the cylinder axis.

PENDULUM: (See CLEVIS and EYE)

RABBET: A mounting device which utilizes matching male and female forms (usually coaxial circular) between the cylinder and its mating element.

REAR END: (See CAP)

REAR HEAD: (See CAP)

ROD END: (See HEAD)

ROD, HEAD: (See HEAD)

SIDE: An envelope surface which is parallel to the piston rod centerline.

TIE ROD: An axial external cylinder element which traverses the length of the cylinder. It is pre-stressed at assembly to hold the ends of the cylinder against the tubing. Tie rod extensions can be a mounting device.

TRUNNION: A mounting device consisting of a pair of opposite projecting cylindrical pivots. The cylindrical pivot pins are at right angle or normal to the piston rod centerline to permit the cylinder to swing in a plane.

SECTION 57

CYLINDER MOUNTINGS

(See Section 56 for detail definitions—wherever the term "piston rod" appears it is intended to also mean "plunger" or "ram").

MOUNTING: A device by which a cylinder is fastened to its mating element.

Mounting, Centerline: A mounting which permits connection on a plane in line with the piston rod centerline.

Mounting, Centerline, Lug: A centerline mounting consisting of two opposite lugs at each end of the cylinder.

Mounting, End: A mounting which permits connection at either or both ends of a cylinder.

Mounting, End, Both: A mounting at both ends of the cylinder.

Mounting, End, Both, Tie Rods Extended: A cylinder mounted at both ends by means of extended tie rods.

Mounting, End, Cap (Back End, Blind End, Blind End Head, Rear End, Rear Head): A mounting which permits connection at the cap end.

Mounting, End, Cap, Circular: A direct circular cap mounting.

Mounting, End, Cap, Circular Flange: A cap mounting consisting of a supplementary circular flange plate.

Mounting, End, Cap, Detachable Clevis: A cap mounting consisting of a clevis which can be removed or rotated.

Mounting, End, Cap, Detachable Eye: A cap mounting consisting of an eye which can be removed or rotated.

Mounting, End, Cap, Fixed Clevis: A cap mounting consisting of a clevis integral with the cap to maintain a fixed clevis-port relationship.

Mounting, End Cap, Fixed Eye: A cap mounting consisting of an eye integral with the cap to maintain a fixed eye-port relationship.

Mounting, End, Cap, Rectangular Flange: A cap mounting consisting of a supplementary rectangular flange plate.

Mounting, End, Cap, Square: A direct square cap mounting.

Mounting, End, Cap, Square Flange: A cap mounting consisting of a supplementary square flange plate.

Mounting, End, Cap, Tie Rods Extended: A cap mounting consisting of extended tie rods.

Mounting, End, Cap, Trunnion: A cap mounting consisting of trunnion pins near or at the cap end of the cylinder.

Mounting, End, Head (Front End, Front Face, Front Head, Rod End, Rod End Head:) A mounting which provides cylinder connection at the head end.

Mounting, End, Head, Circular: A direct circular head mounting.

Mounting, End, Head, Circular Flange: A head mounting consisting of a supplementary circular flange plate.

Mounting, End, Head, Female Rabbet: A head mounting consisting of a female pilot recess.

Mounting, End, Head, Male Rabbet: A head mounting consisting of a male pilot extension.

Mounting, End, Head, Rectangular Flange: A head mounting consisting of a supplementary rectangular flange plate.

Mounting, End, Head, Square: A direct square head mounting.

Mounting, End, Head, Square Flange: A head mounting consisting of a supplementary square flange plate.

Mounting, End, Head, Tie Rods Extended: A head mounting consisting of extended tie rods.

Mounting, End, Head, Trunnion: A head mounting consisting of trunnion pins near or at the head end of a cylinder.

Mounting, Fixed: A mounting which provides rigid connection between the cylinder and the mating element wherein the piston rod reciprocates in a fixed line.

Mounting, Intermediate: A mounting which provides cylinder connection at an intermediate external position along the piston rod centerline between ends.

Mounting, Intermediate, Fixed Trunnion: An intermediate mounting consisting of trunnion pins which cannot be repositioned.

Mounting, Intermediate, Movable, Trunnion: An intermediate mounting consisting of trunnion pins which can be repositioned.

Mounting, Pivot: A mounting which permits a cylinder to change its alignment in a plane.

Mounting, Side (Base): A mounting which provides cylinder connection at one of its sides.

Mounting, Side (Base), End Angles: A side mounting consisting of an angle at each end of the cylinder with the free legs facing a common side.

Mounting, Side (Base), End Lugs: A side mounting consisting of one or more lugs at each end of the cylinder and facing a common side.

Mounting, Side (Base), End Plates: A side mounting consisting of an extending plate at each cylinder end and facing a common side.

Mounting, Side (Base), Lugs (Foot): A side mounting consisting of two opposite lugs at each cylinder end facing a common side.

Mounting, Side (Base), Tapped: A side mounting consisting of one or more tapped holes at each cylinder end facing a common side.

Mounting, Side (Base), Through Holes: A side mounting consisting of holes drilled across both ends to a common side.

Mounting, Universal: A mounting which permits a cylinder to change its alignment in all directions.

SECTION 58

CUSHIONS AND SNUBBERS

CUSHION: A device which provides controlled resistance to motion.

Cushion, Cylinder: A cushion built into a cylinder to restrict flow at the outlet port thereby arresting the motion of the piston rod.

Cushion, Die: A cushion installed with a die on a press to provide controlled resistance against the work. The return motion of the cushion is sometimes used to eject the work.

Cushion, Hydraulic: A cushion in which resistance is developed hydraulically.

Cushion, Hydropneumatic: A cushion in which resistance is developed hydraulically and pneumatically.

Cushion, Pneumatic: A cushion in which resistance is developed pneumatically.

SECTION 61

SEALING DEVICES AND RELATED TERMS

DUROMETER HARDNESS: A comparative indication of elastomer hardness determined by a durometer.

PACK: To install a packing.

PACKING: A sealing device consisting of bulk deformable material or one or more mating deformable elements, reshaped by manually adjustable compression to obtain and maintain effectiveness. It usually uses axial compression to obtain radial sealing.

Packing, Coil: Packing in coil form.

Packing, "U": A packing in which the deformable element has a "U" shaped cross-section.

Packing, "V": A packing in which the deformable element has a "V" shaped cross-section.

Packing, "W": A packing in which the deformable element has a "W" shaped cross-section.

RING, "O": A ring which has a round cross-section.

Ring, Piston: A piston sealing ring. It is usually one of a series and is often split to facilitate expansion or contraction.

Ring, Scraper: A ring which removes material by a scraping action.

Ring, "U": A ring which has a "U" shaped cross-section.

Ring, "V": A ring which has a "V" shaped cross-section.

Ring, "W": A ring which has a "W" shaped cross-section.

Ring, Wiper: A ring which removes material by a wiping action.

SEALING DEVICE: A device which prevents or controls the escape of a fluid or entry of a foreign material.

SEAL: (See SEALING DEVICE)

Seal, Axial: A sealing device which seals by axial contact pressure.

Seal, Cup: A sealing device with a radial base integral with an axial cylindrical projection at its outer diameter.

Seal, Diaphragm: A relatively thin, flat or molded sealing device fastened and sealed at its periphery with its inner portion free to move.

Seal, Dished Diaphragm: A diaphragm in which the central area is depressed in a free state. It permits longer travel than a flat comparable diaphragm.

Seal, Flat Diaphragm: (See Seal, Diaphragm)

Seal, Dynamic: A sealing device used between parts that have relative motion.

Seal, End: (See Seal, Axial)

Seal, Face: (See Seal, Axial)

Seal, Flange: A sealing device with a radial base integral with an axial projection at its inner diameter.

Seal, Gasket: (See Seal, Static)

Seal, Hat: (See Seal, Flange)

Seal, Lip: A sealing device which has a flexible sealing projection.

Seal, Lubricant: A sealing device which uses a lubricant as a sealing barrier.

Seal, Mechanical: A sealing device in which sealing action is aided by mechanical force.

Seal, Oil: A sealing device which retains oil.

Seal, "O" Ring: A sealing ring which has a round cross-section.

Seal, Piston: A sealing device installed on a piston to maintain a sealing fit with a cylinder bore.

Seal, Plunger: (See Seal, Ram)

Seal, Pressure: A sealing device in which sealing action is aided by fluid pressure.

Seal, Radial: A sealing device which seals by radial contact pressure.

Seal, Ram: A sealing device which seals the periphery of a ram.

Seal, Rod: A sealing device which seals the periphery of a piston rod.

Seal, Rotary: A sealing device used between parts that have relative rotary motion.

Seal Shaft: (See Seal, Rod and Seal, Rotary)

Seal, Shoulder: (See Seal, Axial)

Seal, Sliding: A sealing device used between

parts that have relative reciprocating motion.

Seal, Static: A sealing device used between parts that have no relative motion.

Seal, Stem: (See Seal, Rod and Seal, Shaft)

Seal, Water: A sealing device which uses water as a sealing barrier.

Seal, Wiper: A sealing device which operates by a wiping action.

SECTION 63

AUXILIARY DEVICES FOR SEALS

ADAPTER: A seal support shaped to conform with the contour of the seal and the mating element.

Adapter, Female: An adapter with a concave seal support.

Adapter, Male: An adapter with a convex seal support.

Adapter, Pedestal: A male adapter of "T" shaped cross section usually used to support a "U" seal.

ANTI-EXTRUSION RING: A ring which bridges a clearance to minimize seal extrusion.

FILLER RING: A ring which fills the recess of a "V" or "U" seal.

GLAND: The cavity of a stuffing box.

GLAND FOLLOWER: The closure for a stuffing box.

JUNKET RING: (See ANTI-EXTRUSION RING).

LANTERN RING: A ring which provides support and venting for adjacent sealing devices.

SHELL: A structural form to which a sealing element is assembled or bonded.

SPRING, Expander: A spring which produces outward radial sealing force.

Spring, Finger: A spring with flexible fingers which produce sealing force.

Spring, Garter: A compression or tension ring formed from a long helical wire spring with connected ends to produce sealing force.

Spring, Lug: (See Spring, Finger)

Spring, Marcel: (See Spring, Wave)

Spring, Spreader: A spring which produces sealing force against both lips of "U" or "V" seals.

Spring, Wave: A compression spring of waved configuration which produces sealing force.

STUFFING BOX: A cavity and closure with manual adjustment for a sealing device.

SUPPORT RING: (See ADAPTER).

SECTION 71

CONTROLS AND RELATED TERMS

CONTROL: A device used to regulate the function of a component or system.

Control, Automatic: A control which actuates equipment in a predetermined manner.

Control, Combination: A combination of more than one basic control.

Control, Cylinder: A control in which a fluid cylinder is the actuating device.

Control, Electric: A control actuated electrically.

Control, Hydraulic: A control actuated by a liquid.

Control, Liquid-Level: A device which controls the liquid level by a float switch or other means.

Control, Manual: A control actuated by the operator.

Control, Mechanical: A control actuated by linkages, gears, screws, cams or other mechanical elements.

Control, Pneumatic: A control actuated by air or other gas pressure.

Control, Pressure Compensated: A control in which a pressure signal operates a compensating device.

Control, Servo: A control actuated by a feed back system which compares the output with the reference signal and makes corrections to reduce the difference.

CONTROLS, PUMP: Controls applied to positive displacement variable delivery pumps to adjust their volumetric output or direction of flow.

RESTRICTOR: A device which reduces the cross-sectional flow area.

Restrictor, Choke: A restrictor, the length of which is relatively large with respect to its cross-sectional area.

Restrictor, Orifice: A restrictor, the length of which is relatively small with respect to its cross-sectional area. The orifice may be fixed or variable. Variable types are non-compensated, pressure compensated, or pressure and temperature compensated.

SECTION 72

VALVES — FUNCTIONAL TYPES

VALVE: A device which controls fluid flow direction, pressure, or flow rate.

Valve, Air: A valve for controlling air.

Valve, Directional Control: A valve whose primary function is to direct or prevent flow through selected passages.

Valve, Directional Control, Check: A directional control valve which permits flow of fluid in only one direction.

Valve, Directional Control, Diversion: (See Valve, Directional Control, Selector)

Valve, Directional Control, Four Way: A directional control valve whose primary function is to alternately pressurize and exhaust two working ports.

Valve, Directional Control, Selector: A directional control valve whose primary function is to selectively interconnect two or more ports.

Valve, Directional Control, Servo: A directional control valve which modulates flow or pressure as a function of its input signal.

Valve, Directional Control, Straightway: A two port directional control valve.

Valve, Directional Control, Three Way: A directional control valve whose primary function is to alternately pressurize and exhaust a working port.

Valve, Flow Control: A valve whose primary function is to control flow rate.

Valve, Flow Control, Deceleration: A flow control valve which gradually reduces flow rate to provide deceleration.

Valve, Flow Control, Pressure Compensated: A flow control valve which controls the rate of flow independent of system pressure.

Valve, Flow Control, Pressure-Temperature Compensated: A pressure compensated flow control valve which controls the rate of flow independent of fluid temperature.

Valve, Flow Dividing: A valve which divides the flow from a single source into two or more branches.

Valve, Flow Dividing, Pressure Compensated: A flow dividing valve which divides the flow at constant ratio regardless of the difference in the resistances of the branches.

Valve, Flow Metering: (See Valve, Flow Control)

Valve, Hydraulic: A valve for controlling liquid.

Valve, Pilot: A valve applied to operate another valve or control.

Valve, Pneumatic: A valve for controlling gas.

Valve, Prefill: A valve which permits full flow from a tank to a "working" cylinder during the advance portion of a cycle, permits the operating pressure to be applied to the cylinder during the working portion of the cycle, and permits free flow from the cylinder to the tank during the return portion of the cycle.

Valve, Pressure Control: A valve whose primary function is to control pressure.

Valve, Pressure Control, Counterbalance: A pressure control valve which maintains back pressure to prevent a load from falling.

Valve, Pressure Control, Decompression: A pressure control valve that controls the rate at which the contained energy of the compressed fluid is released.

Valve, Pressure Control, Load Dividing: A pressure control valve used to proportion pressure between two pumps in series.

Valve, Pressure Control, Pressure Reducing: A pressure control valve whose primary function is to limit outlet pressure.

Valve, Pressure Control, Relief: A pressure control valve whose primary function is to limit system pressure.

Valve, Pressure Control, Relief, Safety: A relief valve whose primary function is to provide pressure limitation after malfunction.

Valve, Pressure Control, Unloading: A pressure control valve whose primary function is to permit a pump or compressor to operate at minimum load.

Valve, Priority: A valve which directs flow to one operating circuit at a fixed rate and directs excess flow to another operating circuit.

Valve, Sequence: A valve whose primary function is to direct flow in a pre-determined sequence.

Valve, Shutoff: A valve which operates fully open or fully closed.

Valve, Shuttle: A connective valve which selects one of two or more circuits because flow or pressure changes between the circuits.

Valve, Surge Damping: A valve which reduces shock by limiting the rate of acceleration of fluid flow.

Valve, Time Delay: A valve in which the change of flow occurs only after a desired time interval has elapsed.

SECTION 73

VALVES — BASIC DESIGNS

FLAPPER ACTION: A valve design in which output control pressure is regulated by a pivoted flapper in relation to one or two orifices.

JET ACTION: A valve design in which flow effect is controlled by the relative position of a nozzle and a receiver.

SEATING ACTION: A valve design in which flow is stopped by a seated obstruction in the flow path.

Seating Action, Ball: A seating action valve design which utilizes a solid ball to obstruct the flow path.

Seating Action, Diaphragm: A seating action valve design which utilizes a diaphragm to obstruct the flow path.

Seating Action, Disc: A seating action valve design which utilizes a disc to obstruct the flow path.

Seating Action, Disc, Swing: A seating action valve design which utilizes a hinged disc to obstruct the flow path.

Seating Action, Gate: A seating action valve design which utilizes a solid gate to obstruct the flow path.

Seating Action, Gate, Spreader: A gate valve which utilizes two companion discs which are positively seated by common spreaders to obstruct the flow path.

Seating Action, Gate, Wedge: A gate valve which utilizes a solid wedge shaped gate to obstruct the flow path.

Seating Action, Globe: (See Seating Action, Disc)

Seating Action, Needle: A seating action valve design which utilizes an externally adjustable tapered closure to obstruct the flow path.

Seating Action, Plug: A seating action valve design which utilizes a plug to obstruct the flow path.

Seating Action, Poppet: A seating action valve design in which the seating element pops open to obtain free flow in one direction and immediately reseats when flow reverses.

SHEAR ACTION: A valve design in which flow is modulated by an element which slides across the flow path.

Shear Action, Ball: A shear action valve

design which utilizes a ported ball that rotates on an axis normal to the flow path.

Shear Action, Plug: A shear action valve design which utilizes a ported plug that rotates on an axis normal to the flow path.

Shear Action, Plunger: (See Shear Action, Spool)

Shear Action, Spool: A shear action valve design which utilizes a spool that slides through the flow path.

Shear Action, Sliding Plate: A shear action valve design which utilizes a plate that slides across the flow path.

Shear Action, Sliding Plate, Linear: A sliding plate shear action valve design in which the motion of the plate is linear.

Shear Action, Sliding Plate, Rotary: A sliding plate shear action valve design in which the motion of the plate is rotary.

SECTION 74

VALVE POSITIONS

VALVE POSITION: The point at which flow directing elements provide a specific flow condition in a valve.

Valve Position, Center: The selective mid-position in a directional control valve.

Valve Position, Detent: A predetermined position maintained by a holding device acting on the flow-directing elements of a directional control valve.

Valve Position, Normal: The valve position when signal or actuating force is not being applied.

Valve Position, Offset: An off-center position in a directional control valve.

Valve Position, Return: The initial valve position.

Valve, Four Position: A directional control valve having four positions to give four selections of flow conditions.

Valve, Three Position: A directional control valve having three positions to give three selections of flow conditions.

Valve, Two Position: A directional control valve having two positions to give two selections of flow conditions.

SECTION 75

VALVE FLOW CONDITIONS

VALVE FLOW CONDITION: A flow pattern in a directional control valve.

Valve Flow Condition, Closed: All ports are closed.

Valve Flow Condition, Float: Working ports are connected to exhaust or reservoir.

Valve Flow Condition, Hold: Working ports are blocked to hold a powered device in a fixed position.

Valve Flow Condition, Open: All ports are open.

Valve Flow Condition, Regenerative: Working ports are connected to supply.

Valve Flow Condition, Tandem: Working ports are blocked and supply is connected to the reservoir port.

SECTION 76

VALVE ACTUATORS

VALVE ACTUATOR: The valve part(s) through which force is applied to move or position flow-directing elements.

Valve Actuator, Manual: A valve actuator consisting of a hand lever, palm button, foot treadle, or other manual energizing devices.

Valve Actuator, Mechanical: A valve actuator consisting of a cam, lever, roller, screw, spring, stem, or other mechanical energizing devices.

Valve Actuator, Pilot: A valve actuator which utilizes pilot fluid.

Valve Actuator, Pilot, Barrier: A pilot valve actuator wherein the working fluid is isolated from the actuator.

Valve Actuator, Pilot, Differential Area:

A pilot valve actuator wherein pilot fluid acts on unequal areas.

Valve Actuator, Pilot, Differential Pressure: A pilot valve actuator wherein pilot fluid acts at unequal pressure.

Valve Actuator, Pilot, External: A pilot valve actuator wherein fluid is received from an external source.

Valve Actuator, Pilot, Internal: A pilot valve actuator wherein pilot fluid is received from within the valve.

Valve Actuator, Pilot, Solenoid Controlled: A pilot valve actuator wherein pilot fluid is controlled by the action of one or more solenoids.

Valve Actuator, Solenoid: A valve actuaator which utilizes one or more solenoids.

SECTION 77

VALVE MOUNTINGS AND RELATED TERMS

Valve Mounting, Base: The valve is mounted to a plate which has top and side ports.

Valve Mounting, Line: The valve is mounted directly to system lines.

Valve Mounting, Manifold: The valve is mounted to a plate which provides multiple connection ports for two or more valves.

Valve Mounting, Sub-Plate: The valve is mounted to a plate which provides straight-through top and bottom ports.

SECTION 91

INSTRUMENTS

FLOWMETER: A device which indicates either flow rate, total flow, or a combination of both.

GAGE: An instrument or device for measuring, indicating, or comparing a physical characteristic.

Gage, Bellows: A gage in which the sensing element is a convoluted closed cylinder. A pressure differential between outside and inside causes the cylinder to expand or contract axially.

Gage, Bourdon Tube: A pressure gage in which the sensing element is a curved tube that tends to straighten out when subjected to internal fluid pressure.

Gage, Diaphragm: A gage in which the sensing element is relatively thin and its inner portion is free to deflect with respect to its periphery.

Gage, Fluid Level: A gage which indicates the fluid level at all times.

Gage, Piston: A pressure gage in which the sensing element is a piston operating against a spring.

Gage, Pressure: A gage which indicates the pressure in the system to which it is connected.

Gage, Vacuum: A pressure gage for pressures less than atmospheric.

MANOMETER: A differential pressure gage in which pressure is indicated by the height of a liquid column of known density. Pressure is equal to the difference in vertical height between two connected columns multiplied by the density of the manometer liquid. Some forms of manometers are "U" tube, inclined tube, well, and bell types.

PITOT TUBE: A velocity-sensing tubular probe with one end facing fluid flow and the other end connected to a gage.

PRESSURE SWITCH: An electric switch operated by fluid pressure.

Graphic Symbols for Fluid Power Diagrams

1. Introduction

1.1 General

Fluid power systems are those that transmit and control power through use of a pressurized fluid (liquid or gas) within an enclosed circuit.

Types of symbols commonly used in drawing circuit diagrams for fluid power systems are Pictorial, Cutaway, and Graphic. These symbols are fully explained in the USA Standard Drafting Manual (Ref. 2).

1.1.1 Pictorial symbols are very useful for showing the interconnection of components. They are difficult to standardize from a functional basis.

1.1.2 Cutaway symbols emphasize construction. These symbols are complex to draw and the functions are not readily apparent.

1.1.3 Graphic symbols emphasize the function and methods of operation of components. These symbols are simple to draw. Component functions and methods of operation are obvious. Graphic symbols are capable of crossing language barriers, and can promote a universal understanding of fluid power systems.

Graphic symbols for fluid power systems should be used in conjunction with the graphic symbols for other systems published by the USA Standards Institute (Ref. 3–7 inclusive).

1.1.3.1 Complete graphic symbols are those which give symbolic representation of the component and all of its features pertinent to the circuit diagram.

1.1.3.2 Simplified graphic symbols are stylized versions of the complete symbols.

1.1.3.3 Composite graphic symbols are an organization of simplified or complete symbols. Composite symbols usually represent a complex component.

1.2 Scope and Purpose

1.2.1 *Scope*

This standard presents a system of graphic symbols for fluid power diagrams.

1.2.1.1 Elementary forms of symbols are:

Circles	Triangles	Lines
Squares	Arcs	Dots
Rectangles	Arrows	Crosses

1.2.1.2 Symbols using words or their abbreviations are avoided. Symbols capable of crossing language barriers are presented herein.

1.2.1.3 Component function rather than construction is emphasized by the symbol.

1.2.1.4 The means of operating fluid power components are shown as part of the symbol (where applicable).

1.2.1.5 This standard shows the basic symbols, describes the principles on which the symbols are based, and illustrates some representative composite symbols. Composite symbols can be devised for any fluid power component by combining basic symbols.

Simplified symbols are shown for commonly used components.

1.2.1.6 This standard provides basic symbols which differentiate between hydraulic and pneumatic fluid power media.

1.2.2 *Purpose*

1.2.2.1 The purpose of this standard is to provide a system of fluid power graphic symbols for industrial and educational purposes.

1.2.2.2 The purpose of this standard is to simplify design, fabrication, analysis, and service of fluid power circuits.

1.2.2.3 The purpose of this standard is to provide fluid power graphic symbols which are internationally recognized.

1.2.2.4 The purpose of this standard is to promote universal understanding of fluid power systems.

1.3 Terms and Definitions

Terms and corresponding definitions found in this standard are listed in Ref. 8.

2. Symbol Rules (See Section 10)

2.1 Symbols show connections, flow paths, and functions of components represented. They can indicate conditions occurring during transition from one flow path arrangement to another. Symbols do not indicate construction, nor do they indicate values, such as pressure, flow rate, and other component settings.

2.2 Symbols do not indicate locations of ports, direction of shifting of spools, or positions of actuators on actual component.

2.3 Symbols may be rotated or reversed without altering their meaning except in the cases of: a.) Lines to Reservoir, 4.1.1; b.) Vented Manifold, 4.1.2.3; c.) Accumulator, 4.2.

2.4 Line Technique (See Ref. 1)
Keep line widths approximately equal. Line width does not alter meaning of symbols.

2.4.1 Solid Line

(Main line conductor, outline, and shaft)

2.4.2 Dash Line

(Pilot line for control)

2.4.3 Dotted Line

(Exhaust or Drain Line)

2.4.4 Center Line

(Enclosure outline)

2.4.5 Lines Crossing
(The intersection is not necessarily at a 90 deg angle.)

or

IEC

IEC

2.4.6 Lines Joining

IEC

or

IEC

2.5 Basic symbols may be shown any suitable size. Size may be varied for emphasis or clarity. Relative sizes should be maintained. (As in the following example.)

2.5.1 Circle and Semi-Circle

2.5.1.1 Large and small circles may be used to signify that one component is the "main" and the other the auxiliary.

2.5.2 Triangle

2.5.3 Arrow

2.5.4 Square

Rectangle

2.6 Letter combinations used as parts of graphic symbols are not necessarily abbreviations.

2.7 In multiple envelope symbols, the flow condition shown nearest an actuator symbol takes place when that control is caused or permitted to actuate.

2.8 Each symbol is drawn to show normal, at-rest, or neutral condition of component unless multiple diagrams are furnished showing various phases of circuit operation. Show an actuator symbol for each flow path condition possessed by the component.

2.9 An arrow through a symbol at approximately 45 degrees indicates that the component can be adjusted or varied.

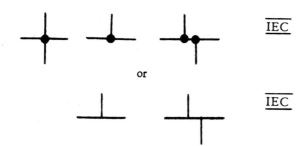

2.10 An arrow parallel to the short side of a symbol, within the symbol, indicates that the component is pressure compensated.

2.11 A line terminating in a dot to represent a thermometer is the symbol for temperature cause or effect.

See Temperature Controls 7.9, Temperature Indicators and Recorders 9.1.2, and Temperature Compensation 10.16.3 and 4.

2.12 External ports are located where flow lines connect to basic symbol, except where component enclosure symbol is used.

External ports are located at intersections of flow lines and component enclosure symbol when enclosure is used, see Section 11.

2.13 Rotating shafts are symbolized by an arrow which indicates direction of rotation (assume arrow on near side of shaft).

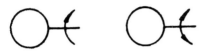

3.   Conductor, Fluid

3.1 Line, Working (main)

3.2 Line, Pilot (for control)

3.3 Line, Exhaust and Liquid Drain

3.4 Line, sensing, etc. such as gage lines shall be drawn the same as the line to which it connects.

3.5 Flow, Direction of

   3.5.1 Pneumatic

   3.5.2 Hydraulic

3.6 Line, Pneumatic
   Outlet to Atmosphere

   3.6.1 Plain orifice, unconnectable

   3.6.2 Connectable orifice (e. g. Thread)

3.7 Line with Fixed Restriction

3.8 Line, Flexible

3.9 Station, Testing, measurement, or power take-off

   3.9.1 Plugged port

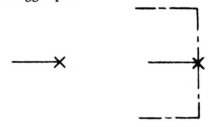

3.10 Quick Disconnect

   3.10.1 Without Checks

   Connected

   Disconnected

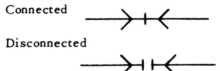

   3.10.2 With Two Checks

   Connected

   Disconnected

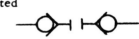

   3.10.3 With One Check

   Connected

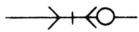

   Disconnected

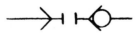

3.11 Rotating Coupling

4. Energy Storage and Fluid Storage

4.1 Reservoir

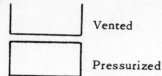

Vented

Pressurized

Note: Reservoirs are conventionally drawn in the horizontal plane. All lines enter and leave from above. Examples:

4.1.1 Reservoir with Connecting Lines

Above Fluid Level

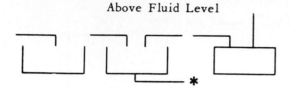

Below Fluid Level

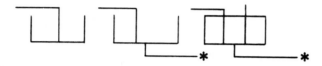

4.1.2 Simplified symbol

The symbols are used as part of a complete circuit. They are analogous to the ground symbol of electrical diagrams. $——\|\cdot$ IEC. Several such symbols may be used in one diagram to represent the same reservoir.

4.1.2.1 Below Fluid Level

4.1.2.2 Above Fluid Level

(The return line is drawn to terminate at the upright legs of the tank symbol.)

4.1.2.3 Vented Manifold

---

* Show line entering or leaving below reservoir only when such bottom connection is essential to circuit function.

4.2 Accumulator

4.2.1 Accumulator, Spring Loaded

4.2.2 Accumulator, Gas Charged

4.2.3 Accumulator, Weighted

4.3 Receiver, for Air or Other Gases

4.4 Energy Source
(Pump, Compressor, Accumulator, etc.)

This symbol may be used to represent a fluid power source which may be a pump, compressor, or another associated system.

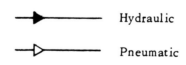

Hydraulic

Pneumatic

Simplified Symbol

Example:

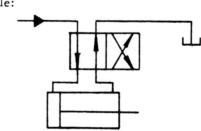

5. Fluid Conditioners

Devices which control the physical characteristics of the fluid.

5.1 Heat Exchanger

5.1.1 Heater

Inside triangles indicate the introduction of heat.

Outside triangles show the heating medium is liquid.

Outside triangles show the heating medium is gaseous.

5.1.2 Cooler

 or

Inside triangles indicate heat dissipation

(Corners may be filled in to represent triangles.)

5.1.3 Temperature Controller
(The temperature is to be maintained between two predetermined limits.)

 or

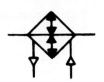

5.2 Filter – Strainer

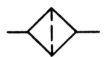

5.3 Separator

5.3.1 With Manual Drain

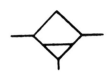

5.3.2 With Automatic Drain

5.4 Filter – Separator

5.4.1 With Manual Drain

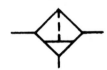

5.4.2 With Automatic Drain

5.5 Dessicator (Chemical Dryer)

5.6 Lubricator

5.6.1 Less Drain

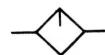

11.4.6 Two Positions, Four Connection Solenoid and Pilot Actuated, with Manual Pilot Override.

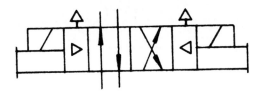

Simplified Symbol

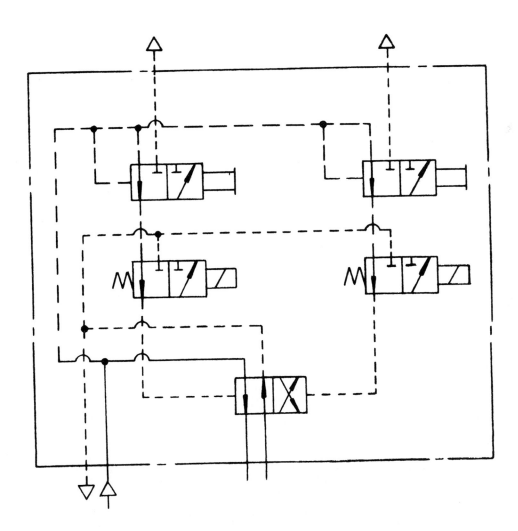

Complete Symbol

11.4.7 Two Position, Five Connection, Solenoid Control Pilot Actuated with Detents and Throttle Exhaust

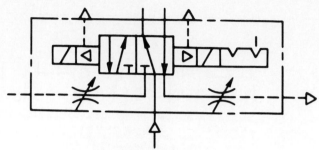

Symplified Symbol

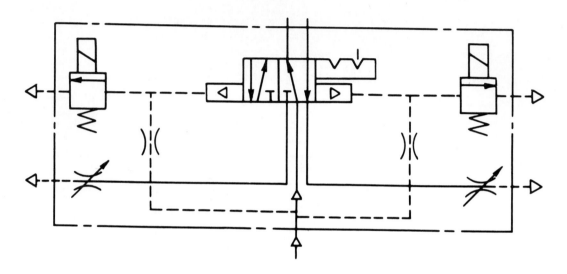

Complete Symbol

11.4.8 Variable Pressure Compensated Flow Control and Overload Relief

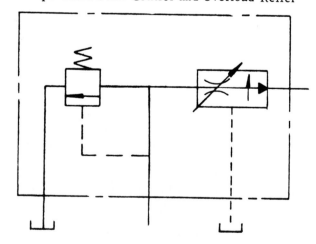

11.4.9 Multiple, Three Position, Manual·Directional Control with Integral Check and Relief Valves

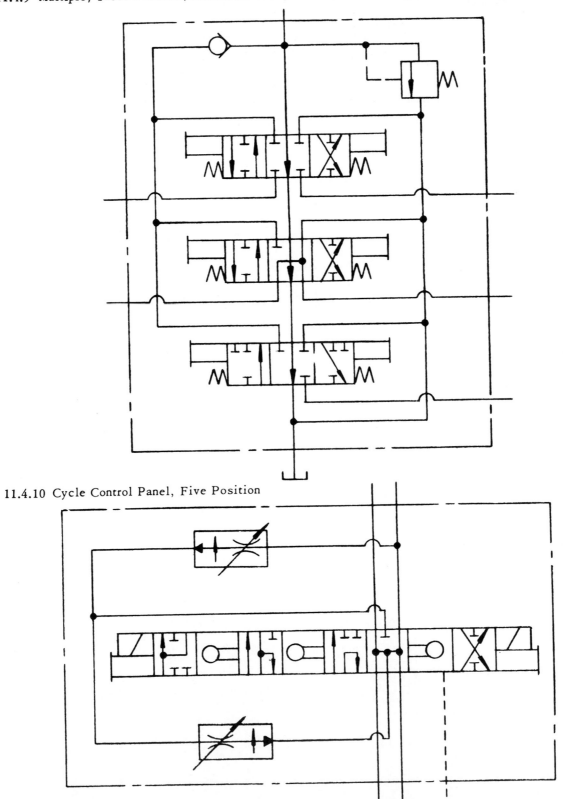

11.4.10 Cycle Control Panel, Five Position

11.4.11 Panel Mounted Separate Units Furnished as a Package (Relief, Two Four-Way, Two Check, and Flow Rate Valves)

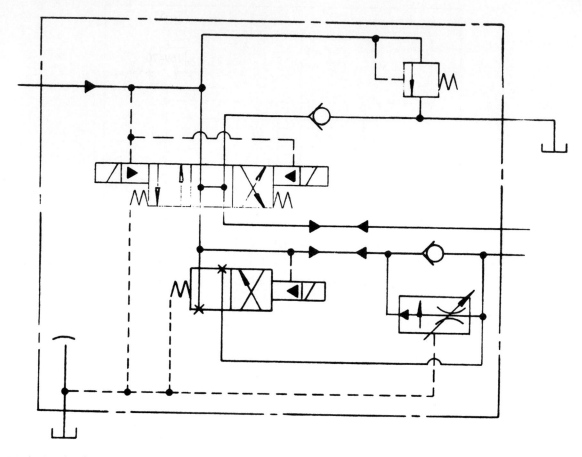

11.4.12 Single Stage Compressor with Electric Motor Drive, Pressure Switch Control of Receiver Tank Pressure

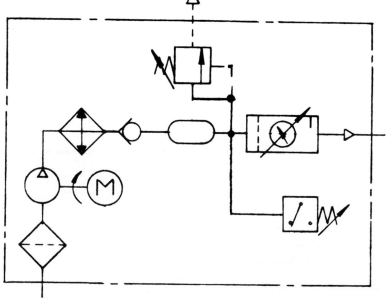

Index